This text is addressed to advanced students in oceanography, meteorology and environmental sciences as well as to professional researchers in these fields. It aims to acquaint them with the state of the art and recent advances in experimental and theoretical investigations of ocean–atmosphere interactions, a rapidly developing field in earth sciences.

Particular attention is paid to the scope and perspectives for satellite measurements and mathematical modelling. Current approaches to the construction of coupled ocean–atmosphere models (from the simplest zero-dimensional to the most comprehensive three-dimensional ones) for the solution of key problems in climate theory are discussed in detail. Field measurements and the results of numerical climate simulations are presented and help to explain climate variability arising from various natural and anthropogenic factors.

T0340050

Cambridge atmospheric and space science series

Ocean–Atmosphere Interaction and Climate Modelling

Cambridge atmospheric and space science series

Editors

Alexander J. Dessler
John T. Houghton
Michael J. Rycroft

Titles in print in this series

M. H. Rees, *Physics and chemistry of the upper atmosphere*
Roger Daley, *Atmospheric data analysis*
Ya. L. Al'pert, *Space plasma, Volumes 1 and 2*
J. R. Garratt, *The atmospheric boundary layer*
J. K. Hargreaves, *The solar–terrestrial environment*
Sergei Sazhin, *Whistler-mode waves in a hot plasma*
S. Peter Gary, *Theory of space plasma microinstabilities*
Ian N. James, *Introduction to circulating atmospheres*
Tamas I. Gombosi, *Gaskinetic theory*
Martin Walt, *Introduction to geomagnetically trapped radiation*
B. A. Kagan, *Ocean–atmosphere interaction and climate modelling*

Ocean–Atmosphere Interaction and Climate Modelling

Boris A. Kagan

P. P. Shirshov Institute of Oceanology, St Petersburg

Translated by Mikhail Hazin

CAMBRIDGE
UNIVERSITY PRESS

CAMBRIDGE UNIVERSITY PRESS
Cambridge, New York, Melbourne, Madrid, Cape Town, Singapore, São Paulo

Cambridge University Press
The Edinburgh Building, Cambridge CB2 2RU, UK

Published in the United States of America by Cambridge University Press, New York

www.cambridge.org
Information on this title: www.cambridge.org/9780521444453

Translation © M.A. Chazin

© Cambridge University Press 1995

First published 1995
This digitally printed first paperback version 2006

A catalogue record for this publication is available from the British Library

Library of Congress Cataloguing in Publication data

Kagan, B. A. (Boris Abramovich)
Ocean–atmosphere interaction and climate modelling / B.A. Kagan.
p.   cm. – (Cambridge atmospheric and space science series)
Includes bibliographical references and index.
ISBN 0 521 44445 4
1. Ocean–atmosphere interaction – Mathematical models.
2. Climatology – Mathematical models.   I. Title.   II. Series.
GC190.5.K34   1995
551.25–dc20   94-14977 CIP

ISBN-13 978-0-521-44445-3 hardback
ISBN-10 0-521-44445-4 hardback

ISBN-13 978-0-521-02593-5 paperback
ISBN-10 0-521-02593-1 paperback

Contents

Preface

In 1963 when the principal concepts of the study of ocean–atmosphere interaction had only been outlined, a group of leading American geophysicists (see Benton *et al.*, 1963) stated: 'We are beginning to realize dimly, although our information on this score is far from complete, that the atmosphere and the oceans which together constitute the fluid portions of the Earth actually function as one huge mechanical and thermodynamical system.' And further: 'A physical understanding of the processes of air–sea interaction should be one of the major objectives of geophysics during the coming decade.' Thirty years have now passed and one of the least developed branches of geophysics has been transformed into an independent discipline with the aim of integrating the varied information on the ocean and atmosphere into a unified and balanced 'picture of the world', in order to provide an explanation as to the natural variability of mutually adjusted fields of climatic characteristics, to detect those common and distinctive features in hydrothermodynamics of both media (ocean and atmosphere) that are important for understanding the evolution of the climate, and to create a climatic theory, on which basis to forecast potential consequences of natural and anthropogenic forcings.

This progress has been achieved as a result of the realization of large experimental programmes on the one hand, and of the introduction of physical models on the other. Both these approaches are mutually complementary: the theoretical approach assumes the utilization of experimental data to test the models; the experimental approach assumes the introduction of conceptual ideas to be confirmed by modelling results.

Among the large experimental programmes performed in the past, and still continuing, mention should be made of the Global Atmospheric Research Programme (GARP), the First GARP Global Experiment (FGGE), the Joint

Air–Sea Interaction Experiment (JASIN), the Tropical Ocean and Global Atmosphere Programme (TOGA), the Global Energy and Water Cycle Experiment (GEWEX), the Atmosphere–Ocean Chemistry Experiment (AEROCE), the International Geosphere–Biosphere Programme (IGBP), the World Ocean Circulation Experiment (WOCE), the Joint Global Ocean Flux Studies (JGOFS), the Soviet programme 'Sections' for studies of energy exchange in the energy active ocean zones, and, finally, the World Climate Research Programme (WCRP).

For the last 30 years the comparatively new technology of field measurements has been tested successfully. This involves the creation of a network of satellite scanned drifting and anchored buoys, as well as satellite and acoustic measurements. In the near future satellite measurements will become the basis of a global network of continuous registration of radiative fluxes, cloudiness, vertical temperature and humidity distribution in the atmosphere, sea surface temperature, wind velocity, wind stress, latent heat flux, wave parameters and ocean level elevations. New perspectives are also being discovered through the use of acoustic tomography of the ocean. This technique measures the travel times of sound pulses between pairs of transducers. Therefore, the difference in the times for travel in opposite directions yields an estimate of the water velocity along the path, averaged along the path, and this estimate, in turn, may be used to evaluate heat and mass transport across the path, heat storage inside the triangles formed by triplets of transducers, and the integral (over the area of the triangles) vertical component of potential vorticity.

The last three decades have been marked by an unprecedented growth of activity in the area of mathematical modelling of ocean–atmosphere interaction and by the creation of an entire hierarchy of models of the ocean–atmosphere system from the simplest zero-dimensional up to high-resolution, three-dimensional models ensuring solution of the key problems of climate theory, among them those which cannot be solved by any other means. Much has been achieved in this field, of which the most important is the realization that oscillations of climate with periods ranging from several years to several decades and more, as well as the climate response to external forcings (e.g. increasing the atmospheric CO_2 concentration, destruction of the vegetation, secular variations of the Earth's orbital elements, etc.), may be simulated only within the framework of models of the ocean–atmosphere system.

Extrapolating the development of activity within the scope of ocean–atmosphere interaction, one may say with confidence that over the next decade the main means of progress will be numerical experiment and satellite monitoring. This, properly speaking, determined the content of this book, whose aims are to expound current representations of the climatic system as a totality of the interacting atmosphere, ocean, lithosphere, cryosphere and biosphere; to introduce the reader to new methods and results of theoretical and experimental researches; to give him or her an opportunity to perceive the extraordinary complexity of real problems which, as a rule, do not have unquestionably final solutions; and thereby to stimulate the reader to independent thinking. By the way, the above-mentioned developments in the science place special demands on the future specialist, who has to be equally well qualified in three closely connected directions: in theory, modelling and observations.

And lastly, this book is written by an oceanologist for oceanologists. This fact, as well as a willingness to emphasize the leading role of the ocean in the long-period variability of the ocean–atmosphere system, has determined the choice of subjects, the arrangement of priorities and even the sequence of the words in the title. This is not to undermine the role of the atmosphere. I have simply shifted accents in an effort to facilitate perception of the subject matter.

When preparing the book I employed the help of, and was influenced by, many people. I am grateful to my teachers, D. L. Laikhtman, A. S. Monin and V. V. Timonov, who gave me a worthy example of world perception. I cannot deny myself the pleasure of thanking my colleagues and friends from the P. P. Shirshov Institute of Oceanology, the Russian Academy of Sciences, and from the Russian State Hydrometeorological Institute: G. I. Barenblatt, A. Yu. Benilov, D. V. Chalikov, L. N. Karlin, N. B. Maslova, A. V. Nekrasov, V. A. Ryabchenko, A. S. Safray, N. P. Smirnov and S. S. Zilitinkevich for their permanent support and helpful advice. I would like to note my special thanks to my students, whose reactions, albeit often completely unexpected, initiated the appearance of this book.

I wish to express my thanks to Conrad Guettler, Michael J. Rycroft and Sheila Shepherd of Cambridge University Press for their courtesy during the publication of this book and for significant refinement of style.

St Petersburg, 1993   *B. A. Kagan*

1

Preliminary information

1.1 Definition of the climatic system

The *climatic system* ($\mathcal{G}$) is the totality of the atmosphere, ($\mathcal{A}$), hydrosphere ($\mathcal{H}$), cryosphere ($\mathcal{C}$), lithosphere ($\mathcal{L}$), and biosphere ($\mathcal{B}$) interacting and exchanging energy and substance with each other, that is,

$$\mathcal{G} = \mathcal{A} \cup \mathcal{H} \cup \mathcal{C} \cup \mathcal{L} \cup \mathcal{B},$$

where by the atmosphere is meant the Earth's gas shell; by the hydrosphere – the World Ocean with all marginal and internal seas; by the cryosphere – the continental ice sheets, sea ice, and land snow cover; by the lithosphere – the active land layer and everything that is on it, including lakes, rivers and underground waters; and, finally, by the biosphere is meant the land and sea vegetation and living organisms, including human beings.

The state of each of the *subsystems* of the climatic system $\mathcal{G}$ is described by a finite number of *determining parameters* $x_1, x_2, \ldots, x_n$. All other variables describing the state of one or another subsystem are functions of these parameters. Just as in thermodynamics, the determining parameters may be divided into *extensive* parameters that are proportional to the size and mass of the relevant subsystems, and *intensive* parameters that do not depend on size and mass. The former relate to volume, internal energy, enthalpy and entropy; the latter relate to temperature, pressure and concentration. *Internal* and *external* determining parameters are differentiated as follows: internal parameters determine the system state, the external ones determine the environment state. This separation of parameters into *internal* and *external* is very conditional in the sense that it depends on the time scales of the processes being investigated. Thus, for a time scale of the order of 10^3 years or less, the area and volume of the continental ice sheets can be identified by external parameters; for longer time scales these can be identified by the internal

1

parameters of the climatic system. Respectively, in the first case the continental ice sheets represent a part of the environment, in the second case they represent the constituents of the climatic system.

Generally, whether one or another subsystem belongs to the environment or climatic system is determined by the ratio between the time scale of the *process* being investigated (changes of the subsystem state because of changes of its parameters) and the relaxation time of the subsystem. If this ratio is small, then the subsystem is defined as belonging to the environment; otherwise it is defined as belonging to the climatic system.

As an illustration of the above we consider the simplest formulation of the problem of the ocean–atmosphere system response to constant external forcing. According to Dickinson (1981), the temperature disturbances in the atmosphere (δT_a), in the upper mixed layer (δT_m) and the deep layer (δT_d) of the ocean, created by the disturbance of heat influx (δQ), can be described by the following equations:

$$c_a \, d\delta T_a/dt + \lambda_{am}(\delta T_a - \delta T_m) + \lambda_a \delta T_a = \delta Q, \tag{1.1.1}$$

$$c_m \, d\delta T_m/dt + \lambda_{am}(\delta T_m - \delta T_a) + \lambda_{md}(\delta T_m - \delta T_d) = 0, \tag{1.1.2}$$

$$c_d \, d\delta T_d/dt + \lambda_{md}(\delta T_d - \delta T_m) = 0, \tag{1.1.3}$$

where Equations (1.1.1)–(1.1.3) describe variations in the heat budget in the atmosphere, in the upper mixed layer (UML) and in the deep layer (DL) of the ocean, and their separate components: mean disturbances of the heat content in separate subsystems (the first terms in (1.1.1)–(1.1.3)), disturbances in the resulting heat flux at the ocean–atmosphere interface (the second terms in (1.1.1) and (1.1.2)), disturbances in heat radiation sources and sinks of heat in the atmosphere (the third term in (1.1.1)) and disturbances in heat exchange between UML and DL (the third term in (1.1.2) and the second term in (1.1.3)). Here c_a, c_m and c_d are the specific heat capacities of the atmosphere, UML and DL; λ_{am} and λ_{md} are the heat exchange coefficients at the atmosphere–UML and UML–DL interfaces; t is time.

Let c_a, c_m and c_d, respectively, be equal to 0.45, 10 and 100 W/m² K/year, that is, equivalent to choosing the following mass values referring to the unit of area surface: 14×10^3 kg/m² for the atmosphere, 81×10^3 kg/m² for the UML and 807×10^3 kg/m² for the DL. The latter restricts the analyses of relatively small (order of decades) time scales where not the whole ocean but, rather, its upper 500–1000 m layer takes part in heat exchange with the atmosphere. We will assign parameters λ_a, λ_{am} and λ_{md} equal to 2.4, 45 and 2 W/m² K and assume also that temperature disturbances are absent at the initial moment, that is,

$$\delta T_a = \delta T_m = \delta T_d = 0 \qquad \text{at } t = 0$$

and the disturbance of heat influx is changed in jumps from 0 up to δQ at $t = 0$ and, further (at $t > 0$), is constant everywhere. Equations (1.1.1)–(1.1.3), forming a linear system of ordinary differential equations, allow an exact solution. But its features can be understood simply by using the asymptotic expansion procedure. Keeping this in mind we first find the stationary solution of Equations (1.1.1)–(1.1.3), representing the equilibrium response of the ocean–atmosphere system to external forcing.

At $t \Rightarrow \infty$ these equations take the form

$$
\left.
\begin{aligned}
\lambda_{am}(\delta T_a - \delta T_m) + \lambda_a \delta T_a &= \delta Q, \\
\lambda_{am}(\delta T_m - \delta T_a) + \lambda_{md}(\delta T_m - \delta T_d) &= 0, \\
\lambda_{md}(\delta T_d - \delta T_m) &= 0,
\end{aligned}
\right\}
\tag{1.1.4}
$$

which results in

$$
\delta T_a = \delta T_m = \delta T_d = \delta Q/\lambda_d. \tag{1.1.5}
$$

Let us assume that the specific heat capacity of the atmosphere is much less than that of the UML, and the specific heat capacity of the UML in its turn is much less than that of the DL, that is, c_a/c_m and c_m/c_d are small parameters, which means that adaptation of the atmosphere to external forcing occurs faster than that of the UML and faster still than that of the DL, and hence, when determining δT_a from (1.1.1) the value δT_m can be assumed to be fixed to a first approximation. Using this fact we fix δT_m in (1.1.1) and rewrite the first equation of the system (1.1.1)–(1.1.3) in the following form:

$$
d\delta T_a/dt + (\lambda_a + \lambda_{am})c_a^{-1}\delta T_a = (\delta Q + \lambda_{am}\delta T_m)/c_a. \tag{1.1.6}
$$

Its integration, with an allowance for the initial condition for δT_a from (1.1.4), yields

$$
\delta T_a = \frac{\delta Q + \lambda_{am}\delta T_m}{\lambda_a + \lambda_{am}}(1 - e^{-t/t_a}), \tag{1.1.7}
$$

where

$$
t_a = c_a/(\lambda_a + \lambda_{am}) \approx 4 \text{ days}.
$$

When $t = O(t_a)$ Equation (1.1.7) takes the form

$$
\delta T_a = \frac{\delta Q + \lambda_{am}\delta T_m}{\lambda_a + \lambda_{am}}, \tag{1.1.8}
$$

where it can be seen that at $\delta T_m = 0$ the atmospheric temperature disturbance and its limiting value defined in (1.1.5) are in the same ratio as λ_a to $(\lambda_a + \lambda_{am})$. In other words, on a time scale of the order of several days the atmospheric response to external forcing does not exceed 4% of its limiting value.

Let us substitute Equation (1.1.8) into Equation (1.1.2) and rewrite (1.1.2) and (1.1.3) as

$$
d\delta T_m/dt + \alpha(\delta T_m - \delta T_d) + \beta\delta T_m = (\beta/\lambda_a)\delta Q, \tag{1.1.9}
$$

$$
d\delta T_d/dt + \varepsilon\alpha(\delta T_d - \delta T_m) = 0, \tag{1.1.10}
$$

where

$$
\alpha = \lambda_{md}/c_m, \qquad \beta = \lambda_a\lambda_{am}/[c_m(\lambda_a + \lambda_{am})], \qquad \varepsilon = c_m/c_d.
$$

Let us take advantage of the smallness of parameter ε and introduce a slow time τ related to a fast time t by the ratio $\tau = \varepsilon t$. Then on a fast time scale Equation (1.1.9) takes the form

$$
d\delta T_m/dt + (\alpha + \beta)\delta T_m = \alpha\delta T_d + (\beta/\alpha)\delta Q, \tag{1.1.11}
$$

and on a slow time scale it takes the form

$$(1 + \alpha/\beta)\delta T_m(\tau) = (\alpha/\beta)\delta T_d(\tau) + \delta Q/\lambda_a$$

or

$$\delta T_m(\tau) = (1 + \alpha/\beta)^{-1}[(\alpha/\beta)\delta T_d(\tau) + \delta Q/\lambda_a]. \tag{1.1.12}$$

Similarly, on a fast time scale Equation (1.1.10) takes the form

$$d\delta T_d/dt = 0 \tag{1.1.13}$$

and on a slow time scale it takes the form

$$d\delta T_d/d\tau + \alpha(\delta T_d - \delta T_m) = 0 \tag{1.1.14}$$

or, after substitution of an expression for δT_m from (1.1.12),

$$\kappa d\delta T_d/d\tau + \delta T_d = \delta Q/\lambda_a, \tag{1.1.15}y$$

where

$$\kappa = \alpha^{-1} + \beta^{-1}.$$

The solutions of Equations (1.1.10) and (1.1.13) with regard to the initial conditions for δT_m and δT_d from (1.1.4) take the form

$$\delta T_m(t) = \frac{\delta Q/\lambda_a}{1 + \alpha/\beta}(1 - e^{-t/t_m}), \tag{1.1.16}$$

$$\delta T_d(t) = 0, \tag{1.1.17}$$

where $t_m = (\alpha + \beta)^{-1} \approx 2.3$ years.

Thus, the quasi-stationary regime of UML is established at a time of the order of several years, and on time scales $t = O(t_m)$ the response of UML to external forcing is less, by $(1 + \alpha/\beta)$, than its limiting value determined by Equation (1.1.5). Rewriting the factor $(1 + \alpha/\beta)$ as $[1 + (\lambda_{md}/\lambda_{am})(1 + \lambda_{am}/\lambda_a)]$ and substituting the above-mentioned values of parameters λ_a, λ_{am} and λ_{md}, we draw the conclusion that on a time scale of the order of t_m the response of UML to external forcing is about half its limiting value.

On the other hand, when $t \gg t_m$ the solution of Equation (1.1.15) after transition from τ to t is written as

$$\delta T_d = (\delta Q/\lambda_a)(1 - e^{-t/t_d}), \tag{1.1.18}$$

where

$$t_d = \varepsilon^{-1}\kappa = \varepsilon^{-1}(\alpha^{-1} + \beta^{-1}).$$

Combining (1.1.18) and (1.1.12) we obtain

$$\delta T_m = \frac{\delta Q/\lambda_a}{1 + \alpha/\beta}\left[1 + \frac{\alpha}{\beta}(1 - e^{-t/t_d})\right], \tag{1.1.19}$$

which holds for $t \gg t_m$.

Let us compare periods of establishing UML and DL. The first, as was found above, is equal to $t_m = (\alpha + \beta)^{-1} \approx 2.3$ years, and the second is $t_d = \varepsilon^{-1}(\alpha^{-1} + \beta^{-1}) \approx 94$ years. Hence, the time required to establish the equilibrium state of UML is about 2.4% of DL.

Let us match asymptotic solutions (1.1.16) and (1.1.19), that is, we sum them and subtract the general term. As a result we obtain

$$\delta T_{\mathrm{m}} = \frac{\delta Q/\lambda_{\mathrm{a}}}{1 + \alpha/\beta}\left[(1 - \mathrm{e}^{-t/t_{\mathrm{m}}}) + \frac{\alpha}{\beta}(1 - \mathrm{e}^{-t/t_{\mathrm{d}}})\right]. \qquad (1.1.20)$$

Equations (1.1.18) and (1.1.20) represent the solution of (1.1.1)–(1.1.4).

We have deduced that the time to establish the equilibrium state of separate subsystems of the atmosphere–UML–DL system is governed by the inequality $t_{\mathrm{a}} \ll t_{\mathrm{m}} \ll t_{\mathrm{d}}$. One can judge the time for setting-in of other subsystems of the climatic system from estimates of the characteristic time for vertical heat diffusion. The latter, together with the geometric and thermophysical parameters of separate subsystems, is presented in Table 1.1. As can be seen, if the characteristic time for vertical heat diffusion is taken as a measure of the relaxation time, then even on time scales of the order of 10^6 s the atmosphere, UML, sea ice, snow cover and active layer of land must be combined in a single climatic system.

Using the terminology of thermodynamics we introduce the concepts of open, closed and isolated systems. A system which exchanges its matter with the environment will be called an *open system*; a system which does not exchange its matter with the environment will be called a *closed system* and a system which does not interact with the environment, that is, does not exchange either energy or matter, will be called an *isolated system*. Thus, in terms of thermodynamics the climatic system represents a non-isolated system consisting of macroscopic subsystems, each of which has an extremely large number of degrees of freedom and interacts with every other and with the environment.

1.2 Scales of temporal variability and its mechanisms

Observational data demonstrate a great diversity of climatic system oscillations. According to Monin (1969), the majority of these fall into one of the following categories:

1. *Small-scale oscillations*, with periods ranging from fractions of a second up to several minutes, governed by turbulence and wave processes of different kinds (such as acoustic and gravitational waves in the atmosphere).
2. *Mesoscale oscillations*, with periods ranging from several minutes to several hours (in particular, inertial oscillations belong to this group). Their intensity is relatively small and, therefore, the energy spectrum in the indicated time range contains wide and deep minima, separating quasi-horizontal synoptic disturbances from three-dimensional small-scale heterogeneities.
3. *Synoptic oscillations*, with periods ranging from several hours up to several days

Table 1.1. *Characteristics of separate subsystems of the climatic system (according to Saltzman, 1983)*

Subsystems	Area (10^{12} m^2)	Characteristic vertical scale (m)	Density (kg/m^3)	Mass (10^{18} kg)	Specific heat capacity (10^3 J/kg/K)	Total heat capacity (10^{21} J/K)	Vertical eddy (or molecular) heat diffusion coefficient (m^2/s)	Typical time scale of vertical heat diffusion (s)
Atmosphere								
free atmosphere	510	10^4	0–1	5	1	5	10^2	10^6
planetary boundary layer	510	10^3	1	0.5	1	0.5	10	10^5
Ocean								
upper mixed layer	334	10^2	10^3	34	4	10^2	10^{-2}	10^6
deep layer	362	4×10^3	10^3	1400	4	5×10^3	10^{-3}	10^{10}
Cryosphere								
sea ice	30	1	10^3	0.5	2	10^{-1}	10^{-6}	10^6
snow-covered land	80	1	5×10^2	10^{-1}	2	10^{-1}	10^{-7}	10^5
mountain glaciers	2	10^2	10^3	10^{-1}	2	10^{-1}	10^{-6}	10^{10}
ice sheets	14	10^3	10^3	10	2	10	10^{-6}	10^{12}
Lithosphere								
lakes and rivers	2	10^2	10^3	0.2	4	1	10^{-3}	10^6
active surface layer	131	2	3×10^3	1	0.8	1	10^{-6}	10^6

in the atmosphere and weeks in the ocean. Among these are diurnal and semidiurnal oscillations arising from diurnal changes of insolation and the gravitational forces of the Moon and the Sun.

4. *Global variations*, with periods ranging from weeks to months. Two-week variations of the *circulation index* (the mean angular velocity of the atmospheric rotation in temperate latitudes with respect to the Earth's surface) and 30- or 60-day oscillations in the Indian and Pacific sectors of the tropical atmosphere may serve as examples.

5. *Seasonal variations*, with annual periods, and their harmonics (including monsoon phenomena).

6. *Interannual variations*, with periods of several years. Among these are the quasi-two-year global variations of meteorological elements, particularly in the equatorial stratosphere (according to Obukhov they occur as the result of *parametric resonance*, that is, resonance amplification of annual oscillations with slowly changing atmospheric parameters over time and with rare changes in phase of oscillations by 180 degrees), quasi-two-year phenomena ENSO (El-Niño – Southern Oscillation), 3.5-year auto-oscillations of the northern branches of the Gulf Stream, and so on.

7. *Intra-centennial variations*, with periods of tens of years; rising temperature during the first half of the current century provides a typical example.

8. *Inter-centennial variations*, with periods of several centuries or tens of ages. They manifested themselves in rising temperatures after the end of the glacial period, (centuries 90–60 BC), the setting-in of the so-called 'climatic optimum' in centuries 40–20 BC, a subsequent fall in temperature (tenth century BC to third century AD), a further rise in temperature during the fifteenth and sixteenth centuries, and, finally, a fall in temperature during the so-called 'small glacial period' (seventeenth to nineteenth centuries).

9. *Long-period oscillations*, with periods of tens of thousands of years (glacial and interglacial epochs of the Pleistocene), related to changes in the parameters of the Earth's orbit and inclination of the Earth's axis. Among oscillations of this kind, which are of astronomical origin, an oscillation with a period of 100 thousand years is the largest. Then, as the amplitude decreases there are oscillations with periods of 22 and 41 thousand years. The first of the above three periods is close to the Earth's orbit eccentricity change; the second period coincides with the period of precession; and the third period coincides with the period of inclination of the Earth's axis.

10. *Geological changes*, with periods of tens and hundreds of millions of years created by orogenesis and tectonics, and by the drift of continents.

All these manifestations of the temporal variability of the climatic system are represented in the idealized spectrum of the surface air temperature shown in Figure 1.1. Depending on the mechanisms of their excitation, variations in temperature, and in any other climatic characteristics, can be divided

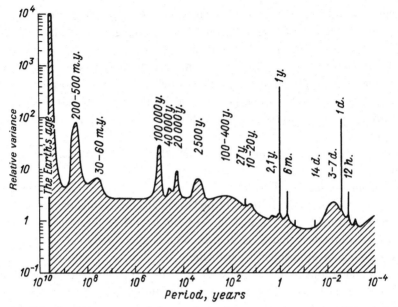

Figure 1.1 Idealized spectrum of the surface air temperature, according to Mitchell (1976).

into *forced* and *free* ones. Oscillations generated by external forcing (say, insolation changes) are termed forced oscillations; oscillations taking place independently of external mechanical or thermal forcing and determined by the internal instability of the climatic system with regard to small disturbances (for example, auto-oscillations in the ocean–atmosphere system) are termed free oscillations.

The basis of this classification can be reduced to the following, according to Lorenz (1970). In any real hydrodynamic system there is always viscous dissipation if the system does not move as a solid body, and thermal dissipation (average rate of degeneration of temperature non-homogeneities) if the system is not isothermal. Hence, in free oscillations the only mechanism which can balance the dissipation effects is energy transfer over the spectrum; in forced oscillations the dissipation effects can be compensated for at the expense of both cascade energy transfer and the work performed by generating forces. In the second case, the mode having the same length scale as the generating force is excited first of all. Then, because of instability, this mode transfers its energy to higher-wave modes. As soon as the energies of these modes approach a critical level the still higher-wave modes are excited. At the same time the high-wave modes have a reverse action on the low-wave-generating modes, resulting in amplification, or stabilization, and even destruction, of the low-wave-generating modes.

From the above it is clear that external forcing is only one of the possible variability mechanisms; the other is the internal stochastic mechanism determined by the existence of *feedbacks* between different internal parameters of the climatic system. Feedbacks can either amplify variations of interacting parameters or attenuate them. In the first case they are called *positive* feedbacks, and in the second case they are called *negative* feedbacks. There are many familiar examples of feedbacks. Let us discuss some of them.

The positive feedback between the greenhouse effect of water vapour and the atmospheric temperature. The amount of water vapour in the air (absolute humidity) is a non-linear function of air temperature, and this in its turn determines the atmospheric transparency to long-wave radiation. Thus, a rise in temperature at constant relative humidity (the assumption about the constancy of relative humidity is justified over a wide range of temperature changes) leads to an increase in the absorption of long-wave radiation in the atmosphere (the greenhouse effect), and thereby contributes to a further rise in temperature in the low atmosphere. In other words, the greenhouse effect of water vapour has a destabilizing effect on the climatic system.

The positive feedback between the albedo of snow and ice cover and the atmospheric temperature. It is common knowledge that ice and snow have higher reflective capacity (albedo) than water or soil. Therefore, an increase in the area of snow and ice cover, or in their lifetimes, has to be accompanied by an increase in the planetary albedo, and this is accompanied by a decrease in the solar radiation assimilated by the climatic system, and a subsequent increase in the snow and ice cover area.

The positive feedback between the greenhouse effect of carbon dioxide and the surface air temperature. An increase in CO_2 in the atmosphere arising from the burning of fossil fuel leads to a rise in temperature as a result of the greenhouse effect of carbon dioxide. In its turn the rise in surface air temperature accompanying an increase in the downward flux of long-wave radiation, and a decrease in sensible and latent heat fluxes (temperature contrast of water–air decreases) contributes to the warming of the ocean surface. This enhances the vertical stability of the upper ocean layers, decreases the absorption of carbon dioxide and eventually favours an increase in CO_2 in the atmosphere.

The negative feedback between the equator-to-pole temperature difference and the meridional heat transport. An increase in the equator-to-pole temperature difference results in enhancement of the meridional heat transport, which leads to a rise in ocean and atmospheric temperatures in high latitudes and, hence, to a decrease in the equator-to-pole temperature difference.

The negative feedback between soil humidity and albedo of land surface. A rise in soil humidity results in a decrease in land surface albedo which, in its turn, leads to an increase in absorption of short-wave solar radiation, a rise in temperature of the underlying surface, an increase in evaporation and, as a result, a decrease in soil moisture.

The negative feedback between air temperature and cloudiness. Radiation properties of clouds are defined by the albedo, height and temperature of their upper and lower boundaries, the amount of cloud and optical thickness. These factors generate simultaneous changes in short-wave and long-wave radiation so that the resulting effect of cloudiness can lead to both a rise and a fall in the temperature of the underlying surface and surface atmospheric layer, that is, it can manifest itself as positive and negative feedbacks. The positive feedback is created by an increase in the greenhouse effect with a preceding increase in the downward long-wave radiation flux, a rise in the underlying surface temperature, an intensificiation of evaporation, an increase in the water vapour content in the atmosphere and cloudiness; the negative feedback is generated by an increase in planetary albedo and a reversal of the short-wave solar radiation assimilated by the climatic system. Increasing amounts of lower-level cloud and decreasing amounts of upper-level cloud contribute to the appearance of negative feedback.

We have discussed two mechanisms of variability of the climatic system: its inherent internal stochasticity and external forcing. The third mechanism comprises the different forms of resonance between internal modes of the climatic system and external forcings of cyclic character. The oscillations of paleoclimatic indicators, with periods of ~100 000 years coinciding with periods of oscillation of the Earth's orbital eccentricity serve as a good example. There are several known explanations for this feature in the spectrum of paleoclimatic indicators, including resonance amplification of forced oscillations that relate to the internal variability of the climatic system.

Following Lorenz (1970, 1979) we assume that the internal stochastic mechanism makes an essential contribution to the variability of the climatic system for all time scales and that this mechanism consists of individual stochastic processes, every one of which has its own time scale and introduces its own contribution (white noise) to the variability, characterized by all time scales exceeding the given one. This defines the existence of the 'background' variability (shaded area in Figure 1.1) and the increase in its intensity with increasing time scale of oscillations. The most important factors controlling the 'background' variability on time scales of the order of one hour and less are small-scale turbulence and convection, and on time scales of the order of one day – inertial waves. On time scales of the order of one month the background variability is determined by the thermal relaxation of the atmosphere. On time scales of the order of one to ten years it is governed by thermal relaxation of the upper mixed layer of the ocean, and on time scales of the order of 10^1–10^3 years it is dictated by processes in the deep ocean. Finally, in the range of timescales from 10^3 to 10^5 years an interaction between the continental ice sheets, deep ocean layer and the atmosphere manifests itself.

External forcing results in the generation of narrow band variability. Hence in the presence of external forcing, and the absence of the 'background' variability the spectrum of climatic variability would resemble the super-position of variations featuring various external factors. Such variations have a high degree of ordering and, therefore, a high degree of predictability. In contrast, the 'background' variability is characterized by a low degree of ordering and predictability.

1.3 Predictability and non-uniqueness

Assume that the initial state of the climatic system is known, and we want to find a solution to the hydrothermodynamics equations describing the evolution of this system, that is, to predict a change in its state. Here is the question: is such prediction possible and, if so, what is the upper bound for prediction? To answer this question let us remember that the initial state of the climatic system is determined by measurements from an irregular network of stations located at some distance from each other, which means that individual motions with horizontal scales of less than the distance between the stations are not recorded at all. Moreover, initial data are distorted by random errors in measurements, interpolation and rounding-off. Thus even exact solutions of initial hydrothermodynamics equations will entail inevitable errors that increase over time due to the non-linearity of the equations.

The above is aggravated by the fact that the initial equations do not describe the state of the climatic system faultlessly due to the limited knowledge of laws governing the evolution of the climatic system. In addition, as is shown by Lorenz (1975), due to the non-linear interaction of motions

of different space and time scales the state of the climatic system can prove
to be unstable, in the sense that two states that are initially close to each
other will not be so forever. All this results in an increase in discrepancies
between predicted and real values.

But any prediction is valid while errors do not exceed the limit of mean
climatic dispersions of the predicted values. The period during which this
condition is valid is referred to as the *limit of predictability* (while using
deterministic models it is called *the limit of deterministic predictability*). Thus,
the causes of limited predictability of climatic system behaviour are: the
instability of the climatic system, inadequate methods of description, and
inaccuracy of initial information.

The limit of deterministic predictability is unambiguously connected with
the growth rate of errors in initial data or, say, with the time of its doubling.
Indeed, with an error of 1 °C in air temperature, a doubling time of three days,
an 8 °C mean climatic dispersion of temperature and a constant growth
rate of error, the limit of deterministic predictability will be equal to nine days.
Such an order of the limit of deterministic predictability is intrinsic to all
current operational numerical models of short-range weather forecasting,
including the ECMWF operational forecasting system (see Figure 1.2). Data
presented in this diagram are interesting in two respects. First, they indicate
that the limit of deterministic predictability depends on the quality of the

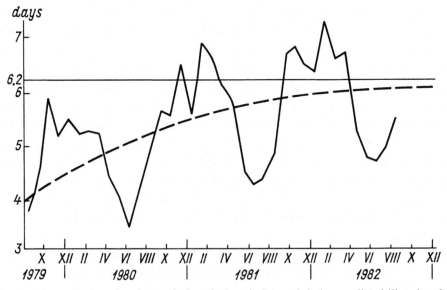

Figure 1.2 Evolution in time of the limit of deterministic predictability in the
ECMWF model, according to Tibaldi (1984).

forecasting model: improvement of the model contributes to an increase in the limit of predictability but only up to a certain point, called *the upper limit of deterministic predictability*. Second, the limit of deterministic predictability undergoes distinctive seasonal variations, taking maximum values in winter and minimum values in summer. This is associated with the fact that the limit of deterministic predictability is determined by both the growth rate of error and the maximum possible magnitude of error which, by definition, is equal to the mean climatic dispersion of the predicted value (Shukla, 1985).

It is a well-known fact that the growth rate of error is less, and dispersion is much less, in summer in the Northern Hemisphere than in winter. As a result the limit of deterministic predictability is higher in winter than in summer. Also, due to the dominance of the intensity of moist-convective instability in the tropics over the intensity of dynamical (generated by horizontal and vertical wind velocity shears) instability in temperate latitudes, and due to the inadequacy of their parametrization, the growth rate of error in the tropics is higher than that in temperate latitudes, while amplitudes of diurnal oscillations of meteorological characteristics are about the same. Therefore, the limit of deterministic predictability in the tropics is lower than that in temperate latitudes.

The influence of the dispersion magnitude on the duration of the predictability period can be illustrated by two more examples. First: owing to increasing ocean area and decreasing planetary scale disturbances and their variability, the dispersion of meteorological characteristics in the Southern Hemisphere is less than that in the Northern Hemisphere and thus the limit of deterministic predictability in the Southern Hemisphere is less than that in the Northern Hemisphere. Second: the intensity of synoptical disturbances is more than that of planetary disturbances. Hence, all other things being equal (the equal growth rates of error in initial data), the limit of deterministic predictability of synoptical disturbances is less than that of planetary disturbances. This is confirmed by the results of analyses of predictability for 500 hPa geopotential height forecast error fields presented by Tibaldi (1984), according to whom the limit of deterministic predictability for disturbances with wave numbers from 5 to 12 (disturbances of the synoptical scale) is about two weeks, and for disturbances with wave numbers from 0 to 4 (disturbances of the planetary scale) it is about four weeks. Thus, the limit of deterministic predictability increases with the spatial scale of disturbances. This last fact indicates the possibility in principle of forecasting large-scale, long-term oscillations of the atmospheric circulation.

The large-scale, long-term oscillations of the atmospheric circulation can be determined by their own instability and interaction with extremely unstable synoptic motions, or by variations of parameters (particularly of the sea surface temperature, soil humidity and area of snow–ice cover) inducing abnormal distribution of sources and sinks of heat and moisture in the atmosphere and, as a result, changes in amplitude and phase of planetary waves. Accordingly, it is appropriate to speak about the *internal* and *external* variability of the climatic system and the corresponding predictabilities of the first and second kind according to Lorenz (1975). By *predictability of the first kind* is meant the prediction of sequential states of the climatic system (the atmosphere, in this case) at fixed values of external parameters and assigned variations of initial conditions; and by *predictability of the second kind* is meant the prediction of an asymptotically equilibrium response (of the limiting state) of the climatic system to prescribed changes in external parameters.

It is clear from general considerations that the internal variability of the climatic system has restricted deterministic predictability beyond the boundaries of which the variability serves as the source of an unpredictable noise, while the external variability can be predicted for a longer term, of course, if variations of external parameters are themselves predictable or if they are sufficiently slow. This agrees with the results of numerical experiments by Shukla (1985), bearing witness to the fact that even for monthly mean characteristics of the atmospheric circulation the limit of deterministic predictability is no more than 30 days. Predictability of external parameter anomalies differs: under appropriate conditions the limit of deterministic predictability can be much higher than the value indicated above. It will be recalled that appropriate conditions are those favouring the initiation of strong disturbances of heat sources and sinks in the atmosphere, and the propagation of their influence at considerable distances from the location of a disturbance. In this respect the location of positive sea surface temperature anomalies in the tropical zone is preferable to their location in temperate latitudes because of the abrupt increase in the saturation mixing ratio and, hence, evaporation and latent heat release. As for possible propagation of the influence of anomalies far from their places of localization, this is confirmed by global-scale modes of variability (so-called *teleconnections*). Quasi-two-year phenomena ENSO serve as good examples of similar kinds of tele-connections (see Section 5.9).

Let us turn back to Figure 1.2 and pay attention to marked fluctuations in the limit of deterministic predictability in the winter time mapping of changes in different types of atmospheric circulation. In particular, the absolute maximum of the limit of deterministic predictability in February 1982 is connected with double (in the Atlantic and Pacific Oceans) *blocking of zonal flow*, that is, with the formation in the high latitudes of the Atlantic and Pacific Oceans of stationary (for about a week) centres of high atmospheric pressure that prevent the normal propagation of cyclones from the west to the east. This poses the obvious question as to whether this specific case is a reflection of general regularity – the existence of several asymptotic equilibrium states of the atmospheric circulation – and if so, what this means in respect of the predictability of long-period oscillations of the climatic system.

Before answering this question let us recall that a dynamic system is termed *ergodic* if the equations describing its evolution at random initial conditions and fixed external parameters have a unique possible stationary solution. If the dynamic system is not ergodic, then its behaviour over an infinitely large time interval will depend on the initial conditions. As applied to the climatic system this is equivalent to the fact that external parameters uniquely determine climate in the first case and non-uniquely in the second case.

The idea of the non-uniqueness of Earth's climate was first put forward by Lorenz (1979), who termed ergodic systems *transitive*, and those systems which do not have the property of transitivity *intransitive*. The real climatic system, according to Lorenz, is *almost intransitive*, that is, it shows signs of transitivity and intransitivity simultaneously. Alternation of glacial and interglacial epochs over the last 3.5 million years of Earth's history testifies to this.

It should be borne in mind that the stationary solution discussed above does not necessarily have to be stable in the sense that the system in question approaches or stays for an infinitely long period in the vicinity of some stationary state oscillating around it. If the stationary state is asymptotically unstable then under the influence of external forces the system begins to move away from it and approach another stationary state, which can also be asymptotically stable or unstable. A vertical rod oscillating in the gravity field around the horizontal axis of rotation serves as the simplest example of systems with asymptotically stable and unstable stationary states. A stable state occurs when the centre of gravity is lower than the axis of rotation, and an unstable state occurs when it is higher than the axis of rotation.

Another example is the motion of a liquid in a one-dimensional narrow canal described by the so-called equation of advection:

$$\partial u/\partial t + u\, \partial u/\partial x = 0, \tag{1.3.1}$$

where u is the current velocity along the x axis.

Let the canal length be L and let it be limited by impermeable walls on both sides where the current velocity vanishes, that is,

$$u = 0 \text{ at } x = 0, L. \tag{1.3.2}$$

In accordance with Wiin Nielsen (1975), we present the solution of Equations (1.3.1) and (1.3.2) as the Fourier series

$$u = \sum_{n=1}^{\infty} u_n(t) \sin(nkx),$$

where $k = \pi/L$. Substitution of this expression into (1.3.1) yields the following set of ordinary differential equations for coefficients u_n of the Fourier series:

$$\frac{du_n}{dt} = \frac{kn}{2} \sum_{m=1}^{\infty} u_m u_{n+m} - \frac{k}{2} \sum_{m=1}^{n-1} m u_m u_{n-m}. \tag{1.3.3}$$

Let us define non-dimensional velocity $\mathcal{V}_n$ and non-dimensional time τ using the formulae $\mathcal{V}_n = u_n/\sqrt{(2E_0)}$, $\tau = \sqrt{(2E_0)}kt$, where $2E_0 = \sum_{n=1}^{\infty} u_n^2(t)$ is twice the kinetic energy; the subscript '0' indicates the initial time instant. Then Equation (1.3.3) can be written as

$$\frac{d\mathcal{V}_n}{d\tau} = \frac{n}{2} \sum_{m=1}^{\infty} \mathcal{V}_m \mathcal{V}_{n+m} - \frac{1}{2} \sum_{m=1}^{n-1} m \mathcal{V}_m \mathcal{V}_{n-m}. \tag{1.3.4}$$

We consider the two-mode ($n = 1, 2$) system. In this case, according to (1.3.4), we have

$$d\mathcal{V}_1/d\tau = \mathcal{V}_1 \mathcal{V}_2/2, \qquad d\mathcal{V}_2/d\tau = -\mathcal{V}_1^2/2, \tag{1.3.5}$$

from which it follows that $dE/d\tau = 0$; here $2E = (\mathcal{V}_1^2 + \mathcal{V}_2^2)$, or, taking into account the fact that $2E = 1$ at $\tau = 0$,

$$\mathcal{V}_1^2 + \mathcal{V}_2^2 = 1. \tag{1.3.6}$$

We substitute this equality into the second equation of the system (1.3.5). As a result we obtain

$$d\mathcal{V}_2/d\tau = -(1 - \mathcal{V}_2^2)/2. \tag{1.3.7}$$

Integrating (1.3.7) and assuming that $\mathcal{V}_2 = \mathcal{V}_2^{(0)}$ at $\tau = 0$, we find

$$\mathcal{V}_2 = \frac{\mathcal{V}_2^{(0)} - \tanh(\tau/2)}{1 - \mathcal{V}_2^{(0)} \tanh(\tau/2)}. \tag{1.3.8}$$

In combination with (1.3.6) this expression yields

$$\mathcal{V}_1 = \frac{\mathcal{V}_1^{(0)}}{\cosh(\tau/2) - \mathcal{V}_2^{(0)} \sinh(\tau/2)}, \tag{1.3.9}$$

where $\mathcal{V}_1^{(0)} = (1 - \mathcal{V}_2^{(0)2})^{1/2}$.

At $\tau \Rightarrow \infty$ it follows from Equations (1.3.8) and (1.3.9) that $\mathcal{V}_1 = 0$ and $\mathcal{V}_2 = -1$, that is, on a phase plane (a plane whose coordinates are the state parameters $\mathcal{V}_1$, $\mathcal{V}_2$) the asymptotic behaviour of the system is presented as a point with coordinates $(0, -1)$. But, as can be seen from Equation (1.3.5), this equilibrium state is non-unique: there is one more point, with coordinates $(0, 1)$.

Let us clarify which of the above is stable. For this purpose we use the perturbation method, and write $\mathcal{V}_1$, $\mathcal{V}_2$ in the vicinity of steady states $(0, \mathcal{V}_2^{(\infty)})$ as $\mathcal{V}_1 = \mathcal{V}_1^{(\infty)} + \mathcal{V}_1'$, $\mathcal{V}_2 = \mathcal{V}_2^{(\infty)} + \mathcal{V}_2'$, where $\mathcal{V}_1^{(\infty)} = 0$ and $\mathcal{V}_2^{(\infty)} = 1$. Then we substitute expressions for $\mathscr{V}_1$ and $\mathscr{V}_2$ into (1.3.5), reject the terms containing the products and squares of perturbations $\mathscr{V}'_1$ and $\mathscr{V}'_2$ and subtract the equations for $\mathscr{V}_1^{(\infty)}$ and $\mathscr{V}_2^{(\infty)}$. As a result we obtain the following equations:

$$\mathrm{d}\mathscr{V}'_1/\mathrm{d}\tau = \mathscr{V}'_1\mathscr{V}_1^{(\infty)}/2, \qquad \mathrm{d}\mathscr{V}'_2/\mathrm{d}\tau = 0.$$

The solution of the first equation takes the form

$$\mathscr{V}'_1 = \mathscr{V}_1^{(0)} \exp(\mathscr{V}_2^{(\infty)}\tau/2), \qquad (1.3.10)$$

where we have taken into account the fact that $\mathscr{V}'_1 = \mathscr{V}_1^{(0)}$ at $\tau = 0$.

From (1.3.10) it follows that deviations from the state $(0, 1)$ increase exponentially, and deviations from the state $(0, -1)$ decrease exponentially, with time. In other words, the first of these two steady states is unstable, and the second is stable.

First indications of the possible existence of the non-uniqueness of the ocean thermohaline circulation appeared in the work of Chamberlin (1906). He proposed that a deep water formation in the absence of permanent ice cover may occur not as the result of intensive heat transfer to the atmosphere and decrease in sea water temperature at high latitudes, but, rather, as the result of intensive evaporation and increase in water salinity in the subtropics. The existence of a similar thermohaline circulation in the Mesozoic, with its characteristic decrease in temperature difference between the equator and the poles, damping of upwelling in the subtropics and with wide propagation of deep water stagnation, is verified by paleontological data. For indirect evidence in favour of fast (on the geological time scale) reorganization or, in effect, weakening of the ocean thermohaline circulation, data from chemical and isotopic analysis of deep water sediment deposition in the Pleistocene may be used.

Stommel (1961) was the first to present a theoretical proof of the possible existence of direct (with downwelling in high, and upwelling in low, latitudes) and reverse (with upwelling in high, and downwelling in low, latitudes) cells of the ocean thermohaline circulation. He ascertained that in the approximation of the ocean by the three-box system (polar boxes in the Northern and Southern Hemispheres and an equatorial box), two-cell thermohaline circulation is formed, the direction of rotation depending on the relationship between the equator-to-pole difference in temperature and salinity. The next step was taken by Rooth (1982), who showed, using a three-box ocean model, that, in the presence of inter-hemispheric water exchange and finite initial salinity perturbation in one of the polar boxes, two-cell circulation may degenerate into one-cell circulation even if forcing factors and ocean–land area ratio are symmetric about the equator. Numerical experiments carried out by Bryan (1986) within the framework of a three-dimensional sectorial ocean model have confirmed this conclusion. According to Bryan (1986), for constant external forcing the ocean thermohaline circulation has at least three steady states, i.e. two-cell (symmetric about the equator) circulation and two one-cell circulations, respectively, with opposite directions of rotation.

A fourth steady state was discovered by Manabe and Stouffer (1988) within the framework of a coupled ocean–atmosphere general circulation model (see the description in Section 5.8). This state features the existence of a sharp halocline at high latitudes of both hemispheres, cessation of deep water formation, and degeneration, or even change of direction, of the ocean thermohaline circulation accompanied by a decrease in the meridional heat and salt transport to the poles, and, as a result, a decrease in the sea surface temperature and salinity of the ocean at high latitudes. The decrease in sea surface temperature and salinity is particularly evident at high latitudes of the North Atlantic. It is not caused by change in precipitation–evaporation difference but, rather, by the deceleration of renewal of surface waters due to downwelling cessation. As a result, surface waters propagating along the European continent evolve into a sub-Arctic gyre where they gradually cool down and desalinate.

Thus, it turns out that the ocean thermohaline circulation may have four stable states: two two-cell and two single-cell states with opposite directions of rotation. In practice, as was established by Welander (1986), within the framework of a three-box ocean model there are nine steady states, but only four of them are stable, the others being unstable. However, if we reject this three-box model and replace it by a more realistic model representing the ocean as a system of three ventilated (participating in heat and moisture exchange with the atmosphere) layers – the surface layer, the intermediate layer and the bottom layer – then, according to Kagan and Maslova (1991), the number of steady states will be equal to 81, including 16 stable states. These steady states differ from each other in the number of cells (one-cell, two-cell and four-cell circulations), direction of rotation (clockwise and anticlockwise), properties of symmetry (symmetric and asymmetric circulations about the equator) and the vertical extension of cells (extension over the whole thickness of the ocean or only over a layer of finite depth). Such a variety of ocean thermohaline circulation forms has far-reaching consequences in respect of ocean climate predictability: even small changes in little-known initial fields of salinity and/or salt flux at the ocean surface may lead to qualitatively different distributions of climatic characteristics.

A similar situation takes place in the atmospheric circulation. According to Lorenz (1967), when using a two-layer quasi-geostrophic model of the general atmospheric circulation and retention for six modes, the increase in the equator-to-pole temperature difference is accompanied by the following change in regimes of the large-scale circulation. Firstly, when the temperature difference does not exceed 58 K there is only one steady solution representing the direct meridional circulation cell (so-called *Hadley cell*) where warm air ascends at the equator and cold air descends at the pole. A distinctive feature of this circulation regime is the absence of zonal flow in the lower layer and, hence, of its interaction with undulations of the underlying surface. When the temperature difference is equal to 58 K a *birfurcation* (replacement of one circulation regime by the other) occurs. The new regime is characterized by the existence of three steady state solutions; one of them (the Hadley cell) is unstable, and the two others, describing wave motions and, respectively, westerly and easterly flows, are stable. The second bifurcation occurs at the equator-to-pole temperature difference of 72 K. The subsequent increase in temperature difference results in the simultaneous existence of five steady solutions, two of them conforming to low and high *zonality indices* (large and small amplitudes of wave oscillations or, in other words, the presence and absence of blocking) respectively. Of these five steady state solutions only one is stable with regard to disturbances of the first meridional mode. It is a solution describing relatively weak zonal flow with superposition of high-amplitude wave constituents.

It should be emphasized that the results mentioned above concerning the change in atmospheric circulation regimes were obtained within the framework of a model based on an indirect assumption of the absence of all wave components with the exception of large-scale forced oscillations, and, hence, of the absence of all forms of non-linear interaction apart from the interaction between large-scale forced waves and zonal flow. In practice, because of a strong baroclinic instability generating free wave motions of synoptic scale, there are always other forms of non-linear interaction in the atmosphere, such as interaction between free waves and stationary flow, between forced and free waves and between free waves. Taking these forms of interaction into account within the framework of a multicomponent spectral model results in only one stable solution and the disappearance of all others. Moreover, it turns out that the unique possible stable solution corresponds to the wave regime of the atmospheric circulation with low zonality index, that is, it coincides with the only steady solution in a small-component model. The main distinction of these stable solutions is the fact that for different initial conditions the trajectories of the solution corresponding to the six-mode model are attracted to one and the same point of *phase space* (space of parameter), while in the multicomponent model they contract to a certain restricted domain inside which the behaviour of the solution becomes so complicated that the solution appears to describe stochastic oscillations of large-scale circulation. From the viewpoint of predictability it means that detailed long-range prediction of large-scale atmospheric circulation is impossible. But since the 'stochastic' oscillations of a large-scale circulation are relatively small (the attractor basin is restricted) then, as was mentioned above, the low-frequency variability of large-scale circulation can be considered as predictable *on the average*, that is, under conditions of preliminary filtration of high-frequency oscillations.

1.4 Methods of experimental research

Depending on the type of measuring instrument carrier, one differentiates between ground-based and satellite-based research of characteristics of the ocean–atmosphere interaction. Let us consider methods of ground-based and satellite measurements, paying particular attention to their fundamental features and possibilities.

1.4.1 Ground-based measurements

Ground-based measurements of characteristics of the ocean–atmosphere interaction are taken out using several types of carriers which serve as a platform for mounting measuring equipment such as commercial, fishing and research vessels, anchor and drifting buoys, fixed foundations (particularly, drilling platforms), pilot balloons, captive balloons, airships and planes. The most commonly used carriers are commercial and fishing ships. There were about 7690 in 1984. Thus, the total number of daily meteorological measurements carried out on these ships is $7690 \times 0.3 \times 4 = 9228$, where the second figure on the left-hand side is the time fraction during which the ships are at sea and the third figure is the number of reports received each day. In practice, Bracknell Regional Centre of meteorological information, for example, receives about 3700 messages every day, that is, 40% of the expected number.

One could have accepted this loss of information if its other part had been distributed more or less uniformly over the ocean. Unfortunately, this is not the case, as is obvious from Table 1.2, which lists information on the percentage of 5-degree squares having more than 100 and those having more than 30 samples per month for the tropical zone of the ocean. As may be seen, the availability of observational data in the Atlantic, Indian and Pacific Oceans leaves much to be desired. Moreover, it turns out that these data are not absolutely independent and are not distributed at random in time and space (they cluster in the vicinity of islands and ship routes). Clearly, even a slight increase in density and/or improvement in the quality of measurements

Table 1.2. *Percentage of 5-degree squares with* > 100 *and* > 30 *samples per month from the tropical zone of the Atlantic, Indian and Pacific Oceans (according to Taylor, 1985)*

Sampling area	> 100			> 30		
	Atlantic Ocean	Indian Ocean	Pacific Ocean	Atlantic Ocean	Indian Ocean	Pacific Ocean
September, 1967	25	10	10	70	45	40
September, 1977	30	15	15	55	40	35
December, 1978	50	40	40	80	60	55
December, 1980	80	90	60	100	95	80

Note: values for December, 1980, include island and buoy reports.

beyond these areas can result in noticeable changes in average values of the characteristics sought.

After these preliminary remarks we give a brief outline of some of the methods available for ground-based measurements of the parameters of the marine surface atmospheric layer, and their accuracy.

Sea surface temperature

For sea surface temperature measurements, use is made of the non-insolated bucket, condenser or cooling system intake, and hull conduit or through hull sensors. Applications of these methods on voluntary observing ships from different countries (a percentage of the total number of ships for each national fleet) are presented in Table 1.3. This suggests that in the mid-1980s the most common method of sea surface temperature measurement was the cooling system intake method. But the bucket method, adopted at the dawn of the industrial epoch, is still widely used.

Both methods yield different temperature values: bucket values are lower by 0.3 °C than intake values. If one takes into account that a mass transition from the first to the second method occurred in the late 1930s, then this might be one cause, among others, of the global rise in sea surface temperature obtained from measurement data in the 1940s. But this is not all. A subsequent check has shown (see Taylor, 1985) that the difference between bucket and non-bucket temperature values varies in time and space. Particularly in a high heat flux situation (e.g. the Gulf Stream, Kuroshio or Agulhas areas in winter) this can mean a difference of 1 °C or more.

Now, as 50 years ago, there is no agreement as to which method is more efficient. Each method has its own disadvantages. The disadvantages of

Table 1.3. *Percentage of ships using each
method of temperature measurement for the
six largest national fleets and for all
voluntary observation ships (WMO, 1984)*

		Method	
Country	Bucket	Intake	Hull
USSR	–	100	–
USA	2	98	–
Japan	4	96	–
FRG	94	4	2
UK	91	5	4
Canada	96	4	–
All ships	26	73	1

the bucket method are related to non-unified measurement techniques and the determination, strictly speaking, not of the sea surface temperature but of the temperature of some near-surface water layer of finite depth. The disadvantages of the non-bucket methods are similar. We note in this connection the strong dependence of temperature measurements on ship size, depth of intake and position of the intake thermometer.

Thus, we have to state that the error of ship sea surface temperature measurements is *a fortiori* more than $\pm 0.1\,°C$.

Air temperature and humidity

Conventional ship measurements of air temperature use either thermometers mounted on screens or take dry-bulb measurements from a portable psychrometer located on the top of the foremast or on the windward side of the top bridge, that is, from a position where the distorting influence of the ship's hull is minimized.

Screen values of air temperature may be biased high due to the radiative heating of the thermometer screen, or due to heat radiation of the hull. The error of such measurements, on average, is $\pm 0.5\,°C$ and not less than $\pm 0.1\,°C$, according to Taylor (1985). Air temperatures from psychrometers subjected to radiation and ventilation errors are necessarily higher than screen values. Moreover, the accuracy of psychrometric measurements depends critically on the location and method of exposure. In this respect well-sited screens are preferable since errors can be more readily assessed. Results of air temperature measurements obtained on USSR ships in GATE confirm this conclusion. As it turns out (see Taylor, 1985), when measurements are produced on the end of a 2.5 m boom extended from the bridge the mean difference between ship and profile buoy measurements is $0.04 \pm 0.12\,°C$.

Near-surface humidity is usually determined by data from dry- and wet-bulb psychrometry. It is clear that distortions created by radiative heating of the ship and screen should be taken into account in this case. But because an increase in air temperature and a decrease in relative humidity partly compensate each other, specific humidity is determined more or less accurately. Obviously, the most serious errors arise from contamination of the wet-bulb wick (say, by salt deposition) and from evaporation of rain or spray from a damp screen. Today it is a matter of common knowledge that ship measurements tend to overestimate air specific humidity and that the mean bias is $0.5 \pm 0.2\,g/kg$ (see Taylor, 1985).

Surface wind velocity

According to WMO data relating to 1984, the anemometric measurements of wind velocity are carried out on only one-third of all ships; on other ships wind velocity is estimated by data from visual observations of sea state. Both methods of wind velocity determination are not without problems. The second method in particular, based on the assumption that there is a one-to-one correspondence between wind force and sea state and also on the assumption that the wind equivalent connecting the Beaufort estimation and wind velocity is known in all cases, has at least two disadvantages: subjectivity and application limited to daytime. Errors of wind velocity determined by this method vary from $\pm 2\,m/s$ when wind force is equal to one point on the Beaufort scale to $\pm 5\,m/s$ when wind force is equal to five points on the Beaufort scale, according to Taylor (1983).

Problems arising from the first method are caused by the distorting influence of the ship's hull. It is usually assumed that the most representative site for anemometer mounting is the top of the foremast or the bow boom. Thorough analysis of the wind velocity field in the vicinity of the ship confirms this conclusion (see Figure 1.3). But even here distortions are still considerable. Moreover, it turns out that errors of anemometric measurements of wind velocity depend significantly on the ratio between the height of anemometer exposure and the height of the superstructures on the top bridge, as well as on the relative wind direction from the ship's course. Data presented in Table 1.4 bear witness to this. As can be seen, for bow winds the accuracy of anemometric measurements on top of a foremast varies from 1 to 8% of wind velocity. For other winds, measured wind velocities can differ from the reference value by 50% and more.

Table 1.4. *Relative error (in %) of anemometric ship measurements of wind velocity*

Author	Location of bench measurements	Wind velocity (m/s)	Relative wind direction		
			from −45 to 45° (bow wind)	from 45 to 115° and from 255 to 315° (beam wind)	from 115 to 255° (stern wind)
Augstein *et al.* (1974)	Meteorological buoy	8	0	–	–
		10	−8	–	–
Ching (1976)	Bow boom	8	from −3 to 5	from −6 to −21	
Kidwall and Seguin (1978)	Bow boom	10	from −2 to −6	from −11 to 45	from −30 to −65
Large and Pond (1982)	Oceanographic buoy	10	0	from −1 to 8	–
				−4	
Romanov *et al.* (1983)	Beyond influence of the domain of ship model	10	from 1 to −4	from −4 to 6	from −6 to 2

Note: Positive values correspond to overestimation; negative values correspond to underestimation of wind velocity.

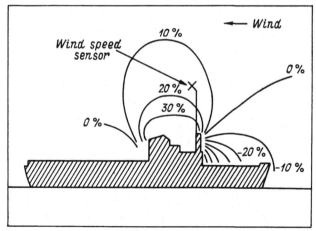

Figure 1.3 Typical relative errors (in %) of anemometer wind velocity measurements taken on a ship, according to Kahma and Leppäranta (1981). The ratio between the height of the anemometer and the height of superstructures on the top bridge is equal to 1.9:1 for a case of bow wind direction.

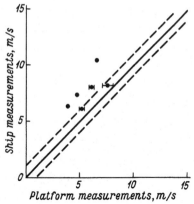

Figure 1.4 Comparison of monthly mean values of wind velocity obtained from standard ship measurement data and measurements on research meteorological platforms, according to Taylor (1983).

The estimates presented above apply to practically instantaneous (averaged over a small time interval) values of wind velocities. It is interesting to determine by how much ship measurements differ from appropriate measurements from buoys and fixed platforms for monthly averaging periods. Results of comparisons are shown in Figure 1.4, from which it is obvious that ship measurements systematically overrate monthly averaged wind velocity. This fact, if confirmed by further research, may imply important and far-reaching climatic consequences.

To determine wind velocity (and temperature, and also humidity) from different heights in the atmosphere radiosonde, airship, airplane and similar measurements are used.

Net radiation flux at the ocean surface

The net radiation flux (radiation balance) at the underlying surface represents the difference between the solar radiation flux assimilated by the underlying surface and the effective emission of the underlying surface, defined as the difference between long-wave upward and downward thermal radiation fluxes.

Note that the Sun radiates energy as a *black body* (body in a state of thermodynamic equilibrium) at a temperature of 6000 K. Its radiation intensity is defined by the Planck formula $I_\lambda = (2hc^2/\lambda^5)[\exp(hc/k\lambda T) - 1]^{-1}$, where $h = 6.62 \times 10^{-20}$ J s is Planck's constant; $k = 1.38 \times 10^{-9}$ J/K is Boltzmann's constant; $c = 3 \times 10^8$ m/s is the velocity of light in a vacuum; λ is the wavelength; T is the absolute temperature. From this formula it follows that $I_\lambda = 0$ at $\lambda = 0$ and $\lambda = \infty$. But since I_λ cannot take on negative values within the wavelength range from $\lambda = 0$ to $\lambda = \infty$, at least one maximum of function I_λ has to exist. Differentiating I_λ over λ and equating the resulting expression to zero we obtain the necessary condition of the extremum, that is, $[(\lambda c/5k\lambda T) + 1 - \exp(hc/k\lambda T)] = 0$.

The above equation has one root at the point $hc/k\lambda t = 4.9651$, from which $\lambda_m T = const = 0.2897$ cm K, where λ_m is the wavelength conforming to the maximum of I_λ. Finding the second derivative and determining its sign we assure ourselves that the function I_λ has to have a maximum at the point indicated above. The equality $\lambda_m T = const$ expressing the Wien law demonstrates that a rise in temperature is accompanied by a shift in the maximum in the direction of shorter wavelengths. As is known, the temperature of the underlying surface and the atmosphere is much less than the temperature of the Sun. Therefore, solar radiation is concentrated in the short-wave (from 0.3 to 5 µm, that is, within the visible) spectral range, and emission of the underlying surface and the atmosphere is concentrated in the long-wave (from 4 to 100 µm) spectral range.

Actinometric measurements are carried out solely on research and ocean weather ships. It is clear that this information is not sufficient to reconstruct the spatial structure of the separate components of radiation balance, and hence there is no other option but to draw on indirect methods. Their practical realization is reduced to searching for empirical relationships associating the components of radiation balance with the parameters determining them (solar radiation at the upper atmospheric boundary, solar altitude, amount of total and low-level cloud, water and air temperature), and consequent reconstruction of information missing from standard ship measurement data. Comparison with independent measurements shows (see Strokina, 1989) that, at least for monthly average periods, the discrepancies between calculated and observed values of radiation balance components in tropical and temperate latitudes of the ocean amount to 2–4% in relative units.

Vertical eddy fluxes of momentum, heat and moisture at the ocean–atmosphere interface

Vertical eddy fluxes of momentum (τ), heat (H), and moisture (E) are defined by equalities $\tau = -\rho\overline{u'w'}$, $H = \rho c_p \overline{w'T'}$, and $E = \rho\overline{w'q'}$, where w' and u' are vertical and horizontal (oriented along the direction of surface wind) velocity fluctuations; T' and q' are fluctuations of temperature and specific humidity; ρ and c_p are the air density and specific heat at constant pressure; the overbar designates time averaging. To estimate eddy fluxes according to these definitions it is necessary, firstly, to measure fluctuations of wind velocity, temperature and humidity; secondly, to obtain instantaneous values of co-variations $u'w'$, $w'T'$, $w'q'$; and, thirdly, to average them within the limits of appropriately selected time intervals.

Measurements of velocity wind fluctuations are made using hot-wire anemometers, acoustic anemometers, anemometers with spherical meters responding to pressure change, cup anemometers, vane anemometers and flag anemometers registering wind load on the axes. Within this group of measuring equipment, acoustic anemometers based on the use of the linear dependence between time of sound propagation along the fixed section of a path guided in a certain direction and wind velocity in the same direction are the most popular. Acoustic anemometers differ favourably from other types of anemometer by the linear dependence of the output signal on wind velocity fluctuations, by the simplicity of detection of the necessary wind velocity component and by the small time lag. These advantages more than compensate for their disadvantages – averaging over the length of the base which is of the order of 10 cm.

Measurements of temperature fluctuations are effected by thermocouples, thermistors and resistance thermometers. All these sensors are characterized by high precision and small time lag. The first attempts to measure humidity fluctuations were undertaken by Dyer in 1961 (see Dyer, 1974). Dry-bulb and wet-bulb thermometers were used for this purpose. Over time, these were replaced by more sensitive optical hygrometers, the general principle of operation being based on the dependence of light absorption in a certain (usually infra-red or ultra-violet) wavelength band on water vapour concentration in the air. Examples of such hygrometers are a hygrometer with an ultra-violet light source (so-called lightman-alpha hygrometer), a microwave refractometer–hygrometer, a gyrometer with a quartz oscillator sensor and an infra-red narrow band hygrometer.

The main requirements to be met by sensors designed to measure fluctuations in wind velocity, temperature and humidity, and their covariances are as follows. First, they should be high-sensitivity (low-inertia) sensors. Otherwise, determination of eddy fluxes will be attended by substantial errors. For example, the use of vane anemometers with a time lag of 0.3 s when measuring vertical velocity fluctuations at a height of four metres from the underlying surface results in an underestimation of eddy fluxes of momentum and heat by 20–30%. The same tendency, but for eddy fluxes of moisture, occurs when using dry-bulb and wet-bulb thermometers (see Kuharets *et al.*, 1980).

Second, the averaging period necessary to obtain statistically stable results should be sufficiently small, on the one hand, to exclude the influence of low-frequency oscillations of synoptic origin, and sufficiently large, on the other hand, to cover the whole spectrum of turbulent fluctuations wherever possible. According to modern concepts the turbulent fluctuations of wind velocity and humidity in the marine surface atmospheric layer have spatial scales ranging from fractions of a millimetre to tens, or even hundreds, of metres, and in the low-frequency range (range of large spatial scales) their spectra essentially differ from each other: temperature and humidity spectra, and especially horizontal velocity spectra, extend to lower frequencies than do vertical velocity spectra. This poses problems when selecting the spectral minimum range separating turbulent fluctuations from disturbances of the synoptic scale.

For vertical eddy fluxes of momentum, heat and moisture the case is simpler because the function $\omega S_{ww}(\omega)$ (where $S_{ww}(\omega)$ is the spectral density of the vertical velocity fluctuations; ω is the frequency) drops to practically zero in the range of spatial scales of the order of several hundred metres in the surface boundary layer and several kilometres in the planetary boundary layer. Accordingly, cospectra $\omega S_{ww}(\omega)$, $\omega S_{wT}(\omega)$, $\omega S_{wq}(\omega)$, in the range of approximately the same spatial scales should also drop down to zero, so that covariances $\overline{u'w'} = \int S_{uw}(\omega)\,\mathrm{d}\omega$, $\overline{w'T'} = \int S_{wT}(\omega)\,\mathrm{d}\omega$, $\overline{w'q'} = \int S_{wq}(\omega)\,\mathrm{d}\omega$ make clear physical sense – these are integrals over the spectrum from scales of about 10 cm (base length of an instrument) to a low-frequency minimum of the function $\omega S_{ww}(\omega)$. Thus, in order to provide reliable estimates of vertical eddy fluxes the frequency characteristics of measuring equipment have to be cut off (reduced to zero) in the range of the minimum of $\omega S_{ww}(\omega)$. This condition is fulfilled if measurements cover the range of $10^{-3} \leq \omega z/u \leq 5$ or even $10^{-3} \leq \omega z/u \leq 10$, where z is the height at which the measurement is taken; ω is frequency in Hz. It follows from this that at $z = 5$ m and $u = 5$ m/s the lowest frequency must be equal to 10^{-3} Hz, and the period of averaging must be not less than 15 minutes.

Finally, the reliability of the information is determined by the degree of mobility of the base on which the equipment is mounted. If the base is not stabilized, its movements will result in distortions of vertical fluctuations of the wind velocity and eddy fluxes on wave frequencies. These distortions can be eliminated by registering equipment movements and amending the data.

We consider the simple procedure to account for the oscillating motion effect proposed by Volkov and Koprov (1974).

Let ξ represent the vertical displacement of a device with respect to its average location and let $w_0' = w' + u \sin\theta + \partial\xi/\partial t$, $u_0' = u' + \xi\, \partial u/\partial z$, $T_0' = T' + \xi\, \partial T/\partial z$, $e_0' = \xi\, \partial e/\partial z$ be fluctuations of wind velocity, temperature and absolute humidity distorted by pitching effects. Here u, w, T and e are average values of wind velocity, temperature and humidity; u', w', T' and e' are turbulent fluctuations of wind velocity, temperature and absolute humidity in the absence of equipment base motion (at $\xi = 0$); the second term on the right-hand side of the first equality describes the effect of slope of the underlying surface; θ is the angle of inclination (angle of pitching).

Let us compile products $u_0' w_0'$, $w_0' T_0'$, $w_0' e_0'$ and then average them over time. Then taking into account that ξ and θ do not correlate with u', w', T' and e', and also that $\overline{\xi\, \partial\xi/\partial t} = 0$ and $\sin\theta \approx \theta$ we have

$$\overline{u'w'} = \overline{u_0'w_0'} - \overline{u\theta\xi}\, \partial u/\partial z,$$

$$\overline{w'T'} = \overline{w_0'T_0'} - \overline{u\theta\xi}\, \partial T/\partial z,$$

$$\overline{w'e'} = \overline{w_0'e_0'} - \overline{u\theta\xi}\, \partial e/\partial z.$$

To estimate the second terms on the right-hand sides of these expressions we assume that $\xi = 1$ m, $\theta = 1°$, $u = 10$ m/s, $\partial T/\partial z = 0.01$ °C/m, $\partial u/\partial z = 0.1$ s^{-1}, $\partial e/\partial z = 0.005$ g/m^4. Substitution of these values yields $\overline{u\theta\xi}\, \partial u/\partial z \approx 150$ cm^2/s^2, $\overline{u\theta\xi}\, \partial T/\partial z \approx 0.20$ °C cm/s, $\overline{u\theta\xi}\, \partial e/\partial z \approx 0.75 \times 10^{-6}$ g/cm^2 s. Typical values of $\overline{u'w'}$, $\overline{w'T'}$ and $\overline{w'e'}$ are equal to 400 cm^2/s^2, 1 °C cm/s and 3×10^{-6} g/cm^2 s, respectively, so that the eddy flux error caused by pitching can be up to 30%. Thus, one more necessary condition for reliability of information on turbulence characteristics in the surface boundary layer is the precise allowance for displacements or stabilization of the height from which measurements are taken.

These requirements are quite restrictive. Perhaps this (plus the sophisticated nature and high cost of the measuring equipment) explains the fact that, nowadays, lack of data on eddy fluxes of momentum, heat and moisture is felt as much as it was 30 years ago when experimental research of ocean–atmosphere interaction was only just beginning.

1.4.2 Satellite measurements

In discussing satellite measurements of surface atmospheric layer characteristics, we make the reservation that equipment mounted on satellites registers
not atmospheric layer characteristics themselves but, rather, different characteristics (such as intensity of thermal radiation or intensity of electromagnetic
wave scattering) that are indirectly related to those in which we are interested.
Moreover, the characteristics registered are functions of several parameters
of the Earth's surface and the atmosphere. Naturally, this complicates the
interpretation of satellite data, resulting in the need to take measurements in
several spectral intervals in order to separate the effects of different parameters,
and, mainly, to enable fall-back on calibration of satellite data by ground-
based measurement data. It is clear that the precision of satellite measurements
is limited by that of ground-based data and, hence, the advantage of satellite
information is not its quality but, rather, its quantity, providing the possibility
of examining ocean–atmosphere interaction on global scales.

 We next consider the principal features of satellite measurements of the
surface layer characteristics and their capacities.

Sea surface temperature

As is well known, the spectrum of terrestrial emission represents the irregular
alternation of bands of transmission (so-called *transparency spectral windows*)
and absorption of thermal radiation. In the former, emission of the ocean
surface, in its passage through the atmosphere, is attenuated to a minimal
degree, and in the latter to a maximal degree. Thus, to indicate the temperature of the ocean surface it seems to be sufficient to measure the intensity of
emission in one or another spectral window and then determine the temperature appropriate to the emission intensity as registered by a radiometer.

 Let us examine, for example, the emission of a black body. According to
Planck's formula (see p. 26), the intensity of emission of such a body is
$I_\lambda = (2hc^2/\lambda^5)[\exp(hc/k\lambda T) - 1]^{-1}$, from which it follows that for small
wavelengths when $\exp(-hc/k\lambda T) \gg 1$ Planck's formula reduces to the Wienn
formula $I_\lambda = (2hc^2/\lambda^5)\exp(-hc/k\lambda T)$ describing an abrupt drop of I_λ in
the violet spectral range; on the other hand, for larger wavelengths, when
$\exp(hc/k\lambda T) \approx 1 + (hc/k\lambda T)$, Planck's formula transforms into the Rayleigh–
Jeans formula $I_\lambda = (2c/\lambda^4)kT$ describing a much slower drop of I_λ in infra-red
and radiowave spectral ranges. Thus, all other things being equal, use of a
transparency window from the near-infra-red range (wavelengths ranging
from 3.4 to 4.2 μm) would be preferable to those from the middle-infra-red
range (wavelengths ranging from 8.1 to 14.0 μm), and especially from the

microwave range[1] (wavelengths ranging from 0.1 to 21.0 cm), if there were no effect from patches of sunlight, which results in an increase in the area of mirror reflection of solar radiation and, hence, the comparability of fluxes of emission and reflected solar radiation. Because of this the selection of a transparency window from the near-IR range means that the ability to receive noiseless information is restricted to night time, while for a transparency window from the medium-IR range there are no such restrictions.

The main disadvantages of radiometric measurements in the IR range (wavelengths ranging from 1 to 15 µm) is the fact that ocean emission is significantly attenuated by the influence of water vapour, aerosols and cloud when passing through the atmosphere. Their influence can be excluded if, instead of thermal emission measurements in the IR range, one takes measurements of radio thermal emission in the microwave range. Here, the absorption of microwave emission occurs only in clouds containing large raindrops. But things are not quite so simple in this case either because of the relatively low sensitivity of the emission intensity to variations in sea surface temperature and the strong dependence of emissivity on the local properties of waves and on the presence of froth, ice and surface-active substances.

Finally, one more source of uncertainty in radiometric measurements is the so-called *skin-effect* (temperature inversion in a thin near-surface ocean layer with a thickness of several millimetres). Note that data on radiometric measurements in the infra-red and microwave ranges describe *radiative* and *radio-brightness temperature*, respectively, defined as the temperature of the black body whose emission is equal to the actual emission. Because the latter refers to the layer with thickness of the order of a millimetre, inversion of temperature in this layer, created mainly by evaporation, should result in underestimation of the sea surface temperature compared with its value as determined by conventional methods in a layer with thickness from several tens of centimetres up to several metres. If we include the fact that the temperature difference in the near-surface inversion layer is not constant (see Table 1.5), then the possibility of establishing a universal dependence between radiative or radio-brightness temperature and sea surface temperature as measured by conventional methods is out of the question since this introduces additional errors.

Nowadays, the root-mean-square error of satellite measurements of the sea surface temperature is no less than 0.6 °C in the IR range, and 1.0 °C in the microwave range.

---

[1] This range is defined according to the classification of radiowaves. Waves belonging to this range have much longer wavelengths than those from the range of long-wave emission.

Table 1.5. *Temperature difference in the near-surface inversion layer*

Source data	Author	Temperature difference (°C)	Comments
Laboratory measurements	Paulson and Parker (1972)	1.14–1.81	Small fetch, no waves
	Hill (1972)	1.0	Dynamic wind velocity less than 35 cm/s
		0.1	Dynamic wind velocity more than 35 cm/s, waves appear and develop
	Katsaros *et al.* (1977)	0.4–2.4	No wind
Natural measurements	Woodcock and Stommel (1947)	0.5–1.0	Nighttime, calm
	Ewing and McAlister (1960)	0.6	Nighttime, wind velocity less than 0.5 m/s
	Malevskii-Malevich (1970)	0.3–0.5	Wind velocity less than 10 m/s, most probable estimations
	Klauss *et al.* (1970)	0.35	Nighttime
		0.5	Daytime
	Bortkovskii *et al.* (1974)	0.22–0.50	Open ocean; wind velocity less than 4 m/s
		0.20–0.30	Wind velocity varies from 6 to 8 m/s
	Grassel (1976)	0.17–0.21	Wind velocity varies from 1 to 10 m/s
	Katsaros *et al.* (1977)	up to 3.4	Wind velocity less than 1 m/s, no waves
	Schooley (1967)	0.2	Wind velocity about 2.5 m/s, cloudless sky
		0.0	Same, but with clouds
	Hunjua and Andreyev (1974)	0.1–0.5	Wind velocity up to 6–8 m/s, waves up to 3 points
	Paulson and Simpson (1981)	0.15–0.30	Wind velocity from 5.5 to 9.2 m/s
	Panin (1985)	∼2.0	Wind velocity less than 3 m/s

Air temperature and humidity

The determination of temperature at any height in the atmosphere (including the surface boundary layer) reduces to registration of the Earth's radiation in different parts of the CO_2 absorption band (wavelengths 4.3 and 15 μm), that is, in the spectral interval where the intensity of emission is controlled mainly by the vertical distribution of temperature and concentration of CO_2, but not by humidity and aerosols. A primary prerequisite is indicated by the well-known physical fact that the emission at various parts of the CO_2 absorption band is formed in different atmospheric layers, and therefore for prescribed and invariable vertical distribution of CO_2 concentration the correspondence between the emission and the temperature of these layers must be single-valued.

In a cloudless sky the dependence of emission intensity I_λ conforming to wavelength λ on temperature T of the atmosphere should have the form (see Malkevich, 1973)

$$I_\lambda(\xi, \theta) = \delta_\lambda B_\lambda(T_s) P_\lambda(1, \theta) - \int_\xi^1 B_\lambda[T(\xi)] \frac{\partial P_\lambda(\xi, \theta)}{\partial \xi} d\xi - 2(1 - \delta_\lambda)$$

$$\times P_\lambda(1, \theta) \int_0^1 B_\lambda[T(\xi)] \frac{\partial \tilde{P}_\lambda(\xi)}{\partial \xi} d\xi, \tag{1.4.1}$$

where θ is the zenith angle; δ_λ is the emissivity of the underlying surface; $B_\lambda(T) = (2hc/\lambda^3)[\exp(hc/kT) - 1]^{-1}$ is Planck's function; $P_\lambda(\xi, \theta)$ is a function of radiation transmission (ratio of radiation passing through the layer with thickness ξ to incoming radiation at the boundary of this layer) in a direction forming an angle θ with the vertical; $\tilde{P}_\lambda(\xi)$ is the transmission function averaged over all θ from the upper half-space of directions; $\xi = p/p_s$ is a vertical variable in the isobaric coordinate system; p is the atmospheric pressure; $c = 3 \times 10^8$ m/s is the velocity of light; $h = 6.62 \times 10^{-20}$ J s is Planck's constant; $k = 1.38 \times 10^{-9}$ J/K is Boltzmann's constant; subscript s indicates belonging to the underlying surface. The first term on the right-hand side of (1.4.1) describes the emission of the underlying surface, the second term describes the emission of the atmospheric layer contained between the underlying surface ($\xi = 1$) and level ξ; and the third term describes the thermal emission reflected from the underlying surface.

If we assume that the ocean to a first approximation is a perfect black body ($\delta_\lambda = 1$), and that measurements of I_λ are carried out from a satellite in the nadir point ($\theta = 0$, $\delta = 0$), then on the basis of (1.4.1) we have

$$I_\lambda = B_\lambda(T_s) P_\lambda(1, 0) - \int_0^1 B_\lambda[T(\xi)] \frac{\partial P_\lambda(\xi, 0)}{\partial \xi} d\xi. \tag{1.4.2}$$

This expression representing the Fredholm equation of first order with respect to $T(\xi)$ allows determination of the vertical profile of temperature if the values of radiation intensity I_λ and transmission function P_λ are given. The same is possible in principle if the angular distribution of $I_\lambda(\theta)$ is known. But in reality this will entail large errors due to horizontal non-uniformity of the temperature field and the presence of cloudiness.

Thus, we realize the necessity of solving the inverse problem of radiation theory – to determine the properties of the environment from data on its emission. This problem is ill-posed in the sense that small errors in signal measurements can entail large errors in estimating environmental parameters.

Let us illustrate this by a simple example. Let us assume that a solution $\bar{x}$ of the equation $f(\bar{x}) = a$ has to be found. It is obvious that $\bar{x} = f^{-1}(a)$, where f^{-1} is the inverse function of f. Using the theorem on derivatives of inverse functions we have $d\bar{x}/\bar{x} = \varphi(a)\,da/a$. This expression sets the relationship between errors at the input (da/a) and at the output $(d\bar{x}/\bar{x})$ of the system. Here $\varphi(a) = a/[f^{-1}(a)f'(a)]$ is an error amplification coefficient. Based on the preceding expression for $d\bar{x}/\bar{x}$ we conclude that output errors will be large at large $\varphi(a)$ or, otherwise, at small $f'(a)$, that is, at those values of a which are close to the roots of the derivative $f'(\bar{x})$.

To achieve acceptable accuracy it is necessary to either minimize errors at the input da/a, or, if this is impossible, to regularize the problem. The latter means that the function $f(\bar{x})$ is replaced by a closer function $f_1(\bar{x})$ such that, in the range of interest, the new error amplification coefficient $\varphi_1(a)$ is not beyond the limits of a certain prescribed value.

Thus, to obtain the solution of an ill-posed problem it is necessary to have *a priori* information about the solution which allows exclusion of large errors in the experimental data. These problems are doubled by the fact that, as applied to the atmosphere, the required and measured parameters are random functions, and so in searching for a solution that is appropriate to an integral equation of type (1.4.2), one has to apply *statistical regularization* with the statistical characteristics of the radiative field of the Earth and fields of atmospheric parameters serving as *a priori* information. Experience in its use indicates that the standard deviation between vertical temperature distributions in the lower 24-kilometre layer of the atmosphere reconstructed from data of satellite and radiosonde measurements is usually about 1.5 °C.

In contrast to the vertical profile of temperature, the vertical profile of humidity is determined by measurements of emission I_λ at different parts of the water vapour absorption band (wavelengths 6.3 and 20–25 μm). In this case, all other things being equal, Equation (1.4.2) is rewritten as

$$I_\lambda = B_\lambda(T_s)P_\lambda(W_A, 0) - \int_0^1 B_\lambda[T(\xi)] \frac{\partial P_\lambda[w_A(\xi), 0]}{\partial \xi}\,d\xi, \qquad (1.4.3)$$

where $w_A(\xi) = (p_s/g)\int_0^{\xi - \kappa si} q(\xi)\,d\xi$ is the moisture content of the atmospheric column between the upper atmospheric boundary $(\xi = 0)$ and level ξ, W_A is the integral moisture content of the atmosphere; q is the air specific humidity.

Equation (1.4.3) can be considered as an integral equation with respect to $w_A(\xi)$, where Planck's function $B_\lambda[T(\xi)]$ is the core of Equation (1.4.2). Because of this the determination of the vertical distribution of specific humidity $q(\xi)$ reduces to the simultaneous recovery of profiles $w_A(\xi)$ and $T(\xi)$ according to measurements of emission in water vapour and CO_2 absorption bands, and to the subsequent transfer from $w_A(\xi)$ to $q(\xi)$. Such a problem, as well as the problem of determination of the vertical temperature distribution, is ill-posed in the sense mentioned above. The difference is that instead of a solution for one integral equation, one has to solve a system of integral equations, which requires the additional use of *a priori* information about the structure of the humidity field in the atmosphere (say, an approximation of the vertical profile of humidity by some analytical expression). Next, because $q(\xi)$ and $w_A(\xi)$ are related to each other by $q(\xi) = (p_s/g)^{-1} \partial w_A(\xi)/\partial \xi$, even small errors in the determination of $w_A(\xi)$ can result in serious distortions of the profile $q(\xi)$. This last-mentioned fact places heavy demands when choosing the transmission function whose accuracy of definition leaves much to be desired because of the influence of aerosol attenuation on the emission transfer in the IR-range.

The obstacle can be overcome if we do not recover the profile $q(\xi)$, but, instead, use the empirical dependence between the integral humidity content of the atmosphere and the specific humidity in the surface layer obtained, for example, from radiosonde measurements. The fact that such dependence really exists can be judged by the following arguments. It is known that the vertical distribution of specific humidity in the atmosphere is sufficiently well described by the relationship $q/q_s = \xi^\kappa$, where κ is an empirical constant. Integrating this relationship over ξ within the whole thickness of the atmosphere and using the definition of the integral water vapour content W_A, we have $q_s = (\kappa + 1)(p_s/g)^{-1} W_A$. The linear relationship between q_s and W_A is confirmed by data from radiosonde measurements at island stations and on ocean weather ships. According to Liu (1983) the standard deviation is only 0.78 g/kg.

It should be emphasized that in the presence of clouds satellite measurements of emission in the IR-range determine profiles $T(\xi)$ and $w_A(\xi)$ only in the atmosphere above the clouds, and that in such a situation there is only one recourse: to start measuring I_λ in the microwave range. The procedure for reconstruction of q_s remains the same: first the vertical profile of the temperature and integral water vapour content of the atmosphere are calculated from I_λ measurement data in CO_2 and water vapour absorption bands; next, using the relationship $q_s = (\kappa + 1)(p_s/g)^{-1} W_A$, one can find q_s. Comparison of q_s values obtained from three-monthly series of SEASAT microwave measurements of I_λ and from standard ship measurements in the Western North Atlantic, which provide better ship measurement data than do other parts of the ocean, show (see Liu, 1983) that the systematic error and root-mean-square error of satellite measurements of specific humidity in the surface atmospheric layer amount to 0.48 and 0.97 g/kg respectively.

Wind velocity

Depending on the type of usable satellite information one can single out the following five methods of determination of wind velocity:

1. by cloud motion;
2. by the brightness and dimensions of patches of sunlight;
3. by the reflected radiation in the microwave band;
4. by the echoes from short-pulse radiolocation signals;
5. by inverse scattering of electromagnetic waves in the centimetric band.

The first method basically reduces to the acquisition of tele- and photo images of fields of cloud, determination of velocity of cloud displacement and wind velocity at a fixed level in the surface layer. This procedure is used at present. Cloud fields serve as the initial information, and these are registered twice per day by a system of geostationary satellites (two American satellites of the GOES series, one European and one Japanese) located along the equator in such a way that the fields of view of cameras mounted on them cover 80% of the ocean area in the tropics.

The disadvantages of this method include, firstly, the fixation of the level of cloud displacement (cumulus clouds in the tropics are assumed to be transported by the wind at the base of the clouds) and, secondly, the necessity of prescribing the vertical shear of wind velocity in the subcloud layer. This last fact suggests a correlative relationship between wind velocities at the base of the cloud and in the surface layer. Considering that such a relationship is not invariable in space and time, then perhaps because of it, or in addition to it, the discrepancies between satellite estimates and data from ground-based observations turn out to be quite large and, according to Wylie and Hinton (1983), amount to ± 2.1 m/s for velocity and $\pm 25°$ for wind direction.

The second method is based on the following simple considerations. If the ocean surface is smooth, then sunbeams are reflected from it at the identical angle so that a patch of sunlight remains small and sharply outlined (contrasting). In contrast, in the presence of waves, sunbeams are reflected from the ocean surface at different angles. This produces scattering of reflected light and, therefore, an increase in patches of sunlight and a decrease in their visibility. Further, because waves are the result of wind forcing, the existence of a relationship between the size and visibility of a patch of sunlight, on the one hand, and wind velocity, on the other, as well as between the shape of the patch of sunlight and wind direction, should not give rise to doubts. Thus, if such a relationship is set, and satellite images of the ocean surface are available, wind velocity and direction can be considered as known. For obvious reasons this method can be used only during the daytime and in the presence of sparse cloud. Its accuracy of estimation of wind velocity is ± 2.0 m/s; the accuracy of determination of wind direction is so low that this is not worth indicating.

We have already discussed the fact that the emission intensity in the microwave spectral range is controlled by the sea surface temperature, the vertical distribution of temperature, humidity and aerosols in the atmosphere, and the emissivity of the ocean. In its turn, the emissivity of the ocean depends on the state of the ocean surface and, hence, on wind velocity. This last fact, having been an obstacle in satellite indication of temperature acquires a new quality, being a prerequisite of the third (radiometric) method of wind velocity determination.

Its use presupposes that the extraneous effects created in this case by the vertical distribution of atmospheric parameters are excluded, and that the dependence between the ocean emission in the wavelength range of ~ 3 cm (here the emission is most sensitive to the state of the ocean surface) and wind velocity is known. The first condition is fulfilled by the simultaneous measurement of the emission intensity in various parts of the microwave band, and for this purpose a multichannel microwave radiometer is used; the second condition is fulfilled by taking account of a specific empirical relationship. Accuracy in determination of wind velocity by this method is ± 1.9 m/s; wind direction cannot be estimated.

The fourth and fifth methods of determination of wind velocity are based on the use of active location of the ocean surface, and on registration of backward scattering of electromagnetic oscillations, presenting a combination of mirror reflection and so-called *Bragg scattering*. The meaning of the mirror reflection is obvious. As for the Bragg scattering, this needs to be explained.

With the above in mind we consider the propagation of a plane wave in a periodical structure, that is, in a structure where properties change in space according to a periodic law.

In the one-dimensional case the equation describing the propagation of a plane wave has the form

$$\frac{\partial^2 u}{\partial x^2} - \frac{1}{c^2}\frac{\partial^2 u}{\partial t^2} = 0, \tag{1.4.4}$$

where u is the surface elevation or current velocity; c is the phase velocity of the wave featuring properties of the environment; x is the spatial coordinate; t is time.

Assume that the environment has a periodic inhomogeneity along the direction of wave propagation, that is, $c^2 = c_0^2(1 - \mu \cos Kx)$, where c_0 is the phase wave velocity in the absence of inhomogeneity; K is the wave number of inhomogeneity defined by the distance between its neighbouring elements; $\mu \ll 1$ is a small parameter (inhomogeneity is considered to be small).

Let us find the solution to Equation (1.4.4) in the form of $u = A(x) \exp(i\omega t)$, where ω is the wave frequency. Then amplitude A must obey the equation

$$\frac{d^2 A}{\partial x^2} + k^2(1 + \mu \cos Kx)A = 0, \tag{1.4.5}$$

where $k = \omega/c_0$ is the wave number of the plane wave in question.

We represent A as an expansion in powers of the small parameter μ and restrict ourselves to the first two terms of the expansion. We treat ω similarly, thereby assuming that wave propagation in an inhomogeneous environment can be accompanied by dispersion. In other words we present A and ω in the form

$$A = A_0 + \mu A_1, \qquad \omega = \omega_0 + \mu \omega_1, \tag{1.4.6}$$

from which, and from the definition of k^2, the equality follows:

$$k^2 = k_0^2 + 2\mu k_0 k_1. \tag{1.4.7}$$

Here we have ignored terms containing the μ parameter to powers higher than one.

Substituting (1.4.6) and (1.4.7) into (1.4.5) and gathering members containing parameter μ to powers of zero and one we find

$$\frac{d^2 A_0}{dx^2} + k_0^2 A_0 = 0, \tag{1.4.8}$$

$$\frac{d^2 A_1}{dx^2} + k_0^2 A_1 = -k_0 A_0(2k_1 + k_0 \cos Kx). \tag{1.4.9}$$

The solution of Equation (1.4.8) has the form

$$A_0 = C_1 \exp(ik_0 x) + C_2 \exp(-ik_0 x). \tag{1.4.10}$$

Its substitution in (1.4.9) yields

$$\frac{d^2 A_1}{dx^2} + k_0^2 A_1 = -k_0 C_1 \{2k_1 \exp(ik_0 x) + \tfrac{1}{2}k_0 \exp[i(K + k_0)x]$$

$$+ \tfrac{1}{2}k_0 \exp[-i(K - k_0)x]\} - k_0 C_2 \{2k_1 \exp(-ik_0 x)$$

$$+ \tfrac{1}{2}k_0 \exp[-i(K + k_0)x] + \tfrac{1}{2}k_0 \exp[i(K - k_0)x]\}. \tag{1.4.11}$$

Let us define two possibilities: non-resonance ($K \neq 2k_0$) and resonance ($K = 2k_0$) wave scattering by environmental inhomogeneities. In the first case the solution of Equation (1.4.11) is written in the form

$$A_1 = c_1 \exp(ik_0 x) + c_2 \exp(-ik_0 x) + ik_1 x[C_1 \exp(ik_0 x) - C_2 \exp(-ik_0 x)]$$

$$+ \frac{k_0^2}{2K(K + 2k_0)} \{C_1 \exp[i(K + k_0)x] + C_2 \exp[-i(K + k_0)x]\}$$

$$+ \frac{k_0^2}{2K(K - 2k_0)} \{C_1 \exp[-i(K - k_0)x] + C_2 \exp[i(K - k_0)x]\}. \tag{1.4.12}$$

As can be seen, there is a term on the right-hand side of (1.4.12) which increases linearly with x. Since we are interested only in waves with restricted amplitudes this term has to be set to zero; this is equivalent to $k_1 = 0$. Thus at $K \neq 2k_0$ wave propagation in an inhomogeneous environment is not accompanied by a change in frequency ($\omega = c_0 k_0$).

But if $K = 2k_0$ then Equation (1.4.11) will take the form

$$\frac{d^2 A_1}{dx^2} + k_0^2 A_1 = -k_0 \left(2k_1 C_1 + \frac{k_0}{2} C_2 \right) \exp(ik_0 x) - k_0 \left(2k_1 C_2 + \frac{k_0}{2} C_1 \right)$$

$$\times \exp(-ik_0 x) - \frac{k_0^2}{2} [C_1 \exp(3ik_0 x) + C_2 \exp(-3ik_0 x)]. \tag{1.4.13}$$

Its integration also results in the appearance of terms which increase linearly with x. For the solution to remain restricted it is necessary that the factors $(2k_1 C_1 + (k_0/2)C_2)$ and $(2k_1 C_2 + (k_0/2)C_1)$ before $\exp(\pm ik_0 x)$ become zero. In other words, it is necessary to demand that

$$2k_1 C_1 + \frac{k_0}{2} C_2 = 0, \qquad 2k_1 C_2 + \frac{k_0}{2} C_1 = 0.$$

These equalities form a system of algebraic equations with reference to C_1 and C_2. It has a non-trivial solution if

$$\begin{vmatrix} 2k_1 & k_0/2 \\ k_0/2 & 2k_1 \end{vmatrix} = 4k_1^2 - \frac{k_0^2}{4} = 0,$$

from which $k_1 = \pm k_0/4$.

Thus at $K = 2k_0$ the frequency of wave propagation in an inhomogeneous environment is determined by the expression

$$\omega = c_0 k_0 \pm \frac{\mu}{4} c_0 k_0.$$

This suggests that the dependence $\omega = \omega(k_0)$ at $k_0 = K/2$ breaks and hence there is a forbidden frequency band of width $\Delta \omega = (\mu/2)c_0 k_0$ such that waves belonging to it are quickly attenuated. This means that at $K = 2k_0$ the incident wave is effectively reflected from inhomogeneities of the environment and its energy is transmitted to the wave propagating backward. The ratio $K = 2k_0$ is called the *condition of Bragg scattering.* In the general case of an arbitrary incidence angle θ calculated from the nadir, this condition is rewritten in the form $K = 2k_0 \sin \theta$.

From the above discussion we draw the following conclusion: if the wavelength of wind waves is much greater than that of the signal, or if the angle of incidence of a radiated signal is close to zero the mirror reflection will dominate over the Bragg scattering; on the other hand, if wavelengths are comparable, or for oblique irradiation (at large angles of incidence), the Bragg scattering will dominate over the mirror reflection. In the fourth method of wind velocity determination (in this case registration of backward scattering is by a radiolocating altimeter) the first circumstance is realized; in the fifth method when a scatterometer is used the second circumstance is realized.

The intensity of backward scattering is characterized by an *effective scattering cross-section* σ, defined as the ratio of backward scattering intensity to density of irradiation flux. The effective scattering cross-section has the dimensions of area. Dividing σ by the geometrical area of the irradiated spot and designating the normalized effective backward scattering cross-section as σ^0, we have that σ^0 has to be a function of the signal incidence, of the azimuthal angle κ (the angle between the beam and wind direction), and of wind velocity u_a in the surface atmospheric layer inducing the appearance of sea surface inhomogeneities (wind waves). When using altimeter measurements of σ^0 from a subsatellite point the dependence $\sigma^0 = \sigma^0(\theta, \kappa, u_a)$ reduces to the form $\sigma = \sigma^0(u_a)$. This facilitates the determination of u_a but excludes the possibility of estimating wind direction. Scatterometric measurements of σ^0 are free from this disadvantage. The accuracy of determination of wind velocity and direction amounts to ± 1.3 m/s and $\pm 16°$ in this case; the accuracy of altimeter measurements is ± 1.6 m/s (see Wylie and Hinton, 1983).

Net radiation flux at the upper atmospheric boundary

This flux is not measured from a satellite. Only its separate components are registered: incident and reflected short-wave solar radiation, and long-wave emission of the atmosphere-underlying surface system into outer space (so-called outgoing emission). Recording of the first components is performed using a pyrheliometer and a spectrobolometer having the same sensitivity for different wavelengths from the short-wave spectrum range, and recording of the second and third components is performed using flat wide-angle and scanning narrow-angle radiometers (the scanning is performed along or across a satellite trajectory).

Flat wide-angle and scanning narrow-angle radiometers provide different information. Radiometers of the first kind measure upward radiation coming from the lower half-space. Therefore, their spatial resolution seems to depend only on the height of the satellite. Actually, not all parts of the visible disc of the Earth contribute equally to the formation of the field of upward radiation; the main contribution is from the vicinity of a subsatellite point. Therefore, the spatial resolution of a flat wide-angle radiometer at the altitude of a satellite orbit, about 600 km, will amount to $\sim 10°$ or ~ 1000 km. Such spatial resolution allows us to gain information on the global distribution of the characteristics based on a minimum amount of data, but at the same time it makes for strong smoothing of the data.

On the other hand, a scanning narrow-angle radiometer measures the upward radiation coming from the lower half-space with a small but finite solid angle. Accordingly, its spatial resolution turns out to be thinner and varies in the range from 50 km in the nadir to 110 km at an angle of 40° from the nadir. But it should be remembered that the radiation recorded by such a radiometer does not meet all angular distributions but, rather, only some of them and, hence, it depends on zenithal and azimuthal angles of the satellite (in the case of short-wave radiation), and on the zenithal angle on the Sun. Because of this the use of scanning narrow-angle radiometers should provide for the assignment of some procedure correcting measurement data in order to take anisotropy (dependence on direction of propagation) of radiation into account. The last circumstance will inevitably entail the introduction of errors.

The current accuracy of satellite estimates of monthly averaged net radiation flux at the upper atmospheric boundary, according to Stephens *et al.* (1981), is of the order of 10 W/m^2, that is, it is approximately equivalent to 3% accuracy in determining the solar constant.

Net radiation flux at the underlying surface

Of all the components of net radiation flux, only the long-wave emission of the underlying surface can be measured directly. Other components are estimated using indirect methods. Let us discuss these methods, paying particular attention to their physical prerequisites. We start with the short-wave solar radiation flux assimilated by the underlying surface.

Its determination using data from satellite measurements is based on a number of assumptions and *a priori* information on the transmission functions of clouds and the atmosphere. One can distinguish two groups of methods. The distinctive feature of the first of them is the presetting of an empirical relation between data from satellite and ground-based measurements of the reflected solar radiation in the visible and near-IR-ranges. The distinctive feature of the second method is the interpretation of satellite measurement data in terms of the characteristics of reflection, absorption and scattering, appearing in the transfer equations for short-wave radiation, followed by the use of the values of these characteristics when integrating the transfer equations.

Both groups of methods have their own disadvantages. The disadvantages of the first group are the strong dependence on accuracy of calibration of the satellite data from ground-based measurements. Therefore, if the calibration relates to short time intervals then its application to other longer periods, strictly speaking, is not well founded. The disadvantages of the second group of methods relate to the restriction in application of transfer equations for the case of monochromatic radiation and, hence, the necessity of integrating over wavelengths and allowing for horizontal inhomogeneity of the characteristics of reflection, absorption and scattering of solar radiation. The relative error of the available methods of determination of short-wave solar radiation flux from satellite measurement data amounts to 10–15% for daily averaged values and 5–10% for monthly averaged values (see *The Report of the Joint Scientific Committee*, 1987). By the way, such large errors result not only from the reasons mentioned above but also from inadequate comparison: a satellite radiometer and ground pyranometer have different viewing areas.

The downward long-wave radiation flux consists of emission created by water vapour, other gases (including CO_2) active in terms of radiation, and also aerosols and cloud. To consider the dependence between the downward flux of long-wave emission and its determining parameters it is necessary to have at least some information about them; information which more often than not is unavailable. In particular, the altitude of the cloud base, under which the downward emission is formed, is not identified by satellite data. The only thing which can be reconstructed by their use is the height of the cloud top and the cloud type. But a problem arises in determining the height of the base, and changing from geometric to optical cloud thickness, or in general in refusing to use the height of the cloud base as the dependent variable. It is clear that in both cases one cannot do without additional assumptions and, hence, this means further deterioration in the quality of satellite information. According to *The Report of the Joint Scientific Committee* (1987), the root-mean-square error in determination of the downward flux of long-wave radiation from satellite measurement data amounts to at least 10–15 W/m^2; in addition, very large deviations from ground-based measurement data will occur when the height of the cloud base undergoes strong variability.

Eddy fluxes of momentum, heat and moisture at the ocean–atmosphere interface

These fluxes, like some components of net radiation flux, cannot be measured directly with the help of satellites. So there is only one recourse: to apply aerodynamic formulae connecting eddy fluxes with marine surface atmospheric layer parameters (temperature, humidity and wind velocity) determined by satellite measurement data. But one should not be under any illusion as to the simplicity and perfect nature of this tool, as explained below.

Meteorological and geophysical satellites functioning at present, and those planned for the future, have either solar-synchronous, precessing or geostationary orbits. Each of them is compiled with its own frequency of information-gathering. Solar-synchronous satellites receive information twice a day at the equator, always at the same local solar time and more often at high latitudes. This excludes the possibility of filtering out daily variations in the examined characteristics, especially where the measurements are obtained by one satellite. Information from precessing satellites is obtained at different times of the day, but it may not cover polar latitudes. Finally, geostationary satellites provide continuous information over time but only within the limits of the restricted part of the equatorial zone whose size depends on the scanning area of the measuring equipment.

Let us recall now that the density of measurements controls the possibility of reconstructing the spatial structure, and their time sampling controls the accuracy of monthly averaged local values of eddy fluxes. To provide support for these statements we refer to the results of the calculation of uncertainty in estimates of monthly averaged sensible and latent heat fluxes received from JASIN data in the North Sea and illustrated in Figure 1.5. In this diagram deviations of monthly averaged fluxes from their 'true' values corresponding to hourly sampling are plotted along the ordinate axis, and the sample interval of measurements in hours is plotted along the abscissa axis. As can be seen for 12-hour sampling an error in estimating latent heat amounts to $\pm 2 \, \text{W/m}^2$, or about 10% of its 'true' value ($21 \, \text{W/m}^2$). The same relative error in the determination of the monthly averaged sensible heat flux is obtained for three-hour sampling. In other words, 12-hour sampling turns out to be sufficient for more or less reliable estimation of latent heat flux but not of sensible heat flux.

Next, if measurements of one or other characteristic are taken using the same device the spatial correlation of measurement errors will be high. This entails a decrease in the amount of independent information, which leads to uncertainty in errors of interpolation and to distortions of the time–space variability of the characteristics examined.

These considerations should be kept in mind when estimating the possibilities of using satellite information. It should also be borne in mind that the specific cost of satellite information (cost of one observation) is not large as compared with the cost of conventional ships' measurements. For

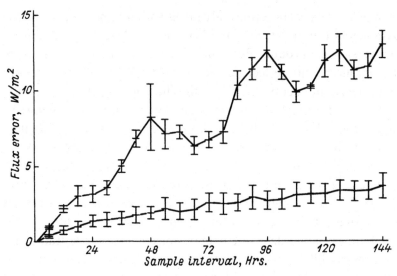

Figure 1.5 Uncertainty in estimating the monthly mean sensible and latent heat fluxes due to sampling at increased intervals as compared to hourly sampling, according to Taylor (1986): upper curve – latent heat flux; lower curve – sensible heat flux.

example, according to WMO data, the maintenance costs of one ocean weather ship, not counting other expenses, amounted to about half a million US dollars per year in the mid-1970s; this is not much less than the maintenance costs for one satellite including the costs of its equipment, launch and centres for receiving and processing the satellite information. But the operation of one satellite equals that of many ocean weather ships, so from the point of view of economy the advisability of using satellite information leaves no room for doubt. However, it is necessary not only to increase the accuracy of satellite information but also to interpret it in the correct way. Results achieved to date are encouraging.

2

Present state of the climatic system

Let us examine the budget of mass, energy, angular momentum and carbon. The presence of the latter in this list is explained by two reasons. First, the carbon budget is interesting in itself. Second, it is closely related to *the atmospheric CO_2 problem*, without solving which it is impossible to estimate the role of anthropogenic factors in climate changes.

In summarizing the factual data we will follow Peixóto and Oort (1984) for the most part.

2.1 Initial information

There are at present about 2000 weather stations for atmospheric sounding where horizontal wind velocity, temperature, humidity and isobaric surface height for 11 levels in the atmosphere are measured twice a day. The amount of information from these stations is enormous. Even if we exclude all monthly series with data gaps exceeding ten days, then during the decade covering May 1963 to April 1973 we obtained $10 \times 365 \times 2 \times 100 \times 11 \times 5 = 4 \times 10^9$ information units. Here the first number specifies the number of years, the second specifies the number of days in a year, the third specifies the number of daily series of measurements, the fourth specifies the number of stations providing qualitative information, the fifth specifies the number of measurement levels in the atmosphere and the sixth specifies the number of measured parameters. The amount of information is thus sufficient to detect large-scale peculiarities in the atmospheric climate. But one should keep in mind that these stations are at very irregular locations: they are concentrated mainly in densely populated areas of the globe while the major part of the atmosphere above the World Ocean, the equatorial belt of South America and Africa, as well as Antarctica, are still not well elucidated by observational data.

45

Almost the same can be said about the network of measurement stations in the World Ocean. Today it has about 5×10^6 records of temperature and salinity and at least 1.0×10^6 soundings carried out by bathythermograph and expendable and free-sinking profilers. The distribution of data over the area of the World Ocean is extremely irregular: in fact there are no data at all for a large part of it, especially in the oceans of the Southern Hemisphere. Fortunately, the problem of shortage of information for the ocean is not as acute as that for the atmosphere because seasonal variability only appears in the upper 200- or 300-metre layer of the ocean, and there are more data for this than for other ocean layers. As regards interannual variability, this is much less distinct in the ocean than in the atmosphere, and that is why all reliable data obtained from the moment the first deep water measurements were taken are suitable for approximate estimates of ocean climatic characteristics (say, meridional heat transfer).

The situation regarding data on current measurements is much poorer. All we have at the present time is information on the drift of ships and a few direct recordings of the current velocity in the surface and deep layers of the ocean. Naturally, this situation does not contribute to the fast progress of knowledge about the global ocean circulation. In addition, a sharp increase in the network of current velocity measurements is hardly possible in the near future due to technical and economic reasons. Thus there is only one thing to do, i.e. make an attempt to utilize non-standard (initially acoustic and satellite) methods of ocean dynamics research. But this is a matter for the future. To date, all necessary information is provided by standard technology.

2.2 Mass budget

Let us start by deriving expressions describing the mass budget in separate subsystems of the climatic system. Define the mass m_A of the atmospheric column with unit cross-section as

$$m_A = \int_\zeta^\infty \rho_A \, dz = \frac{p_s}{g},$$

where ρ_A is the air density, p_s is the surface atmospheric pressure, ζ is the height of the underlying surface elevations, g is gravity and z is the vertical coordinate oriented upward.

Let us integrate the continuity equation over the whole thickness of the atmosphere, and take advantage of the condition by which the vertical mass flux approaches zero at a sufficiently large height in the atmosphere, and also of the kinematic condition: $w = \partial\zeta/\partial t + (\mathbf{v}\nabla)\zeta - (P - E)/\rho_A$ at the underlying surface (here w and $\mathbf{v}$ are the vertical component and horizontal vector of wind velocity, P is the precipitation rate, E is the evaporation rate, ∇ is the gradient operator on the surface of a sphere with radius a and t is time).

As a result we have

$$\frac{\partial m_A}{\partial t} + \nabla \cdot m_A \hat{\mathbf{v}}_A = -(P - E), \qquad (2.2.1)$$

where $\hat{\mathbf{v}}_A$ is the vertically averaged horizontal vector of wind velocity.

Similarly, the conservation equations for water mass in the ocean, as well as those for sea ice and land substance, become

$$\frac{\partial m_O}{\partial t} + \nabla \cdot m_O \hat{\mathbf{v}}_O = (P - E - E_I), \qquad (2.2.2)$$

$$\frac{\partial m_I}{\partial t} + \nabla \cdot m_I \hat{\mathbf{v}}_I = E_I + (P - E), \qquad (2.2.3)$$

$$\frac{\partial m_L}{\partial t} + \nabla \cdot m_L \hat{\mathbf{v}}_L = (P - E), \qquad (2.2.4)$$

where the first terms on the left-hand side represent the rate of change of mass m referring to a unit of surface area; the second terms represent the horizontal divergence of the integral mass transport; $\mathbf{v}_O$, $\mathbf{v}_I$ and $\mathbf{v}_L$ represent the horizontal vectors of current velocity in the ocean, and of sea ice drift, and of river and ground water movement; E_I is the rate of sea ice formation (or melting); the symbol $\hat{}$ signifies mass averaging; and subscripts O, I and L refer to the ocean, sea ice and land.

We note that the integral mass transport $m_L \hat{\mathbf{v}}_L$ within the limits of the land is realized solely by the river and ground waters (river and underground run-off); the remaining part of the land, according to the condition $\hat{\mathbf{v}}_L = 0$, does not contribute to $m_L \hat{\mathbf{v}}_L$.

Let us integrate Equation (2.2.1) over the surface of the entire Earth, Equation (2.2.3) over the surface of the sea ice, and Equation (2.2.4) over the surface of the land. After that we designate the total (river + underground) run-off into the ocean, relating to unit surface area, as $\{E_R\}$. Then, taking into account the continuity of the integral mass transport at the land–ocean boundary, we obtain

$$\frac{\partial}{\partial t} \{m_A\} + \{P - E\} = 0,$$

$$\frac{\partial}{\partial t} \{m_O\} f_O - \{P - E\} f_O + \{E_I\} f_I - \{E_R\} f_L = 0,$$

$$\frac{\partial}{\partial t}\{m_{\mathrm{I}}\}f_{\mathrm{I}} - \{E_{\mathrm{I}}\}f_{\mathrm{I}} + \{P - E\}f_{\mathrm{I}} = 0,$$

$$\frac{\partial}{\partial t}\{m_{\mathrm{L}}\}f_{\mathrm{L}} - \{P - E\}f_{\mathrm{L}} + \{E_{\mathrm{R}}\}f_{\mathrm{L}} = 0,$$

where $f_{\mathrm{O}} = s_{\mathrm{O}}/s$, $f_{\mathrm{I}} = s_{\mathrm{I}}/s$, $f_{\mathrm{L}} = s_{\mathrm{L}}/s$ are the ratios of the ocean area (s_{O}), sea ice area (s_{I}) and land area (s_{L}) to the area s of the Earth's surface (or, otherwise, the ocean, sea ice and land fractions) interrelated as $f_{\mathrm{O}} + f_{\mathrm{I}} + f_{\mathrm{L}} = 1$; braces signify area averaging over the entire Earth's area.

Adding these expressions we obtain

$$\frac{\partial}{\partial t}(\{m_{\mathrm{A}}\} + \{m_{\mathrm{O}}\}f_{\mathrm{O}} + \{m_{\mathrm{I}}\}f_{\mathrm{I}} + \{m_{\mathrm{L}}\}f_{\mathrm{L}}) = 0. \qquad (2.2.5)$$

Equation (2.2.5) describes the obvious fact that for time scales which are much smaller than the lifetime of the continental ice sheets (for such time scales the continental ice sheets may be considered as part of the land) the mass of the atmosphere–ocean–sea-ice–land system remains constant in time.

Let us turn back to Equation (2.2.1) and integrate it over longitude. As a result we obtain

$$2\pi a \cos \varphi \frac{\partial}{\partial t}[m_{\mathrm{A}}] + \frac{1}{a}\frac{\partial}{\partial \varphi} MmT_{\mathrm{A}} = -2\pi a \cos \varphi([P] - [E]), \qquad (2.2.6)$$

where

$$MmT_{\mathrm{A}} = \int_0^{2\pi} m_{\mathrm{A}}\hat{v}_{\mathrm{A}} a \cos \varphi \, \mathrm{d}\lambda$$

is the meridional mass transport in the atmosphere, $\hat{v}_{\mathrm{A}}$ is the vertically averaged meridional component of wind velocity, φ is the latitude, λ is the longitude, square brackets signify zonal averaging.

Let us remember here that atmospheric mass is determined unambiguously by the value of surface atmospheric pressure and that, according to Oort (1983), the annual mean surface atmospheric pressures in the Northern and Southern Hemispheres are 983.6 and 988 hPa (or 2.569×10^{18} and 2.581×10^{18} kg). This suggests that an annual mean mass transport from the Southern to the Northern Hemisphere must exist, and this transport must be controlled by the different intensity of hydrologic cycles in both hemispheres (see Equation (2.2.6)). But the existence of atmospheric mass transport from the Southern Hemisphere to the Northern Hemisphere presupposes a reverse compensational mass transport in other subsystems of the climatic system.

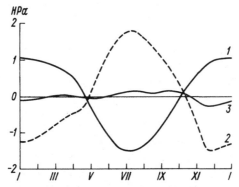

Figure 2.1 Seasonal variations of the difference between actual (not referred to the sea level) surface atmospheric pressure and its annual mean value, according to Oort (1983): (1) the Northern Hemisphere; (2) the Southern Hemisphere; (3) the Earth as a whole.

We postpone further discussion of this subject until Section 2.4 and direct our attention here to Figure 2.1, which shows variations in the difference between actual (not reduced to sea level) surface atmospheric pressure and its annual mean value. One can see that atmospheric mass undergoes distinct seasonal oscillations and these oscillations are out of phase for the Southern and Northern Hemispheres. This last fact is usually associated with mass transport from the summer to the winter hemisphere (see Oort, 1983). Thus, it is implicitly anticipated that sources and sinks of mass on the right-hand side of Equation (2.2.6) counterbalance each other separately in each hemisphere, or because they are too small they can be ignored (excluded from the kinematic condition at the underlying surface, see above). As applied to the annual mean conditions, this assumption is equivalent to an absence of mass transport across the equator, and, hence, to the equality of surface atmospheric pressures in both hemispheres. The controversy surrounding this point is obvious.

2.3 Heat budget

The primary source of energy for the climatic system is an insolation determined by the so-called *solar constant* (flux of solar radiation at an average distance of the Earth from the Sun; the most probable value of this flux is in the range from 1368 to 1377 W/m^2), as well as by the Earth's axis tilt, by the orbit eccentricity and by the longitude of perigee. Variations in the last three astronomical parameters have time scales of the order of tens of thousands of years (see above); that is why they have no real effect on changes in the climatic system on the time scales of decades that we are interested in.

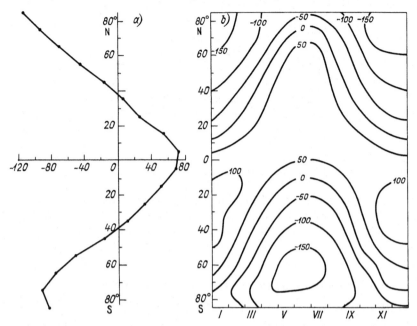

Figure 2.2 Annual mean zonally averaged net radiation flux (W/m²) at the upper atmospheric boundary (*a*), and its seasonal variability (*b*). (After Stephens *et al.*, 1981.)

Figure 2.2 shows the time–space variability of zonally averaged net radiation flux at the upper atmospheric boundary. Note three features: asymmetry of seasonal oscillations which is most distinct in high latitudes in both hemispheres, the local winter maximum in tropical latitudes in the Southern Hemisphere and marked increase in seasonal oscillations in the Southern Hemisphere compared with the Northern Hemisphere. The first and third features are connected to the specific character of the spatial distribution of the outgoing long-wave emission; the second feature is due to an increase in assimilated short-wave solar radiation because of a reduction of the planetary albedo.

The same features are inherent in net radiation flux at the surface of the World Ocean (Figure 2.3), though they manifest themselves less distinctly than in net radiation flux at the upper atmospheric boundary. The cause of this is a reduction in the effective emission of the ocean surface due to the existence of downward long-wave atmospheric radiation. The same cause explains the total increase in net radiation flux at the ocean surface compared with its value at the upper atmospheric boundary.

Denote the radiation heat influx in the atmosphere (the difference in net radiation fluxes at the upper atmospheric boundary and at the underlying

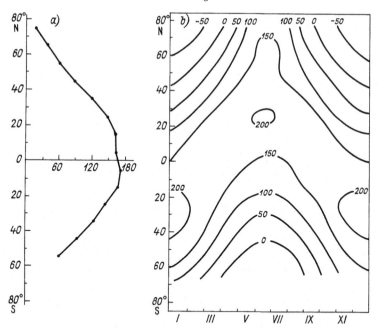

Figure 2.3 Annual mean zonally averaged net radiation flux (W/m²) at the ocean surface (*a*) and its seasonal variability (*b*). (After Strokina, 1989.)

surface) by $(Q_1^A + Q_2^A)$, where Q_1^A and Q_2^A are influxes of short-wave and long-wave radiation respectively. Then $(Q_1^A + Q_2^A)$, together with the incoming sensible heat (Q_3^A) from the underlying surface which is equal and of opposite sign to the sensible heat flux H at the underlying surface, and the heat release $Q_4^A = LP$ due to water vapour phase transitions, as well as the divergence $\nabla \cdot c_p \widehat{m_A v_A T_A}$ of the mass-averaged sensible heat transport and the conversion $C(\Phi, I)$ of internal energy I into potential energy Φ (see Section 2.5), has to be balanced by changes in the internal energy of a unit atmospheric column, that is,

$$\frac{\partial}{\partial t} c_v m_A \widehat{T_A} + \nabla \cdot c_p \widehat{m_A v_A T_A} = Q_1^A + Q_2^A + Q_3^A + Q_4^A - C(\Phi, I), \quad (2.3.1)$$

where here, and above, T_A is the air temperature, L is the heat of condensation, c_p and c_v are the specific air heat for constant pressure and volume; the remaining designations are the same.

Similarly, the net radiation flux $(Q_1^S + Q_2^S)$ at the underlying surface plus the heat release $L_1 E_1$ at the expense of phase transitions of water vapour and minus the fluxes of sensible heat $Q_3^S = H$ and latent heat $Q_4^S = LE$, as well as the divergence of the integral sensible heat transport, have to be balanced by changes in the heat content of a unit column of the ocean or land, that is,

$$\frac{\partial}{\partial t} c_O m_O \hat{T}_O + \nabla \cdot \widehat{c_O m_O v_O T_O} = Q_1^{OS} + Q_2^{OS} - Q_3^{OS} - Q_4^{OS} - L_1 E_1, \quad (2.3.2)$$

$$\frac{\partial}{\partial t} c_L m_L \hat{T}_L + \nabla \cdot \widehat{c_L m_L v_L T_L} = Q_1^{LS} + Q_2^{LS} - Q_3^{LS} - Q_4^{LS}, \quad (2.3.3)$$

where c and T are the specific heat and temperature; m is the mass of a column with unit area; L_1 is the heat of melting (sublimation); subscripts O and L signify belonging to the ocean and land, double superscripts OS and LS signify belonging to the ocean and land surfaces, the symbol $\hat{}$ signifies, as mentioned above, averaging over the mass of the ocean or over the mass of the active land layer.

   The terms appearing on the right-hand sides of Equations (2.3.2) and (2.3.3) completely describe the resulting heat flux at the underlying surface. Note some distinctive features of the time–space distribution of this flux within the World Ocean area not covered by ice ($L_1 E_1 = 0$). Judging from Figure 2.4, for annual mean conditions, the ocean gains heat from the atmosphere in low latitudes and loses it in middle and high latitudes (the secondary maximum in the middle latitudes of the Southern Hemisphere arises from a

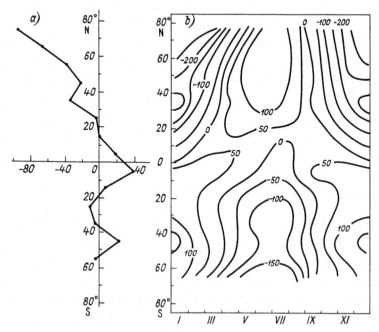

Figure 2.4 Annual mean zonally averaged resulting heat flux (W/m²) at the ocean surface (*a*) and its seasonal variability (*b*). (After Strokina, 1989.)

reduction in the latent heat flux). In the Northern Hemisphere the heat inflow from the atmosphere is lower and the heat transfer from the ocean into the atmosphere is higher than in the Southern Hemisphere. The seasonal variability of the resulting heat flux in the Northern Hemisphere also appears to be greater. This feature can be explained by the difference in the ocean/land area ratio in both hemispheres.

We supplement the system (2.3.1)–(2.3.3) by the budget equations for the latent heat $Lm_A \hat{q}_A$ in the atmosphere and sea ice $L_1 m_1$ (here, q_A is the mass-averaged air specific humidity; m_1 is the sea ice mass referred to a unit of surface area). These equations take the form

$$\frac{\partial}{\partial t} Lm_A \hat{q}_A + \nabla \cdot \widehat{Lm_A v_A q_A} = L(E - P), \tag{2.3.4}$$

$$\frac{\partial}{\partial t} L_1 m_1 + \nabla \cdot L_1 m_1 v_1 = L_1 E_1 + L(P - E), \tag{2.3.5}$$

where v_1 is the horizontal vector of the sea ice drift velocity.

Integrating Equations (2.3.1)–(2.3.5) over the longitude and then summing we obtain the equality

$$2\pi a \cos \varphi \frac{\partial}{\partial t} ([c_v m_A \hat{T}_A] + [c_O m_O \hat{T}_O] f_O' + [c_L m_L \hat{T}_L] f_L'$$

$$+ [Lm_A \hat{q}_A] + [L_1 m_1] f_1') + \frac{1}{a} \frac{\partial}{\partial \varphi} (MHT_A + MHT_O)$$

$$= 2\pi a \cos \varphi([Q_1^\infty] + [Q_2^\infty]) - 2\pi a \cos \varphi[C(\Phi, I)], \tag{2.3.6}$$

where

$$MHT_A = \int_0^{2\pi} m_A (c_P \widehat{v_A T_A} + L\widehat{v_A q_A}) a \cos \varphi \, d\lambda$$

is the meridional sensible and latent heat transport in the atmosphere,

$$MHT_O = \int_0^{2\pi} (f_O' c_O \widehat{m_O v_O} T_O + f_1' L_1 m_1 v_1 + f_L' c_L \widehat{m_L v_L} T_L) a \cos \varphi \, d\varphi$$

is the meridional sensible and latent heat transport in the ocean and active land layer; v_O, v_1 and v_L are the meridional components of current velocity, of sea ice drift and of river and ground water movement; f_O', f_1' and f_L' are fractions of the ocean, sea ice and land in the zonal belt of the unit meridional extent.

Equation (2.3.5) relates the net radiation flux ($[Q_1^\infty] + [Q_2^\infty]$) at the upper atmospheric boundary (here, Q_1^∞ is the flux of the absorbed short-wave solar radiation, Q_2^∞ is the flux of the long-wave emission into space) to the divergence of the meridional heat transport, and energy change in the atmosphere–ocean–sea ice–land system. Let us discuss the time–space variability of the last two components. But first of all we consider the seasonal variability of the global heat budget, which is described by the equality

$$\frac{\partial}{\partial t}(\{c_v m_A \hat{T}_A\} + \{L m_A \hat{q}_A\} + \{c_O m_O \hat{T}_O\} f_O + \{c_L m_L \hat{T}_L\} f_L + \{L_I m_I\} f_I)$$
$$+ \{C(\Phi, I)\} = \{Q_1^\infty\} + \{Q_2^\infty\}.$$

To determine the rate of change in internal energy and its conversion into potential energy in the atmosphere, appearing on the left-hand side of this equality, as well as the rate of sensible heat change in the ocean and latent heat in the atmosphere and sea ice, we use monthly averaged values of air temperature and air specific humidity systematized by Oort (1983), estimates of the rate of change in heat content $\{c_O m_O \hat{T}_O\}$ in a 275-metre layer of ocean, from Levitus (1982), and data on the sea ice area and volume represented in Table 2.1. The calculation results, together with estimates, from Figure 2.2, of the global average net radiation flux at the upper atmospheric boundary are shown in Figure 2.5. They have a lot in common with the results of similar calculations published by Ellis *et al.* (1978). In both cases the amplitude and phase of seasonal variations of the net radiation flux at the upper atmospheric boundary and the rate of sensible heat change in the ocean are in close proximity, which points to the key role of the ocean in the formation of the seasonal variability of the global heat budget. It turns out that the rate of change in internal energy and its conversion into the potential energy of the atmosphere, and the rate of latent heat change in the atmosphere (their combination will be referred to, for the sake of abbreviation, as rate of energy change) are much less than the net radiation flux at the upper atmospheric boundary. Finally, the rate of energy change and the net radiation flux are out of phase with each other. This last fact, as well as the fact that the maximum and minimum of the rate of energy change fall on cold and warm semiannual periods of the Northern Hemisphere, result from the different ocean/land area ratio in the two hemispheres.

It should be emphasized that the rate of energy change in the atmosphere, even amounting to the latent heat change in the sea ice, cannot balance the discrepancy between the net radiation flux at the upper atmospheric boundary and the rate of sensible heat change in the ocean. This is caused by incompleteness and inaccuracy of initial information and by errors in determining the global heat budget components as small differences of large values. But in attempting to define the cause of the discrepancy we

Table 2.1. *Seasonal variability of the sea ice area and volume in the World Ocean and in its separate parts according to different authors*

Characteristic	Region	Author	Month											
			I	II	III	IV	V	VI	VII	VIII	IX	X	XI	XII
Sea ice area (10^{12} m^2)	Arctic Basin	Walsh and Johnson (1979)	13.8	14.1	14.0	13.5	12.5	10.8	8.7	7.0	7.1	9.6	11.6	13.0
		Lemke et al. (1980)	13.0	13.5	13.0	12.5	11.5	10.5	8.5	7.0	8.5	9.5	11.2	13.0
		Lebedev (1986)	11.5	11.5	11.5	11.4	11.1	10.3	8.4	6.7	6.8	8.2	10.3	11.3
	Southern Ocean	Lemke et al. (1980)	5.5	2.5	3.0	6.0	9.0	13.0	15.5	17.5	18.0	17.5	17.0	10.0
		Lebedev (1986)	4.4	2.9	3.1	5.4	9.0	10.9	13.0	14.4	14.8	14.8	11.5	7.3
	World Ocean	Lebedev (1986)	17.8	17.1	17.3	19.2	21.1	21.9	22.0	21.3	21.6	23.1	22.1	19.7
Sea ice volume (10^{12} m^3)	Arctic Basin	Lebedev (1986)	22.5	25.6	28.2	29.9	30.3	28.2	22.4	17.0	14.5	14.8	16.5	19.4
	Southern Ocean	Lebedev (1986)	3.5	2.2	2.2	3.3	5.0	8.0	11.0	14.5	16.8	18.7	15.1	9.1
	World Ocean	Lebedev (1986)	27.9	30.8	34.2	37.0	38.1	37.9	33.2	31.7	31.4	33.6	31.9	29.4

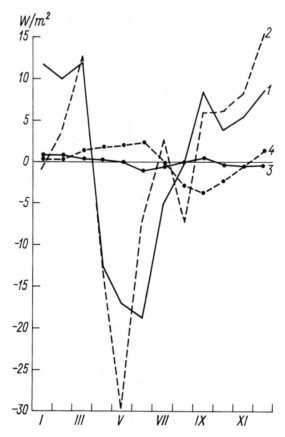

Figure 2.5 Seasonal variability of the global heat budget in the ocean–atmosphere system: (1) the net radiation flux at the upper atmospheric boundary; (2) the rate of change in sensible heat in the ocean; (3) the rate of change in latent heat in sea ice; (4) the rate of change in latent and sensible heat in the atmosphere.

have identified ourselves with the well-known opinion that the active land layer has no marked effect on the seasonal variability of the global heat budget. Let us check this. Let the heat capacity of a unit column of the active land layer be equal to $1.2 \times 10^7 \ \mathrm{J/m^2 \ K}$, let the continent fraction be equal to 0.3 and let the mean temperature change in the active layer be equal to $2 \times 10^{-7} \ \mathrm{K/s}$. Then the rate of heat content change $\partial \{c_L m_L T_L\} f_L / \partial t$ will be approximately $0.6 \ \mathrm{W/m^2}$, that is, it cannot eliminate the imbalance in the global heat budget.

The relation between the components of the heat budget presented above is fulfilled for the planet as a whole but not for its separate parts. Examine, for example, the northern and southern polar regions, bounded by the 70°S and 70°N parallels respectively. In these regions the net radiation flux at the

upper atmospheric boundary during most of the year is much larger than
the rate of energy change in the atmosphere, and this rate is comparable to
the rate of latent heat change in snow and sea ice. The question is: what
balances heat losses? There is an unambiguous answer – they are balanced
only by the meridional energy transport (the transport of sensible and latent
heat, and potential energy) in the ocean–atmosphere system. This can be seen
from Table 2.2. The data in Table 2.2 are interesting in more than one respect:
they allow us to establish the distinctions in the formation of the heat budget
for two polar regions. We discuss the most remarkable distinctions below.

In summer when the net radiation flux at the upper atmospheric boundary
in the northern polar region is small, the meridional energy transport is
basically balanced by the latent heat release due to ice and snow melting,
and by the increase in ocean heat content, whereas in the southern polar
region the meridional energy transport is distributed almost equally between
the latent heat release due to snow and ice melting and the emission
into space. On the other hand, in winter, heat radiation losses at the
upper atmospheric boundary in the northern polar region are large, and 2/3
of these losses are compensated by the meridional energy transport and 1/3
of them are compensated by a reduction in the ocean heat content and by
the heat release due to ice formation. In the southern polar region the
radiation losses at the upper atmospheric boundary are balanced solely by
the meridional energy transport. And again the main cause of distinctions is
the different ratio between ocean–land areas.

Let us discuss the time–space distribution of the rate of heat content change
in the ocean, bearing in mind that this rate (owing to the smallness of its
analogue in the atmosphere) might characterize not only the ocean, but also
the entire ocean–atmosphere system. The most interesting features of this
distribution (Figure 2.6) are the increase in seasonal variations in middle
latitudes in the Northern Hemisphere compared with the same latitudes in the
Southern Hemisphere, the absence of marked seasonal variations in the
equatorial region and in high latitudes of the Northern and Southern
Hemispheres, and, finally, the domination of extreme values in the Northern
Hemisphere compared with analogous values in the Southern Hemisphere.

Turning to an analysis of the annual mean meridional sensible heat transfer
in the ocean–atmosphere system (Figures 2.7(*a*) and 2.8(*a*)), we should note
first of all that in the tropics the meridional sensible (as well as latent) heat
transport in the atmosphere is directed to the equator, and because of this
the tropics are a powerful source of heat for the atmosphere such that
compensation for the meridional sensible and latent heat transport is realized
here mainly by the meridional potential energy transport in the atmosphere,

Table 2.2. *Components of the heat budget in the northern and southern polar regions (according to Nakamura and Oort, 1988)*

Components	I	II	III	IV	V	VI	VII	VIII	IX	X	XI	XII	Annual mean value
Net radiation flux at the upper atmospheric boundary (W/m²)	-164 / -10	-146 / -72	-122 / -106	-78 / -148	-37 / -135	1 / -139	5 / -130	-50 / -123	-126 / -99	-166 / -63	-164 / -38	-162 / -14	-100.7 / -89.6
Rate of energy change in the atmosphere (W/m²)	-2 / -1	4 / -10	16 / -13	25 / -15	25 / -14	21 / -9	3 / -5	-16 / 1	-29 / 8	-26 / 21	-14 / 24	-7 / 11	0.0 / -0.1
Rate of heat content change in the ocean (W/m²)	-29 / 6	-27 / 2	-11 / -2	11 / -4	29 / -5	32 / -5	23 / -4	10 / -4	0 / -1	-6 / 3	-12 / 6	-21 / 8	0.0 / 0.0
Meridional energy transport (W/m²)	113 / 33	94 / 55	94 / 78	96 / 105	82 / 126	81 / 139	85 / 143	88 / 128	102 / 121	116 / 113	114 / 64	116 / 39	98.4 / 95.3
Resulting heat flux at the underlying surface (W/m²)	50 / -42	56 / -13	44 / 11	7 / 17	-20 / 9	-61 / 7	-86 / 7	-55 / 6	-5 / 5	25 / 2	36 / -8	39 / -33	2.4 / -2.7
Rate of latent heat change in snow and sea ice (W/m²)	-21 / 36	-29 / 11	-33 / -9	-18 / -13	-9 / -4	29 / -2	63 / -3	45 / -2	5 / -4	-19 / -5	-24 / 2	-17 / 25	-2.4 / 2.7

Note: values in numerators correspond to the northern polar region, and those in denominators correspond to the southern polar region.

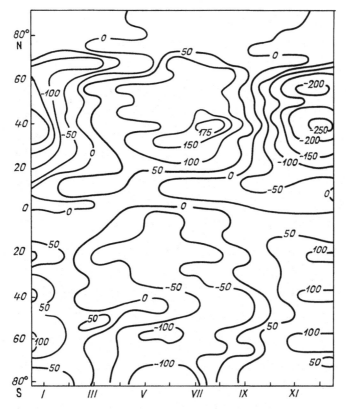

Figure 2.6 Time–space distribution of the rate of change in heat content (W/m²) in the World Ocean, according to Kagan and Tsankova (1986).

that the meridional sensible heat transport in high latitudes of the atmosphere and ocean is directed to the poles, that such orientation is an effect of the domination of the absorbed short-wave radiation over upward long-wave radiation in low latitudes and of the inverse ratio in high latitudes, and, finally, that on the equator the meridional sensible heat transport in the ocean takes on near-zero and positive values, indicating the presence of a small heat transport from the Southern to the Northern Hemisphere.

The seasonal variability of the meridional sensible heat transport in the ocean–atmosphere system can be judged by parts (*b*) in Figures 2.7 and 2.8. The first thing to command our attention is the striking distinctions between time–space distributions of the meridional transport in the atmosphere and the ocean. We mean, firstly, localization of maxima at different latitudes in the atmosphere and ocean (in the atmosphere the maximal variability is coordinated with the equator, and in the ocean it is coordinated with the middle latitudes of both hemispheres), and, secondly, the constancy, during

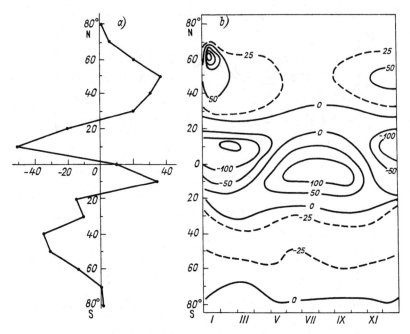

Figure 2.7 Annual mean meridional sensible heat transport (W/m²) in the atmosphere (*a*) and its seasonal variability (*b*). Positive values indicate northward transport; negative values indicate southward transport. (After Carissimo *et al.*, 1985, and Oort and Rasmusson, 1971.)

the annual cycle, of the direction of the meridional transport in middle and high latitudes of the atmosphere and its changing at all latitudes in the ocean.

It is an open secret that the meridional heat transport in the ocean is still the least examined element of the climate, so it is worth dwelling at some length on the analysis of its time–space distribution. Figure 2.8 (part (*b*)) shows the well-known similarity of seasonal variations of MHT_O in the Northern and Southern Hemispheres. In both hemispheres the meridional heat transport reaches a maximum in winter and is directed to the North Pole at this time of the year. In summer, there is a marked reduction and change in direction of MHT_O everywhere with the exception of the tropical region of the ocean. The extreme values of MHT_O occur in the tropics and fall during February and March in the Northern Hemisphere and during October and November in the Southern Hemisphere. The secondary maxima displaced from the primary maxima further from the equator fall during August and September in the Northern Hemisphere and during May and June in the Southern Hemisphere. This last fact points to the presence of the strongly pronounced semi-annual harmonic.

Let us mention two more distinguishing features of the seasonal variability

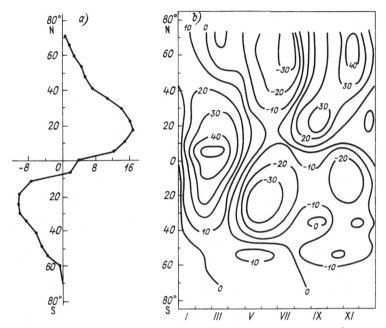

[1]Figure 2.8 Annual mean meridional sensible heat transport (W/m²) in the World Ocean (*a*) and its seasonal variability (*b*), according to Kagan and Tsankova (1987). Positive values indicate northward transport; negative values indicate southward transport.

of MHT_O. The question is, firstly, about the intensification of the poleward heat transport in spring as compared with autumn in the Northern Hemisphere, and in summer as compared with winter in the Southern Hemisphere, and, secondly, about the excess (about twice) of the period with heat transport directed from the Northern Hemisphere to the Southern Hemisphere as compared with the period with reverse direction of meridional heat transport.

And, finally, the comparison of the annual mean distributions in Figures 2.7 and 2.8 is indicative of the commensurability of the meridional sensible heat transports in the atmosphere and the ocean and, mainly, of their absolutely different natures. This is indicated by the fact that the maximum poleward heat transport in the ocean is in low latitudes where meridional gradients of water temperatures are relatively small, and in the atmosphere in middle latitudes, where meridional gradients of air temperature are large by way of contrast.

2.4 Moisture budget

Let us start, as we did in the preceding section, with the derivation of equations for the moisture budget in separate subsystems of the climatic system. For the atmosphere and sea ice these equations are derived by division of (2.3.4) and (2.3.5) by $L(\rho_A/\rho_O)$ and $L_1(\rho_1/\rho_O)$, respectively, where ρ_1 and ρ_O are densities of sea ice and fresh water. Continuing, we obtain the following expressions:

$$\frac{\partial W_A}{\partial t} + \nabla \cdot \widehat{\mathbf{v}_A W_A} = -(P - E)\frac{\rho_O}{\rho_A}, \qquad (2.4.1)$$

$$\frac{\partial W_1}{\partial t} + \nabla \cdot \widehat{\mathbf{v}_1 W_1} = E_1\frac{\rho_O}{\rho_1} + (P - E)\frac{\rho_O}{\rho_A}, \qquad (2.4.2)$$

where $W_A = (\rho_O/\rho_A)m_A\hat{q}_A$ is the moisture content (in water equivalent) of an atmospheric column of unit cross-section, $W_1 = (\rho_O/\rho_1)m_1$ is the fresh water mass in the sea ice normalized to unit area; the remaining designations are the same.

For land the equation for the fresh water mass budget is written as

$$\frac{\partial W_L}{\partial t} + \nabla \cdot \widehat{\mathbf{v}_L W_L} = (P - E)\frac{\rho_O}{\rho_A}, \qquad (2.4.3)$$

where W_L is the moisture content of land, defined as fresh water mass in soil and ground per unit surface area.

The system (2.4.1)–(2.4.3) is closed by the equation of fresh water mass budget in the ocean

$$\frac{\partial W_O}{\partial t} + \nabla \cdot \widehat{\mathbf{v}_O W_O} = (P - E)\frac{\rho_O}{\rho_A} - E_1\frac{\rho_O}{\rho_1}, \qquad (2.4.4)$$

where W_O signifies the fresh water content in a unit column of the ocean.

Equations (2.4.1)–(2.4.4) describe, respectively, atmospheric, cryospheric, continental and ocean components of the hydrologic cycle. After integrating the first of these equations over the whole Earth, the second one – over the sea ice surface, the third one – over the land surface, and the fourth one – over the ocean surface free from ice, and using the continuity condition for the vertically integrated fresh water mass transport at the ocean–land interface, we obtain

$$\frac{\partial}{\partial t}\{W_A\} + \{P - E\}\frac{\rho_O}{\rho} = 0, \qquad (2.4.5)$$

$$\frac{\partial}{\partial t}\{W_1\}f_1 - \{E_1\}f_1\frac{\rho_O}{\rho_1} - \{P - E\}f_1\frac{\rho_O}{\rho_A} = 0. \qquad (2.4.6)$$

$$\frac{\partial}{\partial t}\{W_L\}f_L - \{P - E\}f_L\frac{\rho_O}{\rho_A} + \{E_R\}f_L = 0, \qquad (2.4.7)$$

$$\frac{\partial}{\partial t}\{W_O\}f_O - \{P - E\}f_O\frac{\rho_O}{\rho_A} + \{E_1\}f_1\frac{\rho_O}{\rho_A} - \{E_R\}f_L = 0. \qquad (2.4.8)$$

Adding (2.4.5) and (2.4.8) and considering the equality $(f_O + f_I + f_L) = 1$ results in the obvious relation

$$\frac{\partial}{\partial t}(W_A + \{W_I\}f_I + \{W_L\}f_L + \{W_O\}f_O) = 0, \qquad (2.4.9)$$

expressing the law of conservation for the fresh water mass in the atmosphere–ocean–sea ice–land system.

Let us discuss in turn those elements of the hydrologic cycle on which we have more or less reliable information.

Moisture content in the atmosphere. For annual mean conditions this decreases systematically from the equator to the poles. Such characteristic changing of the moisture content in the atmosphere is explained by the strong dependence of specific air humidity on temperature. Let us note the presence of deviations from zonal symmetry related to the effect of the underlying surface; as a result, atmospheric moisture content is higher over the ocean than over the land. Significant deviations also take place in the vicinity of the east and west boundaries of continents, due to the effect of orography and of the warm and cold ocean currents. There is one more distinguishing feature: the distribution of the moisture content in the Southern Hemisphere is closer to the zonal distribution than in the Northern Hemisphere. This is caused by the different ratio of land to ocean areas in the two hemispheres. The total average moisture content of the atmosphere per year is 13.1×10^{15} kg; this is equivalent to a water layer of 0.025 m thickness.

An idea of seasonal variability can be obtained from Figure 2.9. It is obvious from the diagram that the moisture content of the atmosphere in all latitudes of the Northern Hemisphere is larger in summer than in winter. The opposite situation takes place in the Southern Hemisphere. The maximum seasonal variability of the moisture content in the atmosphere occurs in the zonal belt limited by parallels $20°$ and $30°$N. This feature has its origin in the monsoon circulation over South-East Asia and Africa. In the equatorial region the seasonal variability conditioned by displacement of the moisture content maximum to the summer hemisphere is much less.

Meridional moisture transport in the atmosphere. Let us integrate Equation (2.4.1) over longitude to determine this. As a result we obtain the following expression:

$$2\pi \cos\varphi\, \frac{\partial}{\partial t}[W_A] + \frac{1}{a}\frac{\partial}{\partial \varphi}MWT_A = -2\pi a \cos\varphi([P] - [E])\frac{\rho_O}{\rho_A}, \qquad (2.4.10)$$

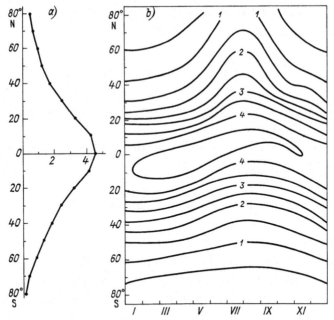

Figure 2.9 Annual mean zonally averaged moisture content ($10 \, \text{kg/m}^2$) in the atmosphere (*a*) and its seasonal variability (*b*). (After Oort, 1983.)

where

$$MWT_A = \int_0^{2\pi} \overbrace{m_A v_A q_A} \, a \cos \varphi \, d\lambda$$

is the meridional transport of fresh water in the atmosphere.

The latter is unambiguously related to the meridional transport of water vapour defined by the division of MWT_A by (ρ_A/ρ_O). The dependence of $(\rho_A/\rho_O)MWT_A$ on latitude is shown in Figure 2.10. Even a quick glance is enough to see two interesting features: first, the strong seasonal variability of the meridional water vapour transport in the tropics, and its almost complete absence in the middle and high latitudes of both hemispheres; second, the existence of considerable water vapour transport from the Southern Hemisphere to the Northern Hemisphere in summer and of the slightly smaller reverse transport in winter. Thus, it turns out that the Southern Hemisphere is a source of water vapour for the Northern Hemisphere.

Water vapour transport across the equator plays an important role in the formation of the water balance of both hemispheres. Suffice it to say that it is this transport that controls the relation between precipitation and evaporation and, in particular, the excess of 39 mm per year of precipitation over evaporation in the Northern Hemisphere (see Peixóto and Oort, 1983).

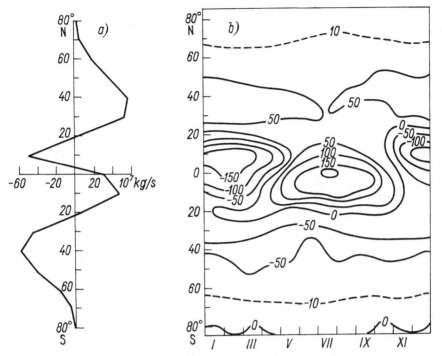

Figure 2.10 Annual mean meridional moisture transport (10^7 kg/s) in the atmosphere (*a*) and its seasonal variability (*b*). Positive values indicate northward transport; negative values indicate southward transport. (After Oort, 1983.)

Next, if we look at this from a more universal point of view then we will see that the presence of water vapour transport in the atmosphere across the equator serves as indirect proof of the presence of a reverse (directed from the Northern Hemisphere to the Southern Hemisphere) fresh water transport in the ocean and/or on the land: for the annual mean meridional moisture transport in all subsystems of the climatic system must vanish. On the basis of this consideration and estimates shown in Figure 2.10 we conclude that the meridional fresh water mass transport in the ocean and continental parts of the hydrologic cycle should be directed to the south in middle latitudes of the Northern Hemisphere as well as on the equator and in the tropics of the Southern Hemisphere, and to the north in all other latitudes.

Evaporation and precipitation. For some latitude zones the difference between precipitation and evaporation is defined not by the meridional water vapour transport but, rather, by its divergence (see Equation (2.4.10)). Say, if the divergence $\partial MWT_A/a\,\partial\varphi$ is positive then $([E]-[P]) > 0$ and, hence, the latitude zone where this inequality is valid serves as a moisture source for the atmosphere. On the other hand, if in some latitude zone $([E] - [P]) < 0$, then a moisture sink takes place there. The annual mean meridional distribution of $\partial MWT_A/a\,\partial\varphi$ signifies the fact that the tropical regions of the ocean serve as a main moisture source for the atmosphere, and the equatorial zone and temperate and high latitudes of both hemispheres serve as a sink. But if that is the case then the interchange of moisture sources and sinks in the atmosphere must have important consequences for the meridional distribution of water salinity in the ocean: where evaporation dominates over precipitation, the salinity of the ocean upper layer has to increase; otherwise it has to decrease. Comparison of meridional distributions of annual mean zonally averaged values of the evaporation minus precipitation rate (Figure 2.11(a)) and the surface salinity in the World Ocean (Figure 2.11(b)) confirms the marked regularity. From this it follows that the determining contribution to the formation of sea surface salinity, at least in equatorial and tropical latitudes, is from evaporation and precipitation. All other factors (including horizontal advection and vertical diffusion) play a secondary role.

Thus, the distribution $([E] - [P])$ is known. Let us examine the behaviour of separate components of this difference (see Figures 2.8 and 2.9). Figures 2.12 and 2.13 do not show anything unexpected. As one would expect, a maximum of evaporation falls on tropical latitudes of the ocean, and also in the tropics of the Northern Hemisphere the evaporation in winter is larger

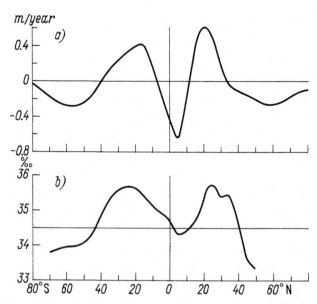

Figure 2.11 Zonal mean profiles of evaporation minus precipitation rate (a) and the salinity at the ocean surface (b), according to Peixóto and Oort (1984).

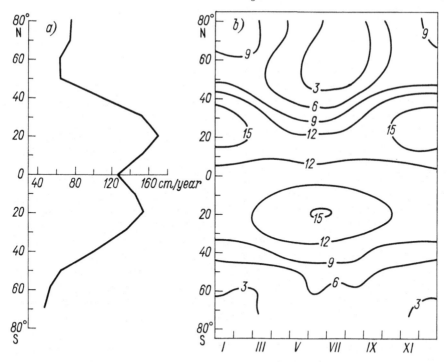

Figure 2.12 Annual mean zonally averaged evaporation rate (cm/year) at the ocean surface (*a*) and its seasonal variation in cm/month (*b*). (After Esbensen and Kushnir, 1981.)

than in the tropics of the Southern Hemisphere in summer. In the equatorial region evaporation is less than in the tropics due to decreasing sea surface temperature and weakening wind velocity. It can be seen from Figures 2.12 and 2.13 that the annual mean meridional distribution of precipitation and evaporation is not symmetric, and their seasonal variations are not antisymmetric about the equator because of differences in land and ocean areas in both hemispheres.

It will be recalled that everywhere other than the tropics the precipitation over the ocean is generally larger than over the land. If we add the fact that evaporation from the ocean surface is larger than that from the land surface, then it follows that evaporation and precipitation are compensated locally. But their compensation can only be partial in the sense that the resulting value of the difference between evaporation and precipitation must be negative on the land and positive over the ocean. Otherwise, the existence of river and underground run-off simply becomes inexplicable.

Thus the atmospheric part of the hydrologic cycle must include the water vapour transport from the ocean to the land, and the continental part must include that from the land to the ocean.

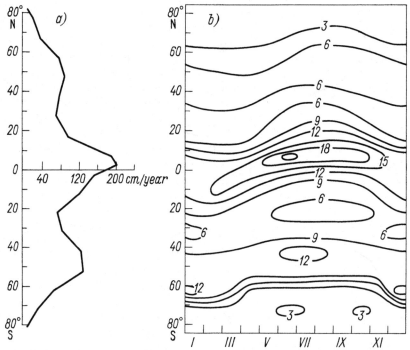

Figure 2.13 Annual mean zonally averaged precipitation rate (cm/year) over ocean and land (*a*) and its seasonal variation in cm/month (*b*). (After Jaeger, 1983.)

Continental run-off. According to Equations (2.4.6)–(2.4.8) the annual mean difference between evaporation and precipitation for the land surface, on the one hand, and evaporation and precipitation for the ocean and sea ice surface, on the other, must be equal to the value of continental (river and underground) run-off from the land into the ocean. This value, together with estimates of global evaporation and precipitation, is given in Table 2.3. As can be seen, all estimations of the run-off, except for the estimation by Bryan and Oort (1984), are in agreement with each other in spite of large systematic errors in precipitation measurements and unavoidable errors in calculations of evaporation, caused, among other things, by the extremely irregular distribution of data of meteorological measurements over the ocean, and by still insurmountable difficulties in determining soil wetness on the land.

Information on the seasonal variability of continental run-off can be found in Korzun (1974), according to which the maximum run-off amounting to 12% of the total for a year falls in June, and the minimal run-off (6% of the total for a year) falls in January. Thereby, one can indicate two features of seasonal variations of continental run-off: the relatively small range of seasonal variations of the global continental run-off and the continental

Table 2.3. *Components of the annual mean water balance of land, ocean and the Earth as a whole according to different researchers*

Area examined	Characteristic	L'vovich (1973)	Budyko (1974)	Korzun (1974)	Baumgartner and Reichel (1975)	Bryan and Oort (1984)	Strokina (1989)	Kagan et al. (1990)
Land	Evaporation (cm/year)	48.4	42.0	48.5	48.1	–	–	44.7
	Precipitation (cm/year)	73.2	73.0	80.0	74.8	–	–	75.6
	Run-off (cm/year)	24.8	31.0	31.5	26.7	4.1	–	30.9
Ocean	Evaporation (cm/year)	124.1	126.0	140.0	117.7	–	150.1	122.5
	Precipitation (cm/year)	113.8	114.0	127.0	106.7	–	133.1	109.9
	Evaporation–precipitation difference (cm/year)	10.3	12.0	13.0	11.0	1.8	17.0	12.6
The Earth as a whole	Evaporation (cm/year)	102.0	102.0	113.0	97.3	–	–	100.0
	Precipitation (cm/year)	102.0	102.0	113.0	97.3	–	–	100.0

Notes: estimations of precipitation presented in the last column of the table are borrowed from Jaeger (1983); estimations of evaporation are obtained from data by Esbensen and Kushnir (1981); bars designate the unavailability of necessary information.

run-off in the Northern Hemisphere, and the fact that these variations are in phase. The first feature is determined by the asynchronism of the continental run-off variations in the Northern and Southern Hemispheres; the second feature is caused by the prevalence of the run-off in the Northern Hemisphere over its value in the Southern Hemisphere.

Let us take advantage of the estimates from Tables 2.1 and 2.3 for the annual mean precipitation rate (1 m/year), river and underground run-off (0.3 m/year) and sea ice melting (0.5 m/year). Dividing the total moisture content in the atmosphere (water equivalent equal to 0.025 m, see above) and fresh water content in the ocean (3700 m) by the first of these estimates, the fresh water content in lakes, rivers, marshes, soils and underground waters (78 m) by the second estimate, and the fresh water content in the sea ice (1.5 m) by the third estimate we have that the times of fresh water renewal in the atmosphere, sea-ice, land and ocean are nine days, three years, 260 years and 3700 years respectively.

2.5 Energy budget

Let us start with the derivation of energy budget equations and then discuss the mechanisms of conversion from one form of energy into the other and the redistribution of energy between the ocean and the atmosphere.

Atmosphere

The state of the atmosphere is determined by the following set of variables: velocity $\mathbf{v}$, pressure p, temperature T and specific moisture q. The evolution equations for these variables, including an allowance for the hydrostatic and conditional (for the Coriolis acceleration) approximations take the form

$$\frac{\partial \mathbf{v}}{\partial t} + (\mathbf{u}\nabla)\mathbf{v} = -2\mathbf{\Omega k} \times \mathbf{v} - \nabla p/\rho + \mathbf{F}, \qquad (2.5.1)$$

$$\frac{\partial p}{\partial z} = -\rho g, \qquad (2.5.2)$$

$$\frac{\partial \rho}{\partial t} + \rho \nabla \cdot \mathbf{u} = 0, \qquad (2.5.3)$$

$$c_p\left(\frac{\partial T}{\partial t} + (\mathbf{u}\nabla)T\right) = F_T + \frac{1}{\rho}\left(\frac{\partial p}{\partial t} + (\mathbf{u}\nabla)p\right), \qquad (2.5.4)$$

$$\frac{\partial q}{\partial t} + (\mathbf{u}\nabla)q = F_q, \qquad (2.5.5)$$

where, as before, $\mathbf{u} = (u, v, w)$ and $\mathbf{v} = (u, v)$ are the three-dimensional and two-dimensional (in the φ, λ-plane) velocity vectors with components $u = a \cos \varphi \, d\lambda/dt$, $v = a \, d\varphi/dt$, $w = dz/dt$; F, F_T and F_q are sources and sinks of momentum, heat and moisture referred to the unit mass; p and ρ are the air pressure and density; Ω is the angular velocity of the Earth's rotation; $\mathbf{k}$ is the unit vector directed along the z-axis (vertically upward); other designations are the same.

Five equations ((2.5.1)–(2.5.5)) involve six unknown functions: $\mathbf{v}$, w, p, ρ, T and q. To close the system we take advantage of the equation of state of unsaturated moist air

$$p = \rho R T, \qquad (2.5.6)$$

where R is the gas constant, and insert definitions of potential energy $\Phi = gz$, internal energy $I = c_v T$, latent heat Lq, kinetic energy $K = |u|^2/2$, and total energy $\mathscr{E} = K + \Phi + I + Lq$ refer to the unit mass.

The budget equation for the potential energy,

$$\frac{\partial \Phi}{\partial t} + (\mathbf{u}\nabla)\Phi = gw, \qquad (2.5.7)$$

results from the definition of Φ, if it is affected by the operator $(\partial/\partial t + \mathbf{u}\nabla)$ and the identity $w \equiv dz/dt$ is taken into consideration.

The budget equation for the internal energy is found by a combination of Equations (2.5.3), (2.5.4) and (2.5.6). Namely, combining (2.5.3) and (2.5.6) and substituting in (2.5.4) the resulting expression

$$\frac{1}{\rho}\left(\frac{\partial p}{\partial t} + (\mathbf{u}\nabla)p\right) = -\frac{p}{\rho}\nabla \cdot \mathbf{u} + R\left(\frac{\partial T}{\partial t} + (\mathbf{u}\nabla)T\right),$$

we obtain

$$\frac{\partial I}{\partial t} + (\mathbf{u}\nabla)I = F_T - \frac{p}{\rho}\nabla \cdot \mathbf{u}$$

or

$$\frac{\partial I}{\partial t} + (\mathbf{u}\nabla)I = F_T - \frac{1}{\rho}\nabla p\mathbf{u} + \mathbf{v}\frac{\nabla p}{\rho} - gw. \qquad (2.5.8)$$

In hydrostatic approximation the integral (over the atmospheric mass) internal energy $\int_0^\infty \rho I \, dz$ is proportional to the integral potential energy $\int_0^\infty \rho\Phi \, dz$ with a proportionality factor c_v/R. Really, for the unsaturated moist air

$$\int_0^\infty \rho\Phi \, \partial z = -\int_0^{p_s} z \, dp = \int_0^\infty p \, dz = \int_0^\infty \rho R T \, dz,$$

where p_s is the surface atmospheric pressure.

This suggests

$$\int_0^\infty \rho I \, dz \Big/ \int_0^\infty \rho \Phi \, dz = c_v/R.$$

Considering this fact we can combine these two forms of energy into one, referred to as the total potential energy $\Pi = I + \Phi = c_p T$, where $c_p = c_v + R$. The corresponding budget equation takes the form

$$\frac{\partial \Pi}{\partial t} + (\mathbf{u}\nabla)\Pi = F_T - \frac{1}{\rho}\nabla p\mathbf{u} + \mathbf{v}\frac{\nabla p}{\rho}. \tag{2.5.9}$$

The kinetic energy budget equation is derived by means of scalar multiplication of (2.5.1) by $\mathbf{v}$, and taking into account that in hydrostatic approximation $dw/dt = 0$, we have

$$\frac{\partial K}{\partial t} + (\mathbf{u}\nabla)K = -\mathbf{v}\frac{\nabla p}{\rho} + \mathbf{v}\cdot F. \tag{2.5.10}$$

Equations (2.5.7)–(2.5.10) include similar (but opposite in sign) terms describing the conversion from one form of energy into the other. Thus, the term gw appearing in Equations (2.5.7) and (2.5.8) and describing the work performed by the buoyancy force features the conversion of internal energy into potential energy and back, and the term $\mathbf{v}\nabla p/\rho$, appearing in Equations (2.5.9) and (2.5.10) and describing the work performed by the pressure force describes the conversion of total potential energy into kinetic energy and back. Not all total potential energy can be converted into kinetic energy, but only the part $A = \int \rho(\Pi - \Pi_*) \, dV$ designated as *available potential energy*. Here, $\int \rho\Pi_* \, dV$ is the *unavailable potential energy*, that is, that part of the total potential energy which remains if the atmosphere is adjusted to a state with constant pressure over any isentropic surface (surface with equal values of specific entropy $\eta = c_p \ln T - R \ln p + const$) retaining the stable stratification; $dV = a^2 \cos \varphi \, d\lambda \, d\varphi \, dz$ is an element of the volume; the integration extends to the whole volume of the atmosphere.

To represent A in terms of the parameters to be measured, let us carry out the following chain of transformations:

$$\int \rho\Pi \, dV = \int \rho(I + \Phi) \, dV = \frac{c_p}{R}\int \rho\Phi \, dV = -\frac{c_p}{g}\int ds \int_{p_0}^0 T \, dp$$

$$= -\frac{c_p}{g}\int ds \int_{p_0}^0 \theta\left(\frac{p}{p_0}\right)^{R/c_p} dp = -\frac{c_p}{g}\left(1 + \frac{R}{c_p}\right)^{-1} p_0^{-R/c_p}\int ds \int_{p_0}^0 \theta \, dp^{1+R/c_p}$$

$$= \frac{c_p}{g}\left(1 + \frac{R}{c_p}\right)^{-1} p_0^{-R/c_p}\int ds \int_0^\infty p^{1+R/c_p} \, d\theta.$$

Here, it is anticipated that the potential temperature θ is equal to zero at $p = p_0$ ($p_0 = 1000$ HPa is the standard atmospheric pressure at the sea surface); the integration over s extends to the whole surface of the sphere.

Let us take into account the fact that in adiabatic processes the air mass over any isentropic surface is constant. Then the mean pressure p_* and the value

$$\int \rho \Pi_* \, dV = \frac{c_p}{g}\left(1 + \frac{R}{c_p}\right)^{-1} p_0^{-R/c_p} \int ds \int_0^\infty p_*^{1+R/c_p} \, d\theta$$

also remain constant above this surface. Substituting these expressions for $\int \rho \Pi \, dV$ and $\int \rho \Pi_* \, dV$ in definition of A we have

$$A = \frac{c_p}{g}\left(1 + \frac{R}{c_p}\right)^{-1} p_0^{-R/c_p} \int ds \int_0^\infty (p^{1+R/c_p} - p_*^{1+R/c_p}) \, d\theta. \quad (2.5.11)$$

But inasmuch as $p > 0$ and p_*, by definition, is the mean value of p, then, according to Lorenz (1967), the integral of the difference $(p^{1+R/c_p} - p_*^{1+R/c_p})$ will be positive for each isentropic surface, and therefore it can be expressed in terms of a pressure dispersion at this surface. Further, when it is considered (see Lorenz, 1967) that inclination of the isentropic surface with respect to the horizontal plane is small, then the pressure dispersion at the isentropic surface can be approximated in terms of the potential temperature dispersion or of absolute temperature dispersion at the isobaric surface. At present, there are several approximations to the relationship (2.5.11). They are all based on the approximated Lorenz formula (1967):

$$A = \tfrac{1}{2}c_p \int \frac{\gamma}{(\gamma_a - \gamma)} \frac{T^2 - \tilde{T}^2}{\tilde{T}} a^2 \cos \varphi \, d\varphi \, d\lambda \, dp/g,$$

where $\tilde{T}$ is the global mean air temperature on the fixed isobaric surface, $\gamma_a = g/c_p$ and $\gamma = -\partial \tilde{T}/\partial z$ are adiabatic and real vertical gradients of temperature in the atmosphere.

Turning back to the derivation of the energy budget equation, we multiply (2.5.5) by L and add it to (2.5.9) and (2.5.10). As a result, we have the total energy budget equation

$$\frac{\partial \mathscr{E}}{\partial t} + (\mathbf{u}\nabla)\mathscr{E} = -\frac{1}{\rho}\nabla p\mathbf{u} + \mathbf{v}\cdot\mathbf{F} + F_T + LF_q, \quad (2.5.12)$$

or, in divergent form,

$$\frac{\partial \rho \mathscr{E}}{\partial t} + \nabla\cdot\mathbf{u}(p + \rho\mathscr{E}) = \rho(\mathbf{v}\cdot\mathbf{F} + F_T + LF_q). \quad (2.5.12')$$

Conversion from one form of energy to the other has already been mentioned above but nothing was said about the mechanism of this conversion. To analyse it we use the quasi-zonality of the atmospheric circulation resulting from the zonality of the insolation, and extract zonal average and eddy (related to deviations of zonal averages) components in the fields of climatic characteristics. Then any climatic characteristics (say, components u, v and w of wind velocity or air temperature T) can be presented as

$$u = [u] + u^*, \; v = [v] + v^*, \; w = [w] + w^*, \atop T = [T] + T^* + \tilde{T}, \qquad \qquad \qquad \qquad \qquad (2.5.13)$$

where, as before, square brackets signify zonal averaging, the asterisks signify departure from zonal average values and the tilde signifies a global average value.

Accordingly, the kinetic energy K referring to the unit mass can be expressed as the sum of the kinetic energy K_M of the zonal mean motion and the kinetic energy K_E of eddy disturbances, that is,

$$K = K_M + K_E, \quad K_M = \tfrac{1}{2}([u]^2 + [v]^2), \quad K_E = \tfrac{1}{2}([u^{*2}] + [v^{*2}]). \quad (2.5.14)$$

Similarly, the available potential energy is defined as

$$A = A_M + A_E,$$

$$A_M = \tfrac{1}{2}c_P \int \frac{\gamma_a}{(\gamma_a - \gamma)} \frac{[T]}{\tilde{T}} a^2 \cos \varphi \, d\varphi \, d\lambda \, dp/g,$$

$$A_E = \tfrac{1}{2}c_P \int \frac{\gamma_a}{(\gamma_a - \gamma)} \frac{[T^{*2}]}{\tilde{T}} a^2 \cos \varphi \, d\varphi \, d\lambda \, dp/g. \qquad (2.5.15)$$

Let us obtain the budget equations for K_M and K_E. At first, we move from coordinates λ, φ, z and t to isobaric coordinates λ, φ, p and t. Then the atmospheric dynamics equations, Equations (2.5.1)–(2.5.3), are written in the form

$$\frac{\partial u}{\partial t} + \frac{u}{a \cos \varphi} \frac{\partial u}{\partial \lambda} + \frac{v}{a} \frac{\partial u}{\partial \varphi} + \omega \frac{\partial u}{\partial p} - \left(f + \frac{u}{a} \tan \varphi \right) v = - \frac{1}{a \cos \varphi} \frac{\partial \Phi}{\partial \lambda} + F_\lambda,$$

$$(2.5.16)$$

$$\frac{\partial v}{\partial t} + \frac{u}{a \cos \varphi} \frac{\partial v}{\partial \lambda} + \frac{v}{a} \frac{\partial v}{\partial \varphi} + \omega \frac{\partial v}{\partial p} + \left(f + \frac{u}{a} \tan \varphi \right) u = - \frac{1}{a} \frac{\partial \Phi}{\partial \varphi} + F_\varphi,$$

$$(2.5.17)$$

$$\frac{\partial \Phi}{\partial p} = -\alpha, \qquad \qquad (2.5.18)$$

$$\frac{1}{a\cos\varphi}\left(\frac{\partial u}{\partial\lambda}+\frac{\partial}{\partial\varphi}v\cos\varphi\right)+\frac{\partial\omega}{\partial p}=0, \tag{2.5.19}$$

where $\omega=dp/dt$ is the isobaric vertical velocity; $\alpha=1/\rho$ is the specific air volume; F_λ and F_φ are the components of the vector $\mathbf{F}$ in the direction of axes λ, φ; $f=2\Omega\sin\varphi$ is the Coriolis parameter; the remaining designations are the same.

Rewriting Equations (2.5.16) and (2.5.17) in divergent form and averaging them over longitude we have

$$\frac{\partial}{\partial t}[u]+\frac{1}{a\cos\varphi}\frac{\partial}{\partial\varphi}[uv]\cos\varphi+\frac{\partial}{\partial p}[u\omega]-f[v]-[uv]\frac{\tan\varphi}{a}=[F_\lambda],$$

$$\frac{\partial}{\partial t}[v]+\frac{1}{a\cos\varphi}\frac{\partial}{\partial\varphi}[v^2]\cos\varphi+\frac{\partial}{\partial p}[v\omega]+f[u]+[u^2]\frac{\tan\varphi}{a}$$

$$=-\frac{1}{a}\frac{\partial}{\partial\varphi}[\Phi]+[F_\varphi].$$

Let us substitute the definitions of u, v and ω from (2.5.1) into these equations and then multiply the first by $[u]$, the second by $[v]$ and add the resulting expressions. After these transformations we have

$$\frac{\partial}{\partial t}\tfrac{1}{2}([u]^2+[v]^2)+\frac{1}{a\cos\varphi}\left\{[u]\frac{\partial}{\partial\varphi}(([u][v]+[u^*v^*])\cos\varphi)\right.$$

$$+[v]\frac{\partial}{\partial\varphi}(([v]^2+[v^{*2}])\cos\varphi)\Big\}+\Big\{[u]\frac{\partial}{\partial p}([u][\omega]+[u^*\omega^*])$$

$$+[v]\frac{\partial}{\partial p}([v][\omega]+[v^*\omega^*])\Big\}-([u^*v^*][u]-[u^{*2}][v])\frac{\tan\varphi}{a}$$

$$=-\frac{[v]}{a}\frac{\partial}{\partial\varphi}[\Phi]+([u][F_\lambda]+[v][F_\varphi]). \tag{2.5.20}$$

Let us present the second and third terms on the left-hand side and the first term on the right-hand side of this equation in the form

$$\frac{1}{a\cos\varphi}\Big\{[u]\frac{\partial}{\partial\varphi}(([u][v]+[u^*v^*])\cos\varphi)$$

$$+[v]\frac{\partial}{\partial\varphi}(([v]^2+[v^{*2}])\cos\varphi)\Big\}=\frac{1}{a\cos\varphi}\Big\{\frac{\partial}{\partial\varphi}\left(\left(\frac{[u]^2+[v]^2}{2}[v]\right.\right.$$

$$+[u^*v^*][u]+[v^{*2}][v]\Big)\cos\varphi\Big)-\cos\varphi\left([u^*v^*]\frac{\partial[u]}{\partial\varphi}+[v^{*2}]\frac{\partial[v]}{\partial\varphi}\right)\Big\};$$

$$\left\{ [u] \frac{\partial}{\partial p} ([u][\omega] + [u^*\omega^*]) + [v] \frac{\partial}{\partial p} ([v][\omega] + [v^*\omega^*]) \right\}$$

$$= \frac{\partial}{\partial p} \left(\frac{[u]^2 + [v]^2}{2} [\omega] + [u^*\omega^*][u] + [v^*\omega^*][v] \right)$$

$$- \left([u^*\omega^*] \frac{\partial[u]}{\partial p} + [v^*\omega^*] \frac{\partial[v]}{\partial p} \right);$$

$$- \frac{[v]}{a} \frac{\partial[\Phi]}{\partial \varphi} = - \frac{[v] \cos \varphi}{a \cos \varphi} \frac{\partial[\Phi]}{\partial \varphi} = - \frac{1}{a \cos \varphi} \frac{\partial}{\partial \varphi} [\Phi][v] \cos \varphi$$

$$+ \frac{[\Phi]}{a \cos \varphi} \frac{\partial}{\partial \varphi} [v] \cos \varphi;$$

or according to (2.5.18) and (2.5.19)

$$- \frac{[v]}{a} \frac{\partial[\Phi]}{\partial \varphi} = - \frac{1}{a \cos \varphi} \frac{\partial}{\partial \varphi} [\Phi][v] \cos \varphi - [\Phi] \frac{\partial[\omega]}{\partial p}$$

$$= - \frac{1}{a \cos \varphi} \frac{\partial}{\partial \varphi} [\Phi][v] \cos \varphi - \frac{\partial}{\partial p} [\Phi][\omega] - [\alpha][\omega].$$

Substitution of these equations into (2.5.20) yields

$$\frac{\partial}{\partial t} \frac{[u]^2 + [v]^2}{2} + \frac{1}{a \cos \varphi} \frac{\partial}{\partial \varphi} \left(\frac{[u]^2 + [v]^2}{2} [v] \cos \varphi \right)$$

$$+ \frac{\partial}{\partial p} \left(\frac{[u]^2 + [v]^2}{2} [\omega] \right) + \frac{1}{a \cos \varphi} \frac{\partial}{\partial \varphi} (([u^*v^*][u] + [v^{*2}][v]) \cos \varphi)$$

$$- \frac{1}{a} \left([u^*v^*] \frac{\partial[u]}{\partial \varphi} + [v^{*2}] \frac{\partial[v]}{\partial \varphi} \right) + \frac{\partial}{\partial p} ([u^*\omega^*][u] + [v^*\omega^*][v])$$

$$- \left([u^*\omega^*] \frac{\partial[u]}{\partial p} + [v^*\omega^*] \frac{\partial[v]}{\partial p} \right) - ([u^*v^*][u] - [u^{*2}][v]) \frac{\tan \varphi}{a}$$

$$= - \frac{1}{a \cos \varphi} \frac{\partial}{\partial \varphi} [\Phi][v] \cos \varphi - \frac{\partial}{\partial p} [\Phi][\omega] - [\alpha][\omega]$$

$$+ ([u][F_\lambda] + [v][F_\varphi]). \tag{2.5.21}$$

Integrating (2.5.21) over the entire mass of the atmosphere and taking into account appropriate boundary conditions, we have

$$\frac{\partial}{\partial t} \int K_M a^2 \cos \varphi \, d\lambda \, d\varphi \, dp/g = C(A_M, K_M) + C(K_E, K_M) - D(K_M), \tag{2.5.22}$$

where

$$C(A_M, K_M) = - \int [\alpha][\omega] a^2 \cos \varphi \, d\lambda \, dp/g; \qquad (2.5.23)$$

$$C(K_E, K_M) = \int a \cos \varphi \left\{ \left([u^*v^*] \frac{\partial}{a \, \partial \varphi} + [u^*\omega^*] \frac{\partial}{\partial p} \right) \left(\frac{[u]}{a \cos \varphi} \right) \right.$$

$$+ \left. \left([v^{*2}] \frac{\partial}{a \, \partial \varphi} + [v^*\omega^*] \frac{\partial}{\partial p} - \frac{\tan \varphi}{a} ([u^{*2}] + [v^{*2}]) \right) \left(\frac{[v]}{a \cos \varphi} \right) \right\}$$

$$\times a^2 \cos \varphi \, d\lambda \, d\varphi \, dp/g; \qquad (2.5.24)$$

$$D(K_M) = - \int ([u][F_\lambda] + [v][F_\phi]) a^2 \cos \varphi \, d\lambda \, d\varphi \, dp/g. \qquad (2.5.25)$$

To derive the budget equation for K_E we direct our atttention again to the primitive atmospheric dynamics equations, Equations (2.5.16)–(2.5.19). Multiplying the first of these equations by u, and the second by v, and then adding the resulting expressions, and using the two remaining equations of the system, Equations (2.5.16)–(2.5.19), we obtain the local equation of the kinetic energy budget for the total (zonal mean plus eddy) motion. This takes the form

$$\frac{\partial}{\partial t} \frac{u^2 + v^2}{2} + \frac{1}{a \cos \varphi} \left(\frac{\partial}{\partial \lambda} \left(\frac{u^2 + v^2}{2} u \right) + \frac{\partial}{\partial \varphi} \left(\frac{u^2 + v^2}{2} v \cos \varphi \right) \right) + \frac{\partial}{\partial p} \left(\frac{u^2 + v^2}{2} \omega \right)$$

$$= - \frac{1}{a \cos \varphi} \left(\frac{\partial}{\partial \lambda} \Phi u + \frac{\partial}{\partial \varphi} \Phi v \cos \varphi \right) - \frac{\partial}{\partial p} \Phi \omega - \omega \alpha + (u F_\lambda + v F_\varphi).$$

Let us substitute Equations (2.5.13) into this equation and average it over the longitude. As a result we have

$$\frac{\partial}{\partial t} \left(\frac{[u]^2 + [v]^2}{2} + \frac{[u^{*2}] + [v^{*2}]}{2} \right) + \frac{1}{a \cos \varphi} \frac{\partial}{\partial \varphi} \left\{ \frac{[u]^2 + [v]^2}{2} [v] \right.$$

$$+ \frac{[u^{*2}] + [v^{*2}]}{2} [v] + ([u][u^*v^*] + [v][v^{*2}]) \right\} \cos \varphi + \frac{\partial}{\partial p} \left\{ \frac{[u]^2 + [v]^2}{2} [\omega] \right.$$

$$+ \frac{[u^{*2}] + [v^{*2}]}{2} [\omega] + ([u][u^*\omega^*] + [v][v^*\omega^*]) \right\}$$

$$= - \frac{1}{a \cos \varphi} \frac{\partial}{\partial \varphi} ([\Phi][v] + [\Phi^*v^*]) \cos \varphi - \frac{\partial}{\partial p} ([\Phi][\omega] + [\Phi^*\omega^*])$$

$$- ([\omega][\alpha] + [\omega^*\alpha^*]) + ([u][F_\lambda] + [v][F_\varphi]) + ([u^*F_\lambda^*] + [v^*F_\varphi^*]). \qquad (2.5.26)$$

Subtracting (2.5.21) from (2.5.26) we have the desired equation for K_E:

$$\frac{\partial}{\partial t} \frac{[u^{*2}] + [v^{*2}]}{2} + \frac{1}{a \cos \varphi} \frac{\partial}{\partial \varphi} \left\{ \frac{[u^{*2}] + [v^{*2}]}{2} [v] + ([u^* v^*][u] \right.$$

$$+ [v^{*2}][v] \Bigg\} \cos \varphi + \frac{\partial}{\partial p} \left\{ \frac{[u^{*2}] + [v^{*2}]}{2} [\omega] + ([u^* \omega^*][u] \right.$$

$$+ [v^* \omega^*][v]) \Bigg\} + \frac{1}{a} \left([u^* v^*] \frac{\partial [u]}{\partial \varphi} + [v^{*2}] \frac{\partial [v]}{\partial \varphi} \right)$$

$$+ \left([u^* \omega^*] \frac{\partial [u]}{\partial p} + [v^* \omega^*] \frac{\partial [v]}{\partial p} \right) + \frac{\tan \varphi}{a} ([u^* v^*][u] - [u^{*2}][v])$$

$$= -\frac{1}{a \cos \varphi} \frac{\partial}{\partial \varphi} [\Phi^* v^*] \cos \varphi - \frac{\partial}{\partial p} [\Phi^* \omega^*] - [\omega^* \alpha^*]$$

$$+ ([u^* F_\lambda^*] + [v^* F_\varphi^*]). \tag{2.5.27}$$

Integrating (2.5.27) over the whole mass of the atmosphere and taking into account the boundary conditions we finally have

$$\frac{\partial}{\partial t} \int K_E a^2 \cos \varphi \, d\lambda \, d\varphi \, dp/g = C(A_E, K_E) - C(K_E, K_M) - D(K_E), \tag{2.5.28}$$

where

$$C(A_E, K_E) = -\int [\omega^* \alpha^*] a^2 \cos \varphi \, d\varphi \, d\lambda \, dp/g; \tag{2.5.29}$$

$$D(K_E) = -\int ([u^* F_\lambda^*] + [v^* F_\varphi^*]) a^2 \cos \varphi \, d\lambda \, d\varphi \, dp/g. \tag{2.5.30}$$

In Equations (2.5.22) and (2.5.28) the constituents $C(K_E, K_M)$ describe the mutual transformations of kinetic energy of zonal mean (K_M) and eddy (K_E) motions. It is obvious from Equation (2.5.24) that these constituents represent the work of Reynolds stresses on the gradients of zonal velocity. In addition, the first and third terms in the integrand describe the so-called *barotropic instability* of zonal circulation. The remaining terms in (2.5.22) and (2.5.27) describe the dissipation of kinetic energy of zonal mean and eddy motions (components $D(K_M)$ and $D(K_E)$) and mutual transformations of kinetic and available potential energy (components $C(A_M, K_M)$ and $C(A_E, K_E)$).

The budget equations for the available potential energy can be obtained in much the same manner as we have just shown for K_M and K_E. Therefore, without repeating the derivation, one might restrict oneself to a discussion of the final expressions. But to avoid misunderstanding, which often occurs when presenting this subject, we will proceed in the opposite direction. So, we will start from the heat budget equation, written in terms of potential temperature. This equation, after transition to the isobaric system of co-ordinates, takes the form

$$\frac{\partial \theta}{\partial t} + \frac{u}{a \cos \varphi} \frac{\partial \theta}{\partial \lambda} + \frac{v}{a} \frac{\partial \theta}{\partial \varphi} + \omega \frac{\partial \theta}{\partial p} = \frac{\theta}{T} \frac{F_T}{c_p}. \tag{2.5.31}$$

Let us define θ as the sum of the average (over the sphere surface) value $\tilde{\theta}$ and of its deviation θ'. We define velocity components as well as heat sources and sinks in the atmosphere similarly. Substitution of these relationships into (2.5.31) and subsequent subtraction of the equation for $\tilde{\theta}$

$$\frac{\partial \tilde{\theta}}{\partial t} = \frac{\theta}{T} \frac{\tilde{F}_T}{c_p} \tag{2.5.32}$$

from the resulting equality yields

$$\frac{\partial \theta'}{\partial t} + \frac{u}{a \cos \varphi} \frac{\partial \theta'}{\partial \lambda} + \frac{v}{a} \frac{\partial \theta'}{\partial \varphi} + \omega \frac{\partial \theta'}{\partial p} + \omega \frac{\partial \tilde{\theta}}{\partial p} = \frac{\theta}{T} \frac{F'_T}{c_p}. \tag{2.5.33}$$

Here we have assumed that the multiplier θ/T on the right-hand sides of Equations (2.5.31) and (2.5.32) does not depend on horizontal coordinates, and have also taken into consideration the fact that in accordance with the continuity equation integrated over the surface of a sphere the isobaric vertical velocity $\tilde{\omega}$ obeys the equation $\partial \tilde{\omega}/\partial p = 0$; from here, with $\tilde{\omega}$ equal to zero at the upper atmosphere boundary ($p = 0$), it follows that $\tilde{\omega} \equiv 0$.

We introduce definitions

$$\begin{aligned}
\theta' &= [\theta] + \theta^*, & u &= [u] + u^*, \\
v &= [v] + v^*, & \omega &= [\omega] + \omega^*, \\
F'_T &= [F_T\} + F^*_T,
\end{aligned} \tag{2.5.34}$$

where, as before, the square brackets signify an average over the longitude, the asterisk is a deviation from the zonal mean, and then we substitute (2.5.34) into (2.5.33), rewritten in divergent form, and average the equation obtained over the longitude. As a result we have

$$\frac{\partial [\theta]}{\partial t} + \frac{1}{a \cos \varphi} \frac{\partial}{\partial \varphi} (([\theta][v]) + [\theta^* v^*]) \cos \varphi)$$

$$+ \frac{\partial}{\partial p} ([\theta][\omega] + [\theta^* \omega^*]) = \frac{\theta}{T} \frac{[F_T]}{c_p} - [\omega] \frac{\partial \tilde{\theta}}{\partial p},$$

or, after multiplying by $[\theta]$,

$$\frac{\partial}{\partial t}\frac{[\theta]^2}{2} + \frac{1}{a\cos\varphi}\frac{\partial}{\partial\varphi}\left(\frac{[\theta]^2}{2}[v] + [\theta][\theta^*v^*]\right)\cos\varphi$$

$$+ \frac{\partial}{\partial p}\left(\frac{[\theta]^2}{2}[\omega] + [\theta][\theta^*\omega^*]\right) - \left([\theta^*v^*]\frac{\partial}{a\,\partial\varphi} + [\theta^*\omega^*]\frac{\partial}{\partial p}\right)[\theta]$$

$$= \frac{\theta}{T}[\theta]\frac{[F_T]}{c_p} - [\theta][\omega]\frac{\partial\tilde\theta}{\partial p}. \quad (2.5.35)$$

We transform the expression for the last term on the right-hand side of (2.5.35). But first we note that

$$\frac{\partial\tilde\theta}{\partial p} = \left(\frac{p_0}{p}\right)^{R/c_p}\left(\frac{\partial\tilde T}{\partial p} - \frac{R}{c_p}\frac{\tilde T}{p}\right) = \left(\frac{p_0}{p}\right)^{R/c_p}\left(\frac{\partial\tilde T}{\partial z}\frac{\partial z}{\partial p} - \frac{1}{c_p\tilde\rho}\right)$$

$$= -\frac{1}{g\tilde\rho}\left(\frac{p_0}{p}\right)^{R/c_p}\left(\frac{\partial\tilde T}{\partial z} + \frac{g}{c_p}\right) = -\frac{1}{g\tilde\rho}\left(\frac{p_0}{p}\right)^{R/c_p}(\gamma_a - \gamma),$$

where, as before, $\gamma_a = g/c_p$ and $\gamma = -\partial\tilde T/\partial z$ are adiabatic and real vertical temperature gradients.

Thus,

$$[\theta][\omega]\frac{\partial\tilde\theta}{\partial p} = -\frac{\tilde T}{g}\left(\frac{p_0}{p}\right)^{2R/c_p}(\gamma_a - \gamma)[\alpha][\omega]$$

$$= -\frac{\tilde T}{c_p}\left(\frac{p_0}{p}\right)^{2R/c_p}\frac{(\gamma_a - \gamma)}{\gamma_a}[\alpha][\omega]$$

$$= -\frac{\tilde T}{c_p}\left(\frac{\theta}{T}\right)^2\frac{(\gamma_a - \gamma)}{\gamma_a}[\alpha][\omega].$$

Let us substitute this expression into (2.5.35), and then multiply the resulting equations by $(c_p/\tilde T)(T/\theta)^2\gamma_a/(\gamma_a - \gamma)$, represent $[\theta]$ in terms of $[T]$ and integrate over the whole mass of the atmosphere. Then, finally, we have

$$\frac{\partial A_m}{\partial t} = G(A_M) - C(A_M, K_M) - C(A_M, A_E), \quad (2.5.36)$$

where

$$A_M = \tfrac{1}{2}c_p\int\frac{\gamma_a}{(\gamma_a - \gamma)}\frac{[T]^2}{\tilde T}a^2\cos\varphi\,d\lambda\,d\varphi\,dp/g, \quad (2.5.37)$$

$$G(A_M) = \int\frac{\gamma_a}{(\gamma_a - \gamma)}\frac{[T]}{\tilde T}[F_T]a^2\cos\varphi\,d\lambda\,d\varphi\,dp/g, \quad (2.5.38)$$

$$C(A_M, A_E) = -c_p \int \frac{\gamma_a}{(\gamma_a - \gamma)} \left([T^*v^*] \frac{\partial}{a\,\partial\varphi} + [T^*\omega^*] \frac{\partial}{\partial p} \right)$$

$$\times \frac{[T]}{\tilde{T}} a^2 \cos\varphi \, d\lambda \, d\varphi \, dp/g. \tag{2.5.39}$$

The definition of $C(A_M, K_M)$ was given above.

Equation (2.5.36) is the equation for the available potential energy budget. The terms on the right-hand side describe, respectively, the generation of available potential energy by zonal mean motion, mutual conversions of available potential and kinetic energy of zonal mean motion and mutual conversions of available potential energy of zonal mean motion and eddy disturbances. The last named process has its origin in the so-called *baroclinic instability* of zonal motion.

Let us turn to Equation (2.5.33) to derive the budget equation for the available potential energy A_E of eddy disturbances. We rewrite it in divergent form, then multiply by θ', average over the longitude and use Equations (2.5.34). As a result we obtain

$$\frac{\partial}{\partial t} \tfrac{1}{2}([\theta]^2 + [\theta^{*2}]) + \frac{1}{a\cos\varphi} \frac{\partial}{\partial\varphi} (\tfrac{1}{2}[\theta]^2[v] + [\theta^{*2}][v]$$

$$+ 2[\theta][\theta^*v^*]) \cos\varphi + \frac{\partial}{\partial p} (\tfrac{1}{2}[\theta]^2[\omega] + [\theta^{*2}][\omega] + 2[\theta][\theta^*\omega^*])$$

$$= \frac{1}{c_p} \frac{\theta}{T} ([\theta][F_T] + [\theta^*][F_T^*]) - ([\theta][\omega] + [\theta^*\omega^*]) \frac{\partial\tilde\theta}{\partial p}. \tag{2.5.40}$$

Subtracting (2.5.33) from (2.5.40) we obtain

$$\frac{\partial}{\partial t} \frac{[\theta^{*2}]}{2} + \frac{1}{a\cos\varphi} \frac{\partial}{\partial\varphi} (\tfrac{1}{2}([\theta^{*2}][v] + [\theta][\theta^*v^*]) \cos\varphi)$$

$$+ \frac{\partial}{\partial p} \tfrac{1}{2}([\theta^{*2}][\omega] + [\theta][\theta^*\omega^*]) + \left([\theta^*v^*] \frac{\partial}{a\,\partial\varphi} + [\theta^*\omega^*] \frac{\partial}{\partial p} \right)[\theta]$$

$$= \frac{1}{c_p} \frac{\theta}{T} [\theta^*F_T^*] - [\theta^*\omega^*] \frac{\partial\tilde\theta}{\partial p}. \tag{2.5.41}$$

We turn in (2.5.41) from $[\theta^{*2}]$ to $[T^{*2}]$ and multiply the resulting equation by $c_p\gamma_a/\tilde{T}(\gamma_a - \gamma)$, integrating it over the whole mass of the atmosphere.

If we introduce designations

$$A_E = \tfrac{1}{2}c_p \int \frac{\gamma_a}{(\gamma - \gamma)} \frac{[T^{*2}]}{\tilde{T}} a^2 \cos \varphi \, d\lambda \, d\varphi \, dp/g, \qquad (2.5.42)$$

$$G(A_E) = \int \frac{\gamma_a \tilde{T}^{-1}}{(\gamma_a - \gamma)} [T^*F_T^*]a^2 \cos \varphi \, d\lambda \, d\varphi \, dp/g, \qquad (2.5.43)$$

for the available potential energy of eddy disturbance and its generation, respectively, and take into account the definitions of mutual energy conversions (see (2.5.29) and (2.5.39)), then the budget equation for A_E takes the form

$$\frac{\partial A_E}{\partial t} = G(A_E) - C(A_E, K_E) + C(A_M, A_E). \qquad (2.5.44)$$

The energy cycle of the atmosphere described by Equations (2.5.22), (2.5.28), (2.5.36) and (2.5.44) is shown in Figure 2.14. The arrows here signify the directions of energy transitions; the numerical estimates are based on factual data throughout the decade 1963–1973 and relate to the total years' average conditions. An analysis of this figure leads to the following conclusions. The annual mean radiational heating of the atmosphere in the tropics and cooling in high latitudes result in the generation of the available potential energy of zonal mean motion. Affected by baroclinic instability, it is converted into available potential energy of eddy disturbances with a rate of 1.27 W/m². This conversion, together with the generation of the available

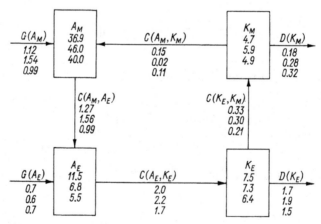

Figure 2.14 Energy cycle of the global atmosphere, according to Oort and Peixóto (1983). The upper numbers indicate annual mean conditions; the middle numbers indicate winter conditions; and the lower numbers indicate summer conditions. Different forms of energy are in 10^5 J/m²; generation, energy transitions and dissipation are in W/m².

potential energy of eddy disturbances (0.74 W/m^2), determined by release of latent heat, is balanced (due to the mechanism of barotropic instability) by the conversion of available potential energy into kinetic energy of eddy disturbances. A small part (0.33 W/m^2) of the incoming kinetic energy of eddy disturbances goes towards maintaining the zonal average circulation that is equivalent to reverse energy transfer over the spectrum (i.e. the transfer from motions with small horizontal scales to large ones) – the phenomenon known as *negative viscosity*. The main part (1.7 W/m^2) is involved in direct cascade energy transfer from the interval of energy supply to the viscous range, where the kinetic energy of eddy disturbances dissipates into heat. Finally, of the kinetic energy of the zonal mean circulation, part (0.18 W/m^2) dissipates, and part (0.15 W/m^2) transforms into the available potential energy of zonal mean circulation under the influence of ordered meridional motions. This is the energy cycle of the atmosphere in general terms. Its schematic representation can be given in the form $A_M \Rightarrow A_E \Rightarrow K_E \Rightarrow K_M \Rightarrow A_M$.

The above-mentioned features of the atmospheric energy cycle are qualitatively similar not only in different seasons of the year, but in different hemispheres. From the quantitative perspective, the differences reduce to amplification of the generation of available potential energy of zonal mean motions and eddy disturbances in summer, to the appearance of the powerful transport of available potential energy of the zonal mean circulation from a summer hemisphere into a winter one, to the disappearance of the reverse energy transfer over the spectrum and its replacement by direct transfer, and, finally, to the intensification of kinetic energy dissipation in winter. Let us note one more interesting detail: there is almost a doubling of amplitudes of seasonal variations for different forms of energy in the Northern Hemisphere as compared to their values in the Southern Hemisphere, which is caused by the different ocean–land area ratio in both hemispheres.

Ocean

Before embarking upon a discussion of ocean energetics, let us note that the compexity of the equation of sea water state relating density to temperature, salinity and pressure excludes the possibility of representing the total potential energy as the sum of internal and potential energies. But it does not prevent the introduction of the concept of available gravitational potential energy.

We define the potential energy Φ so that it takes on zero value at a reference depth $z = -z_r$, that is, $\Phi = g\rho(z + z_r)/\rho_0$, where ρ is the sea water density and ρ_0 is its constant value, and apply the operator d/dt to both parts of this equation. As a result we obtain

$$\partial\Phi/\partial t + (\mathbf{u}\nabla)\Phi = g(\rho/\rho_0)w + g(z + z_r)(\partial/\partial t + \mathbf{u}\nabla)\rho/\rho_0. \quad (2.5.45)$$

We find the second term on the right-hand side of (2.5.45) using the evolution equation for density

$$\partial\rho/\partial t + (\mathbf{u}\nabla)\rho = F_\rho, \quad (2.5.46)$$

where F_ρ are sources and sinks of mass in the ocean.

Substitution into (2.5.45) yields the budget equation for potential energy

$$\partial\Phi/\partial t + (\mathbf{u}\nabla)\Phi = g(\rho/\rho_0)w + g(z + z_r)F_\rho/\rho_0. \quad (2.5.47)$$

The budget equation for kinetic energy $K = |u|^2/2$ can be written similarly to (2.5.10) as

$$\partial K/\partial t + (\mathbf{u}\nabla)K = -(1/\rho_0)\nabla p\mathbf{u} - g(\rho/\rho_0)w + \mathbf{v}\cdot\mathbf{F}. \quad (2.5.48)$$

We now introduce the definition of the *available gravitational potential energy*:

$$A = -\frac{g}{2}\int \frac{(\rho - \tilde{\rho})^2}{\partial\tilde{\rho}/\partial z} a^2 \cos\varphi \, d\varphi \, d\lambda \, dz, \quad (2.5.49)$$

where $\tilde{\rho}$ is the density value averaged over the ocean area; the integration extends to the whole ocean volume and we then represent the climatic characteristics as the sum of average (in the above-mentioned sense) values and departures from them. We designate the former values by tildes, and the latter values by asterisks. After substitution of these expressions into (2.5.46) we obtain

$$\partial\rho^*/\partial t + (\mathbf{u}\nabla)\rho^* + w^* \, \partial\tilde{\rho}/\partial z = F_\rho^*, \quad (2.5.50)$$

where we have taken into account the fact that $\tilde{w} = 0$ and $\tilde{F}_\rho = 0$, and, hence, $\partial\tilde{\rho}/\partial t = 0$.

We multiply Equation (2.5.50) by $-g\rho^*(\partial\tilde{\rho}/\partial z)^{-1}$ and then integrate it over the ocean volume and use the appropriate boundary conditions. Then, discarding negligible terms describing the effect of river and underground run-off, we arrive at the following equation:

$$\partial A/\partial t = g \int \rho^* w^* a^2 \cos\varphi \, d\varphi \, d\lambda \, dz - g \int \frac{\rho^* F_\rho^*}{\partial\tilde{\rho}/\partial z} a^2 \cos\varphi \, d\varphi \, d\lambda \, dz. \quad (2.5.51)$$

This is the budget equation for the available gravitational potential energy in the ocean. The first term on the right-hand side in this equation represents the mutual conversions of the kinetic and available gravitational potential energy; the second term represents the production and degeneration of the available gravitational potential energy. Let us note that, because $\tilde{w} = 0$,

$\widetilde{\rho^*\rho^*} = \widetilde{\rho w}$, and therefore not all potential energy but, rather, only that part defined by Equation (2.5.49) is involved in the exchange with the kinetic energy (see (2.5.48)).

Separation of K and A into zonal and eddy components and the derivation of the corresponding budget equations are carried out formally in the same manner as for the atmosphere. Without repeating this procedure we turn to a discussion of empirical estimates of ocean energy cycle characteristics. According to Oort *et al.* (1989) the kinetic energy density of the zonally averaged circulation and eddy disturbances in the World Ocean for the annual mean conditions amounts to, on average, 0.006 and 0.075 J/m^2, and the available gravitational potential energy density is 4.40 and 1.68 J/m^2 respectively. For comparison: in the atmosphere these are equal to 4.5 and 7.3 J/m^2, and 33.3 and 11.1 J/m^2, respectively.

The above suggests that the kinetic energy in the ocean is much less than the available gravitational potential energy. This, in turn, means that the conversion of the available gravitational potential energy into kinetic energy of large-scale ocean circulation is small, and hence the time scales of processes in the ocean are much larger than in the atmosphere. Indeed, the rates of generation and dissipation of kinetic energy in the ocean are of the order of 0.1 W/m^2 and the rate of generation of available gravitational potential energy and its conversion into kinetic energy is of the order of 2×10^{-4} W/m^2 (see Lueck and Reid, 1984). In other words, the ocean is characterized by the approximate parity between the production of available gravitational potential energy and its conversion into kinetic energy, on the one hand, and the generation and dissipation of kinetic energy, on the other.

Moreover, the kinetic and available gravitational potential energy in the ocean are much less than in the atmosphere, which results from the fact that in the ocean the thickness of the layer in which the main density disturbances are concentrated is no more than several hundred metres, while in the atmosphere the temperature disturbances extend to the limits of its total thickness. The fact that the ocean stratification is more stable than that of the atmosphere is of no small importance. Thus, in interaction with the atmosphere the ocean behaves as a passive, inertly responding partner.

To complete this discussion, let us enumerate the most distinguishing features of the energetics of the climatic system as a whole. According to empirical data classified by Monin (1982) the incoming short-wave solar radiation flux per unit area of the upper atmospheric boundary amounts to 1356 W/m^2. Part of this radiation (the *planetary albedo*) which is reflected back to space is equal, on the average, to 0.30, so that the short-wave solar radiation flux assimilated by the climatic system amounts to

$244 \, W/m^2$. The latter is redistributed in the following way: from the 70% of the incoming solar radiation remaining after reflection, 20% is absorbed by the atmosphere and 50% is absorbed by the ocean and the land. In its turn, the short-wave solar radiation absorbed by the ocean and land is expended on evaporation from the underlying surface (24%) and on sensible heat exchange with the atmosphere (8%). The remaining part (20%) is expended on ocean and land heating and hence on the conversion of short-wave solar radiation into long-wave thermal radiation from which 14% is absorbed by the atmosphere and 6% is emitted into space. Only a very small part (about $4 \, W/m^2$, or slightly more than 1%) of the assimilated solar radiation is converted into the kinetic energy of atmospheric and ocean motions. The rate of kinetic energy dissipation must be of the same order. Finally, the heat energy, absorbed by the atmosphere and equal to 64% of incoming short-wave solar radiation (20% of this is determined by the absorption of short-wave radiation, 16% is due to the absorption of long-wave radiation and 30% is determined by the exchange of sensible and latent heat with the underlying surface), is returned to space by means of long-wave radiation, thus completing the energy cycle. This is illustrated in Figure 2.15.

2.6 Angular momentum budget

The absolute angular momentum of a unit mass relative to rotation of the Earth's axis is defined by the expression

$$M = \Omega a^2 \cos^2 \varphi + ua \cos \varphi, \tag{2.6.1}$$

where the first term on the right-hand side is called the *planetary momentum* (it describes the angular momentum of the unit mass rotating together with the Earth as a solid body); the second term is called the *relative momentum* (it has its origin in motion relative to the rotating Earth). The relative momentum is thought to be positive when moving from west to east, and negative when moving backwards.

We apply the individual derivative operator for (2.6.1) and write down the expression for dM/dt, which takes the form

$$\frac{\partial M}{\partial t} = \left(\frac{\partial u}{\partial t} + \left(f + \frac{u}{a} \tan \varphi \right) v \right) a \cos \varphi.$$

Combining this relationship with the equation of motion for the zonal velocity component:

$$\frac{\partial u}{\partial t} + \left(f + \frac{u}{a} \tan \varphi \right) v = -\frac{1}{\rho} \frac{\partial p}{a \cos \varphi \, \partial \lambda} + F_\lambda,$$

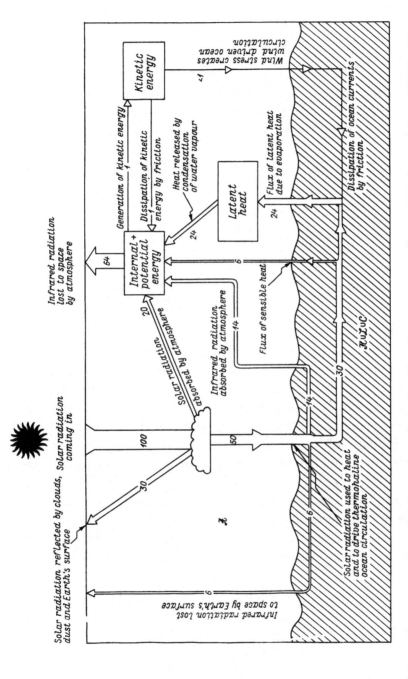

Figure 2.15 Schematic representation of the energy cycle in the climatic system, according to Peixóto and Oort (1984). All energy transitions are given in % of the incoming solar radiation.

we obtain

$$\rho \frac{\partial M}{\partial t} = -\frac{\partial p}{\partial \lambda} + \rho F_\lambda a \cos \varphi, \qquad (2.6.2)$$

from which it follows that a change in the absolute angular momentum is determined only by the torque of pressure and friction forces.

We integrate (2.6.2) over the longitude and then over the vertical within the limits of the atmospheric thickness. Then, instead of (2.6.2), we obtain

$$2\pi a \cos \varphi \, \frac{\partial}{\partial t} \, [m_A \hat{M}_A] + \frac{1}{a} \frac{\partial}{\partial \varphi} \, MMT_A = \int_0^\infty \sum_i (p^i_{AE} - p^i_{AW}) a \cos \varphi \, dz$$

$$- 2\pi [\tau_\lambda] a^2 \cos^2 \varphi. \qquad (2.6.3)$$

where

$$MMT_A = \int_0^{2\pi} m_A v_A \widehat{M}_A a \cos \varphi \, d\lambda$$

$$= \int_0^{2\pi} m_A (\Omega \hat{v}_A a^2 \cos \varphi + \widehat{u_A v_A} a \cos \varphi) a \cos \varphi \, d\lambda$$

is the meridional transport of the absolute angular momentum in the atmosphere which is equal to the sum of the meridional transports of the planetary momentum (the first term on the right-hand side) and the relative momentum (the second term); $(p^i_{AE} - p^i_{AW})$ is the pressure difference on the west and east slopes of the ith mountain ridge; summation is carried out for all mountain ridges and other irregularities of the Earth's surface that cross the latitude belt in question; τ_λ is the zonal component of tangential wind stress at the underlying surface; the remaining designations are the same.

Equation (2.6.3) describes the budget of the angular momentum in the zonal belt of unit meridional extension. To estimate the components of the budget we use observational data of the annual mean zonal wind velocity (Figure 2.16). The distinguishing feature of its distribution in the meridional

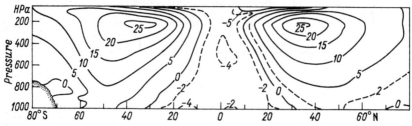

Figure 2.16 Annual mean latitude–altitude distribution of the zonal component of wind velocity (m/s), according to Peixóto and Oort (1984).

plane is the three-cell structure of the circulation in each hemisphere with two direct cells in the tropical and polar latitudes and one reverse cell in the temperate latitudes. It will be recalled that the *direct cell* of the circulation is the one with upwelling of warmer air and with downwelling of cooler air. On the other hand, in the *reverse cell* there is upwelling of cooler air and downwelling of warmer air. The direct cell of the circulation in low latitudes is usually referred to as the Hadley cell, and the reverse cell as the Ferrel cell.

It can be seen from Figure 2.16 that the zonal component of the wind velocity at low atmospheric levels is directed to the west (easterly trade winds) in low latitudes, and to the east (the westerlies) in temperate latitudes. Accordingly, the surface friction stress directed against the wind will be positive in the first case and negative in the second case. From this it follows that the atmosphere passes momentum to the Earth in low latitudes and obtains it from the Earth in temperate latitudes. But since, in accordance with Newton's third law, the Earth renders an equal and opposite (in direction) effect on the atmosphere, one can suggest that it is as a source of momentum for the atmosphere in low latitudes and a sink of momentum in temperature latitudes. The same can be said with regard to relative angular momentum. To confirm this we refer to Figure 2.17, which presents the results of calculations of the meridional distribution of the total surface torque due to both pressure and friction forces. This torque is thought to be positive if it tends to increase the eastward angular momentum of the atmosphere, that is, if, in the last term in Equation (2.6.3), the tangential wind stress is replaced by an equal, but opposite in direction, tangential stress of the underlying surface. According to Figure 2.16 the surface friction contributes to the transport of relative angular momentum from the underlying surface to the atmosphere in the tropics, and from the atmosphere to the underlying surface in temperate latitudes, and this direction of the transport probably remains constant throughout the year.

We note two more sources of relative angular momentum for the atmosphere: the regions of localization of the north and south polar cells of circulation. The appropriate radius of the latitude circle (the arm $a \cos \varphi$) is small and this is why the relative angular momentum transferred from the underlying surface to the atmosphere is also small.

It has been found that the relative angular momentum is transferred from the underlying surface to the atmosphere in the tropics. Its subsequent fate is as follows: first it is carried by the ascending branches of Hadley cells into the upper tropospheric layers, then it is transported by the transient eddies into the midlatitudes, and here it is reduced to compensate for losses in the

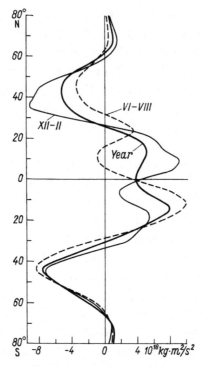

Figure 2.17 Zonal mean profile of the total surface torque due to both friction and pressure differences across mountains, according to Peixóto and Oort (1984).

relative angular momentum in the atmospheric planetary boundary layer due to friction.

It is known that the velocity Ωa of solid rotation is equal to 464 m/s and is much larger than the relative wind velocity u. One might think that because of this the meridional transport of absolute angular momentum has to be determined mainly by the planetary momentum transport. But in the steady state the integral (over the atmospheric mass) meridional transport of absolute angular momentum does not depend on the planetary momentum transport, for, according to the law of conservation of mass, the integral zonal component of the wind velocity is identically equal to zero. In other words, the integral meridional transport of absolute angular momentum is determined solely by the transport of relative angular momentum, and this in turn depends solely on the correlation $[\widehat{u_A v_A}]$ between the zonal and meridional components of the wind velocity. But because infinite accumulation of absolute angular momentum in temperate latitudes, as well as its infinite subtraction in the low latitudes, does not occur, the reverse (compensatory) transport of absolute angular momentum must exist. It is natural to assume that such transport is concentrated in the ocean. Let us check this possibility.

First we write down the budget equation for absolute angular momentum in the zonal belt of ocean with unit meridional extension. By analogy with (2.6.3) we obtain

$$2\pi a \cos \varphi \frac{\partial}{\partial t}[m_O \hat{M}_O] + \frac{1}{a}\frac{\partial}{\partial \varphi} MMT_O = -\int_{-H}^{0} \sum_i (p_{OE}^i - p_{OW}^i) a \cos \varphi \, dz$$

$$+ 2\pi[\tau_\lambda^0] a^2 \cos^2 \varphi, \qquad (2.6.4)$$

where

$$MMT_O = \int_0^{2\pi} \widehat{m_O v_O M_O} a \cos \varphi \, d\lambda$$

is the meridional transport of absolute angular momentum in the ocean; $(p_{OE}^i - p_{OW}^i)$ is the pressure difference at the east and west edges of continents, of mid-ocean ridges and other bottom undulations crossing the zonal belt in question; the tangential stress at the ocean bottom is assumed to be negligible compared to the tangential stress τ_λ^0 at the free ocean surface.

We compare estimates of the meridional transport of absolute angular momentum in the atmosphere and the ocean. Let us assume that typical values of ratios of correlation terms $\widehat{[u_A v_A]}$ and $\widehat{[u_O v_O]}$, of the mass of unit columns of the atmosphere and the ocean (m_A and m_O), and of the extension of the atmosphere and the ocean in the zonal direction, are equal to 10^4, 3×10^{-3}, and 2 respectively. Then $O(MMT_A/MMT_O) \approx 10^2$, that is, the meridional transport of the absolute angular momentum in the ocean is only 1% of that in the atmosphere, and, hence, it cannot compensate for the accumulation of the absolute angular momentum in the midlatitudes of the atmosphere.

Considering this fact, Equation (2.6.4) for the annual mean conditions is rewritten in the form

$$0 \approx -\int_{-H}^{0} \sum_i (p_{OE}^i - p_{OW}^i) a \cos \varphi \, dz + 2\pi[\tau_\lambda^0] a^2 \cos^2 \varphi. \qquad (2.6.5)$$

This suggests that the torque of the frictional force at the ocean surface is balanced by the torque created by the pressure difference (difference of mean ocean level) at the eastern and western edges of the continents, and, therefore, in contrast to the atmosphere the redistribution of the absolute angular momentum in the ocean occurs in the zonal rather than in the meridional direction (Figure 2.17).

Thus, we have only one possibility: to close the cycle of absolute angular momentum by its meridional transport in the Earth's solid body. This accords with Oort (1985), who proposed the diagram (Figure 2.18) describing the

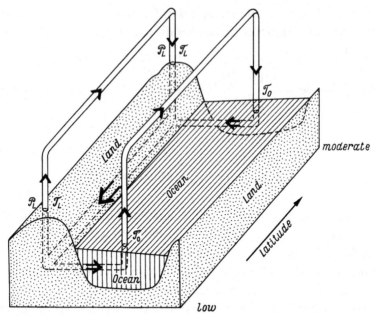

Figure 2.18 Schematic representation of the absolute angular momentum cycle in the ocean–atmosphere–lithosphere system, according to Oort (1985). The pressure and friction torques are designated by symbols $\mathscr{P}$, $\mathscr{T}$ respectively; index L indicates belonging to the land, index O indicates belonging to the ocean.

cycle of absolute angular momentum. The clarity of this diagram barely hides its shortcomings, related to difficulties of interpretation of the mechanism of absolute angular momentum transport in the Earth's solid body.

2.7 Carbon budget

The distinguished place of carbon among other chemical elements is determined first by the fact that carbon atoms, due to their electrical neutrality, easily interact with each other and with atoms of other elements, creating stable organic compounds and thereby producing a crucial effect on biological processes and on the life evolution on Earth. Moreover, carbon is included in the composition of such important (in the climatic sense) compounds as carbon dioxide whose role in the formation of the Earth's radiation is difficult to overestimate.

Carbon content. The carbon in the atmosphere is mainly present in the form of two stable compounds (carbon dioxide CO_2 and methane CH_4) and of one unstable compound (carbon monoxide CO) which quickly oxidizes to CO_2.

The volume concentration of the first compound in January 1988 was 351 ppm, of the second one it was about 1.6 ppm, and of the third one it was about 0.1 ppm. It is clear that the carbon content in the atmosphere is almost completely determined by carbon dioxide.

Regular precise measurements of CO_2 concentration in the atmosphere were initiated during the International Geophysical Year (1957–1958) on Mauna Loa (Hawaii). During 1960–1963 the South Pole station (Antarctica) began operations, and today there are more than 30 stations for background monitoring of CO_2 – these stations being located far from industrial centres. Analyses of data from these stations indicate that atmospheric CO_2 concentration has marked seasonal variability arising from a phase shift in the process of formation (*production*) and disintegration (*destruction*) of the organic matter substance in terrestrial ecosystems. The maximum amplitude of seasonal variations (~ 8.2 ppm) have been recorded at Gold Bay (Alaska) located in the region of the boreal forests of North America, and in Eurasia. The maximum amplitude decreases to the north to about 3 ppm on Mauna Loa and 0.5–1.5 ppm in the Southern Hemisphere where the phase shift between production and destruction of the organic matter localized mainly in the forests of the tropical belt is low.

Large-scale spatial variations in the volume concentration of atmospheric CO_2 are less than local seasonal variations. Thus, the average values for eight years (1976 through 1983) of volume concentrations at two neighbouring stations, Mauna Loa and Kumutahi, located at 3397 and 3 metre heights respectively, differ from each other by only 0.4 ± 0.3 ppm, and the average values for the same period at stations located in the northern and southern polar regions of the Earth differ by 3.2 ppm. Such relatively small spatial variability of the volume concentrations of atmospheric CO_2 is explained by rapid (compared with the time of renewal, see below) mixing of the atmosphere whose characteristic time scale (as shown in Table 1.1) does not exceed 10^6 s.

The carbon content in a unit atmospheric column is determined by the expression $C_A = (\mu_C/\mu_{CO_2})m_A\hat{c}_A$, where $c_A = 10^{-6}(\mu_{CO_2}/\mu)q_{CO_2}$ is the specific concentration of CO_2; $q_{CO_2} = 10^6 \times p_{CO_2}^A/p_s$ is the volume concentration of CO_2; and p_{CO_2} and p_s are the partial pressure of CO_2 and the atmospheric pressure at the sea level respectively; m_A is the mass of a unit atmospheric column with unit cross-section; μ_C, μ_{CO_2} and μ are the molecular weights of carbon, carbon dioxide and air; the symbol $\hat{}$ designates, as before, the operation of averaging over mass. According to this expression the carbon content in the atmosphere in January 1988 was 745 GtC (1 Gt $= 10^9$ t). For comparison, before the industrial revolution, that is, before 1860, it was equal to 594 GtC.

The dissolving of CO_2 in water is described by two expressions: $CO_2 \leqslant CO_2$ (diss) and CO_2 (diss) $+ H_2O \leqslant H_2CO_3$ defining the hydration phenomenon and condition for the chemical equilibrium of carbonic acid H_2CO_3. The dissolving takes several minutes and is determined by the equilibrium constant $K_0 = [H_2CO_3]/[CO_2]$, where $[CO_2]$ and $[H_2CO_3]$ are concentrations of dissolved carbon dioxide and dissolved carbonic acid. If the inorganic carbon content in the ocean were controlled only by this process then it would amount to 1% of its present value. In practice, carbon dioxide is not an inert gas like oxygen, nitrogen or like the noble gases such as argon, and this is why the dissolving of CO_2 in water is accompanied by a reaction with the substances of the Earth's crust and by the formation of other compounds which gradually accumulate in the ocean and, as a result, their concentrations become many times larger than the concentration of carbonic acid.

The major carbonic compound in the ocean is the bicarbonate ion HCO_3^-. Its content as a percentage of the total amount of inorganic carbon in the ocean is about 95%. The bicarbonate ion is formed by the dissociation of carbonic acid when the proton is lost. This stage proceeds instantly and is described by the condition of chemical equilibrium $H_2CO_3 \leqslant HCO_3^- + H^+$ with dissociation constant $K_1 = [HCO_3^-][H^+]/[H_2CO_3]$, where $[HCO_3^-]$ and $[H^+]$ are the concentrations of bicarbonate and hydrogen ions.

The next stage is the dissociation of the bicarbonate ion and the formation of a carbonate ion CO_3^{2-}. The appropriate condition for chemical equilibrium takes the form $HCO_3^- \leqslant CO_3^{2-} + H^+$ and is determined by the dissociation constant $K_2 = [CO_3^{2-}][H^+]/[HCO_3^-]$, where $[CO_3^{2-}]$ is the concentration of carbonate ions. The carbonate ion is the final product of the reaction of dissolved carbon dioxide with bases. Its concentration increases in the ocean until the solubility limit of calcium carbonate $CaCO_3$ is reached. The dissolving of $CaCO_3$ is described by the expression $CaCO_3 \leqslant Ca^{2+} + CO_3^{2-}$, and the dissociation constant K_S is determined by the product of concentration of calcium ions $[Ca^{2+}]$ and carbonate ions $[CO_3^{2-}]$ coexisting in equilibrium with one of two crystal forms of calcium carbonate: calcite and aragonite.

Let us define the total concentration c_0 of inorganic carbon in the ocean as $c_0 = [CO_2] + [HCO_3^-] + [CO_3^{2-}]$. Available data obtained by gas chromatography indicate (see Takahashi *et al.*, 1981) that the total concentration of inorganic carbon at the ocean surface increases with latitude from 1900 μmol/kg in the equatorial zone up to 2150 μmol/kg at latitude 55°N and up to 2250 μmol/kg at latitudes 55–60°S. It also increases with depth, but only within the limits of the upper two-kilometre layer of the ocean. A decrease in the total concentration of inorganic carbon in the surface layer is explained by inorganic carbon consumption during the process of photo-synthetic activity of phytoplankton, and to some extent by CO_2 exchange with the atmosphere, while an increase in total concentration below 1000 metres is explained by the decomposition of falling residues of phyto- and zooplankton and by the oxidation of soft organics. According to Kobak (1988) the total content of inorganic carbon in the ocean amounts to 38 200 GtC. And 864 GtC of this, that is, slightly more than in the atmosphere, is in the upper 100-metre layer of the ocean.

Data for the last few years obtained from satellite images, aerophotography and ground-based studies have allowed us to establish (see Kobak, 1988) the fact that the total content of organic carbon in the terrestrial biota does not exceed 560 GtC. Note that 20 years ago it was estimated as 827 GtC. The causes of these discrepancies are the limited nature and inaccuracy of initial information on the specific (referred to unit area) content of carbon in forest ecosystems and the areas which they occupy.

At present, the variety of species of life on Earth is determined by its terrestrial inhabitants: angiosperms and insects. The last-mentioned group accounts for one million species. In the ocean the specific variety of fauna and flora is much less (about 30 000 species of plants and about 160 000 species of animals) and this is explained by the lower (than on the land) variability of ecologic and climatic conditions controlling species composition and density of the biota. The seaweeds are the major producers of organic matter in the ocean. There are about 30 000–35 000 known water-plant species and more than half of these dwell in the ocean.

The main mass of organic carbon in the ocean is contained in *dissolved organic matter* (DOM) representing the intermediate substances between living organisms and biogenous inorganic matter, and combining true solutions and colloids of organic carbon as well as particles with sizes from 0.45 through 1 μm. According to experimental data the mean concentration of DOM in the ocean is equal to 1.36 ± 0.20 mg/l, which is equivalent to 1800–2000 GtC in the entire ocean volume.

Another form of existence of organic carbon in the ocean is *suspended organic matter* (SOM), including living cells of phyto- and zooplankton, residues of organisms, organic matter of skeleton formations, terrigenous and aeolian influx as well as precipitated, adsorbed from a solution and aggregated organic matter with particle sizes larger than 1 μm. The concentration of SOM in the ocean is much less than that of DOM and accounts for an average 53 ± 26 μm/l. Its maximum values fall in high-productive regions, and its minimal values fall in low-productive regions and deep ocean layers. The total content of SOM in the ocean is estimated as 25.5 ± 1.5 GtC.

About 50% of this is in depths of over 1000 m where bacterial activity is low and oxidation of organic matter is very slow. About the same amount of SOM (15–20 GtC) is not mineralized and gradually descends to the ocean bottom.

Finally, the third form of existence of carbon in the ocean is *living organic matter*. All present-day estimates of phyto- and zoomass are close to 3 GtC. Thus, estimates of the total content of DOM, SOM and living organics are related to each other as 100:1.4:0.15. This 'pyramid of mass' is caused by equilibrium of the production and destruction of organic matter in the ocean. In other words, it reflects the different stability (from the viewpoint of decomposition) of separate forms of organic matter.

The organic matter created in the process of vital functions by living organisms (primarily by plants and photosynthesizing organisms) is involved in rapid biological transformations and is also accumulated in relatively stable (in the sense of destruction) complexes forming soil organics on the land and water humus in the ocean. Affected by biochemical and chemical processes the organic residues are subjected to destruction with the result that part of the organic matter is transformed into mineral compounds (carbon dioxide, ammonia, nitrites and nitrates, etc.); the remaining part is converted into more stable (slowly oxidized) organic forms. This is why, for example, it is customary to classify all organic residues on the land in three groups: vegetable deciduosity and organic underlay (fallen leaves, dead parts of trees, bushes and grass cover); unstable biochemical compounds (incompletely modified vegetable residues, products of metabolism and newly formed humus matter); and stable biochemical compounds (humus, peat, sapropel, etc.) containing 84.2, 673 and 1346 GtC respectively. Thus, the total content of organic carbon in soil is equal to 2104 GtC, that is, it practically coincides with its value in the dissolved organic matter of the ocean.

The largest amount of soil carbon is concentrated in the boreal belt, and the least amount is concentrated in the polar zone of the land (35.1% and 6.5% of the total content respectively). Note, however, that the minimum content of the organic carbon in the polar belt is explained not by its low concentration, which is even higher than in the tropical, subtropical and subboreal belts but, rather, by the small area of the polar belt.

The organic matter in the ocean is also classified into stable and unstable groups, which, for the most part, have *autochthonic* (formed by living organisms) origin. The vertical variability for the unstable fraction is greater than for the stable fraction. Suffice it to say that the concentration of water humus in the ocean surface layer is equal to $\sim 2.2 \times 10^{-3}$ gC/l, and at a scale of 3000 m it is equal to $\sim 1.5 \times 10^{-3}$ gC/l, while the concentration of the unstable fraction, determined only by the vital function processes of zooplankton and bacteria, can differ by a hundred times within the limits of the ocean thickness.

Estimates of the carbon content in sedimentary deposits of the ocean and continents are rather approximate. The most recent estimates (see Budyko *et al.*, 1985) attest that the sedimentary shell of the Earth contains 97.8×10^6 GtC, including 86×10^6 Gt of carbonate and 11.8×10^6 Gt of organic carbon.

Carbon sources and sinks. We have already said that the carbon content in the atmosphere is closely related to the vital function processes of the terrestrial biota and that carbon dioxide is practically the only (at present at least) source of it. The first experimental data confirming the existence of the relation between the atmospheric CO_2 content and the assimilating activity of plants were obtained in the early 1920s. The advent of optical-acoustic gas analysers in the 1950s and increased measurement accuracy opened up the possibility of studying the physical mechanism of CO_2 assimilation in vegetable communities and creating a quantitative theory of photosynthesis.

The intensity of carbon exchange between the atmosphere and vegetable communities is characterized by the *pure primary production* defined as the difference between the *total primary production* (the rate of organic matter formation) and *losses by respiration* (the rate of organic matter destruction under respiration) of autotrophs and heterotrophs. It will be recalled that *autotrophs* are organisms that assimilate the carbon of inorganic compounds, and *heterotrophs* are organisms that use organic matter. If the total primary production is higher than losses by respiration, the ecosystem serves as a carbon sink for the atmosphere; otherwise it represents a carbon source for the atmosphere. According to Kobak (1988) the total primary production of all forest, steppe and tundra vegetable communities is equal to 118.2 GtC/year.

Let us define losses by autotroph respiration as the difference between the total primary production and the rate of carbon photosynthetic assimilation, and losses by heterotroph respiration as the sum of the rate of organic matter mineralization in the upper soil layer and the rate of organic matter transformation into humus. According to Kobak (1988) the rate of carbon photosynthetic assimilation is 58.2 GtC/year, while the rate of organic matter mineralization and its transformation into humus is 41.4 and 2.5 GtC/year. The rate of organic matter transformation into humus consists of two components characterizing the rate of formation of unstable (1.44 GtC/year) and stable (1.04 GtC/year) fractions of the soil humus. Using these estimates and those mentioned above we find that the *pure primary production of the terrestrial biota*, which is equal to the rate of photosynthetic assimilation minus the rate of organic matter mineralization, and of its transformation into humus must constitute $58.2 - 41.4 - 2.5 = 14.3$ GtC/year. Hence, the terrestrial biota serves as the carbon sink for the atmosphere.

The flux of CO_2 into the atmosphere is not only due to respiration of the terrestrial vegetation, but also to so-called *soil breathing* (CO_2 emission from soil) representing the result of oxidation of soil organics by microorganisms and of respiration of vegetation roots. But because in the absence of other sources and sinks of carbon its withdrawal from the atmosphere through the process of photosynthesis can be balanced only by soil breathing, it has to be equal to the carbon photosynthetic assimilation, that is, 58.2 GtC/year. Part ($41.4 + 2.5 = 43.9$ GtC/year) of this value is determined by the destruction of mineralized organic matter and humus; the other part is determined by respiration of vegetation roots.

In contrast to terrestrial vegetable communities, in ocean ecosystems the carbon is not a factor which limits photosynthesis. One such factor is the intensity of short-wave solar radiation, which is not less than the limiting value of 2.08 W/m^2 when photosynthesis stops. The depth appropriate for this value is called the *light compensation depth*. Above this depth the intensity of photosynthetic assimilation is greater than losses for respiration, and below this depth the reverse situation takes place where phytoplankton can exist only due to the organic matter which has been formed previously, that is, due to conversion to heterotrophic nutrition.

The other factor limiting photosynthesis is the presence of biogenous elements in sea water – nitrate NO_3^- and phosphate HPO_4^{2-}. Indeed, according to the Redfield formula

$$106CO_2 + 16NO_3^- + HPO_4^{2-} + 18H^+ + 122H_2O \gtrless$$

$$(CH_2O)_{106}(NH_3)_{16}(H_3PO_4) + 138O_2,$$

describing the process of photosynthesis and its reverse process of destruction of organic matter, for every 106 moles of CO_2 expended for photosynthesis 16 moles of nitrate and 1 mole of phosphate are consumed. It follows, therefore, that the content of carbon, nitrogen and phosphorus in phytoplankton cells is in the ratio 106:16:1, and thus lack of nitrogen and phosphorus limits the intensity of carbon assimilation.

If we attribute the heterotrophic nutrition of phytoplankton to secondary production and attribute the same for *chemosynthesis*, which uses the energy of chemical reactions as a source, then the photosynthetic assimilation of carbon will represent the *pure primary production of the ocean biota*. There are many of its estimates that have mainly been obtained using the radiocarbon method. The most reliable of these are within the limits 23–46 GtC/year. It has also been shown that 75% of the pure primary production occurs in the open ocean, 17.5% occurs on the continental shelf, 4% occurs in estuaries, and only 0.5% occurs in local upwelling zones. Information about the seasonal variability of the ocean biota production can be classified as very approximate. Everything we know about this reduces to the following: at low latitudes phytoplankton production does not change throughout most of the year, but at temperate latitudes it undergoes distinct seasonal variations, increasing in spring and decreasing in summer. The summer decrease is related to the facts that phytoplankton is grazed by herbivorous zooplankton, and particularly to the lack of biogenous elements whose influx from the deep layers is blocked by the abrupt thermocline in the base of the upper mixed layer. In autumn, when vertical mixing is enhanced and the seasonal thermocline degenerates, phytoplankton productivity increases again. At high latitudes one (summer) peak only of productivity is recorded. These presentations are supported by data from colour scanning of the ocean surface by satellites that allow recovery of the chlorophyll concentration in water.

Apart from the autochthonic mechanism of organic matter formation in the ocean (photosynthesis), there is another mechanism – the allochthonic mechanism characterized by the inflow of organic matter with river and underground run-off, and by the removal of suspended particles and aeolian matter containing organic carbon from the land. According to Kobak (1988) the intensity of such a source of organic matter in the ocean amounts to ≈ 1 GtC/year. In addition, the contribution of river run-off is 0.21 GtC/year, that of underground run-off is 0.06 GtC/year, that of suspended particles is 0.4 GtC/year and that of aeolian matter is 0.3 GtC/year.

Autochthonic and allochthonic inflows of organic matter are compensated by lifetime secretions of plants and animals, as well as by decomposition of secretions and remains of plants and animals due to heterotrophic organisms (bacteria). The resulting production of dissolved and suspended organic matter amounts to 1.08 and 1.0–3.0 GtC/year, respectively (see Kobak, 1988), which signifies that 2–5% of the pure primary production goes into solution and about the same amount is precipitated. The precipitating organic matter formed during the process of biochemical and chemical reactions is subjected to destruction and is then dissolved at a rate of 0.9–2.9 GtC/year, so that the accumulation of organic matter in sedimentary deposits does not exceed 0.1 GtC/year.

Carbon exchange at the ocean–atmosphere interface. As previously mentioned, among all carbon-containing atmospheric gases only CO_2 has a sufficiently high concentration. Therefore, carbon exchange at the ocean–atmosphere interface is determined by the carbon dioxide flux. The first estimates of time–space variability of CO_2 flux on the basis of direct measurements of CO_2 partial pressure difference in water and air were obtained in 1986 at the Lamont Doherty Geological Observatory of Columbia University (US). They were then revised by Ariel *et al.* (1991). As a result it was found that the equatorial zone of the ocean (10°N–10°S) is a carbon source for the atmosphere. Here, due to the strong upwelling which ensures export to the surface of deep waters rich in carbon and biogenes, and due to the low solubility of CO_2 in water (the latter is due to the high temperature), carbon transfer from the ocean to the atmosphere amounts to 0.14 GtC in the Atlantic Ocean, 0.05 GtC/year in the Indian Ocean, and 0.54 GtC/year in the Pacific Ocean. These differences are caused by the different intensity of the equatorial upwelling.

In the subtropical gyres of the Northern (10°N–40°N) and Southern (10°S–40°S) Hemispheres the picture is more varied: in the Northern Hemisphere the Pacific and Atlantic Oceans serve as a sink of carbon for the atmosphere, and the Indian Ocean serves as a source. The respective values are: -0.01, -0.19 and 0.05 GtC/year. In the subtropical gyres of the Southern Hemisphere the carbon transfer from the ocean into the atmosphere takes place in the Atlantic (0.04 GtC/year), that from the atmosphere into the ocean takes place in the Pacific (-0.20 GtC/year) and Indian (-0.21 GtC/year) Oceans. Similarly, carbon transfer from the ocean into the atmosphere occurs in the northern subpolar area of the Pacific Ocean (0.17 GtC/year), that from the atmosphere into the ocean occurs in the northern (-0.38 GtC/year) and southern (-1.39 GtC/year) subpolar areas and in the southern polar area (-0.83 GtC/year) of the Atlantic Ocean.

In general, as shown by Ariel *et al.* (1991), the ocean absorbs atmospheric carbon at a rate of -2.21 GtC/year. Here, as before, the negative values of the flux conform to carbon transfer from the atmosphere into the ocean, and positive values do so in the reverse direction. Note that the annual mean global average carbon flux at the ocean–atmosphere interface turned out to be different from zero, which it would be in the absence of any long-term disturbances. Some of the arguments put forward by Ariel *et al.* (1991) favour the anthropogenic origin of such an imbalance.

Time of carbon renewal. The features, mentioned above, of the natural (not subjected to anthropogenic impacts) carbon cycle, together with the times of carbon renewal, are summarized in Table 2.4. Let us turn our attention to the following three facts. Firstly, the natural carbon cycle is closed in the atmosphere, ocean and biosphere, but it is not closed in the lithosphere, and thus in the climatic system as a whole. This is due to disregarding the interaction between the sedimentary shell, formed by sedimentary and volcanic rocks, and deep earth layers: during this process organic carbon accumulation on the ocean bottom in the form of carbonate sediments, that is, a net flux carbon from the biosphere to the lithosphere has to be compensated for by emission of volcanic carbon dioxide from the deep earth layers into the atmosphere and ocean.

Secondly, in accordance with the data presented in Table 2.4, the age of most ancient sediments and deposits of the ocean and continents, if estimated as the ratio between the carbon content and the characteristic rate of accumulation or erosion, should not exceed 100 million years. In other words, throughout the geological history of the Earth the sedimentary rocks must have been renewed repeatedly.

At first sight, this result contradicts the known facts and, in particular, the discovery of sedimentary rocks at the entrance to the Ameralik-Fjord (western Greenland) aged 3.8 billion years. This contradiction is explained by the variability of rates of sedimentary accumulation in the ocean and of erosion on the continents and thereby by the impossibility of applying present-day values to other geological periods. Let us recall in this connection the decrease in accumulation of carbonate sediments in the Medium and Late Carboniferous (346–232 million years ago), or the so-called *interruptions in the sedimentary deposition sequence* (intervals when there was no sediment accumulation) with recurrence changing from 40–60% at the beginning to 70–76% at the end of the Eocene Period (58–37 million years ago). As to the rate of continental erosion, this can be judged from variations in the mean ocean level characterizing the base of continental erosion. The most significant increase of 300–350 m in erosion over the last 570 million years, accompanied by the abrupt decrease in the export of terrigenous material into the ocean, occurred in the late Cretaceous Period (100–67 million years ago). This event and the ocean transgression which caused it had their origins in tectonic processes giving rise to the convergence of the African and Eurasian continents and the degeneration of the Tethys Ocean.

One should keep in mind the rather approximate character of estimates of the carbon accumulation rate in sedimentary depositions, and, most of all, the fact that the age of the most ancient sedimentary depositions in the ocean is determined by plate tectonics with characteristic time scales of ≈ 150 million years, and on the continents it is determined by the duration of

Table 2.4. *Characteristics of the natural carbon cycle in the atmosphere–ocean–biosphere–lithosphere system*

Subsystem	Component	Carbon content (GtC)	Carbon sources and sinks		Time of renewal (years)
			Process	Intensity (GtC year^{-1})	
Atmosphere		594	Soil breathing	58.2	10.2
			Photosynthetic assimilation	−58.2	
			Exchange with the ocean	0	
Ocean		38 200	Destruction of unstable organics and dissolution of part of the water humus	from 23 to 46	from 830 to 1660
			Photosynthetic assimilation	from −23 to −46	
			Exchange with the atmosphere	0	
Biosphere	Continental phyto- and zoomass	560.5	Photosynthetic assimilation	58.2	9.6
			Formation of the organic sublayer and humus	−43.9	
			Plant roots breathing	−14.3	
	Ocean phyto- and zoomass	3	Pure primary production	from 23 to 46	from 0.06 to 0.13
			Destruction of unstable organics	from −20 to −43	
			Formation of dissolved and suspended matter	from −2.1 to 4.1	
	Dissolved organic matter	from 1800 to 2000	Allochthonous formation of organic matter	1.0	from 1636 to 1818
			Production	1.1	
			Mineralization	−1.1	

Suspended organic matter	from 24 to 27	Production	from 1.0 to 3.0	from 8 to 27
		Destruction	from −0.9 to −2.9	
		Accumulation in ocean sedimentary deposits	0.1	
Lithosphere Vegetative falling and organic sublayer	84.2	Formation	41.4	2.0
		Mineralization of organic matter	−41.4	
Unstable fraction of soil humus	673	Production	1.4	481
		Mineralization of organic matter	−1.4	
Stable fraction of soil humus	1346	Production	1.0	1346
		Accumulation in continental sedimentary deposits	−1.0	
Sedimentary deposits of the ocean and continents	97.8×10^6	Accumulation in continental and ocean sedimentary deposits	1.1	10^8
		Export of suspended particles and Aeolian matter into the ocean	−1.0	

Note: the positive values correspond to carbon sources, the negative values correspond to carbon sinks; the carbon content in the atmosphere relates to the beginning of the industrial revolution.

existence of granite sheets with characteristic time scales of ≈ 4 billion years. Therefore, the data in Table 2.4 do not contradict those for silt sediments discovered in the composition of ancient formations.

The third circumstance which should be particularly emphasized is the different renewal times for separate components of the carbon cycle. Indeed, as one can see from Table 2.4, the carbon renewal time is 0.1 year, and less for the ocean biota; about one year for plant deciduosity and organic underlayer; about ten years for the atmosphere, terrestrial biota and suspended organic matter in the ocean; hundreds, or even thousands, of years for the inorganic carbon content in the ocean, dissolved organic matter and soil humus; and, finally, hundreds of millions of years for sediments in the ocean and on the continents. Such a diversity of renewal times serves as the basis for the separation of three subcycles in the carbon cycle: the *mobile* subcycle, describing the processes of organic matter transformation and carbon circulation in the ocean–atmosphere system with time scales of about 10^3 years and less; the *geochemical* subcycle describing the processes of interaction between sea-water and carbonate sediments with time scales of about 10^4–10^5 years; and the *geological* subcycle, describing processes of organic matter burial and metamorphism, and of mantle outgassing with time scales of about 10^6 years and more.

We are interested in processes with time scales under 10^3 years. This restriction is equivalent to fixing slow (with time scales over 10^3 years) carbon subcycle characteristics. In this case the budget equation for the carbon in the atmosphere–ocean–biosphere–lithosphere system takes the form

$$\partial C_A/\partial t + \nabla \cdot \widehat{m_A v_A c_A} = -Q_C^{OS} - Q_C^{AB}, \tag{2.7.1}$$

$$\partial C_O/\partial t + \nabla \cdot \widehat{m_O v_O c_O} = Q_C^{OS} - Q_C^{OB}, \tag{2.7.2}$$

$$\partial C_{OB}/\partial t + \nabla \cdot \widehat{m_A v_A c_{OB}} = Q_C^{OB} - Q_C^{O}, \tag{2.7.3}$$

$$\partial C_{LB}/\partial t = Q_C^{AB} - Q_C^{LB}, \tag{2.7.4}$$

$$\partial C_L/\partial t + \nabla \cdot \widehat{m_L v_L c_L} = Q_C^{LB} \, Q_C^{l}, \tag{2.7.5}$$

where c_A and c_O are specific concentrations of inorganic carbon in the atmosphere and ocean; c_{LB} (see below), c_{OB}, and c_L are specific concentrations of organic carbon in the terrestrial and ocean biota and in the lithosphere; $C_A = m_A \hat{c}_A$ and $C_O = m_O \hat{c}_O$ are the inorganic carbon content in the atmospheric and ocean columns with unit cross-section; $C_{OB} = m_O \hat{c}_{OB}$, $C_{LB} = m_L \hat{c}_{LB}$ and $C_L = m_L \hat{c}_L$ are the same but for inorganic carbon in the ocean and terrestrial biota and in the lithosphere; Q_C^{OS} is the carbon flux (gas exchange) at the ocean–atmosphere interface; Q_C^{AB} and Q_C^{OB} are the intensities of inorganic carbon sources and sinks in the atmosphere and ocean referred to the unit area; Q_C^{LB} is the intensity of carbon exchange between fast components of the terrestrial biota and the lithosphere normalized to the unit area; Q_C^L and Q_C^O are the intensities of organic carbon exchange between fast and slow components of the carbon cycle in the lithosphere and the ocean normalized in an analogous way; all other designations are the same.

Let us integrate Equation (2.7.1) over the whole of the Earth's surface, Equations (2.7.2) and (2.7.3) over the ocean surface and Equations (2.7.4) and (2.7.5) over the land surface, and then sum them up. As a result we obtain the equality

$$\partial(\{C_A\} + \{C_O\}f_O + \{C_{OB}\}f_O + \{C_{LB}\}f_L + \{C_L\}f_L)/\partial t = -\{Q_C^O\}f_O - \{Q_C^L\}f_L,$$

$$(2.7.6)$$

representing the law of carbon conservation on time scales of the order of 10^3 years or less.

Similarly, integration of Equations (2.7.1)–(2.7.5) over longitude and their consequent addition yields

$$2\pi a \cos\varphi \, \partial([C_A] + [C_O]f_O' + [C_{OB}]f_O' + [C_{LB}]f_L' + [C_L]f_L')/\partial t$$

$$+ \partial(MCT_A + MCT_O + MCT_L)/a \, \partial\varphi = -2\pi a \cos\varphi([Q_C^O]f_O' + [Q_C^L]f_L'),$$

$$(2.7.7)$$

where

$$MCT_O = \int_0^{2\pi} m_O(\widehat{v_O c_O} + \widehat{v_O c_{OB}})a \cos\varphi \, d\lambda,$$

$$MCT_A = \int_0^{2\pi} m_A \widehat{v_A c_A} a \cos\varphi \, d\lambda,$$

$$MCT_L = \int_0^{2\pi} m_L \widehat{v_L c_L} a \cos\varphi \, d\lambda,$$

are the meridional carbon transports in the ocean, atmosphere and lithosphere.

A rough estimate of MCT_A and MCT_O can be obtained if we turn to (2.7.7) and assume that $([Q_C^O]f_O' + [Q_C^L]f_L') = 0$, $MCT_L = 0$. Considering that the meridional carbon transport in the atmosphere and ocean has to vanish at the pole, we obtain, on average for the year $(MCT_A + MCT_O) = 0$, that is, the annual mean meridional carbon transport in the atmosphere and ocean must balance each other. Let us recollect, then, that the ocean serves as a carbon source for the atmosphere at low latitudes and as a carbon absorber at high latitudes, and that the intensity of the annual mean carbon exchange between the ocean and the atmosphere in the equatorial zone accounts for

~ 0.8 GtC/year. Hence, the annual mean meridional carbon transport is directed from low latitudes to high latitudes in the atmosphere, and in the reverse direction in the ocean. In addition, since the ratios of masses and zonal extensions of the atmosphere and ocean are equal to 3×10^{-3} and 2 respectively, then, all things being equal, the typical value of the meridional carbon transport across the unit length of a latitude circle in the atmosphere has to be two orders greater than in the ocean. To convert relative units into absolute units we note that in the pre-industrial epoch, when sources and sinks of atmospheric CO_2 counterbalanced each other on average for the year, the meridional transport of CO_2 in the atmosphere amounted to 3–6 GtC/year.

3

Small-scale ocean–atmosphere interaction

3.1 Surface atmospheric layer

The surface atmospheric layer is a layer within the limits of which the vertical fluxes of momentum, heat and moisture remain approximately constant in height. Let us expand this definition.

We examine the averaged (in terms of the Reynolds conditions) equation of motion for the horizontal velocity vector v. With the statistically homogeneous (in the horizontal direction) velocity field it takes the form

$$\frac{d\mathbf{v}}{dt} = 2\Omega\mathbf{k} \times \mathbf{v} - \frac{\nabla p}{\rho} + \frac{\partial}{\partial z}\frac{\tau}{\rho},\tag{3.1.1}$$

where p is the atmospheric pressure; τ is the horizontal vector of the tangential wind stress; ρ is the air density; d/dt and ∇ are the operators of the total derivative and of the horizontal gradient; the axis z is directed vertically upward; $\mathbf{k}$ is the unit vector directed along the z-axis.

The first two terms on the right-hand side of Equation (3.1.1) describing, respectively, the effects of the Coriolis force and of the force of the horizontal pressure gradient, are the main terms. Considering that they balance each other, then assuming that fu_0 (here u_0 is the characteristic wind velocity scale; f is the Coriolis parameter) is an upper estimate of the third term on the right-hand side of (3.1.1) and that the change of vector τ/ρ in the vertical direction is $0.2|\tau|/\rho$ we obtain the following inequality for the thickness h_1 of the layer of approximate constancy of vertical momentum flux: $h_1 > 0.2$ $(|\tau|/\rho)(fu_0)^{-1}$, from which with $|\tau|/\rho = 0.1\text{ m}^2/\text{s}^2$, $u_0 = 10\text{ m/s}$, and $f = 10^{-4}\text{ s}^{-1}$ the estimate of $h_1 > 20\text{ m}$ is derived.

The estimate of the thickness h_2 of the layer of approximate constancy of vertical heat flux H can be obtained from the averaged heat budget equation which, in the absence of water vapour transitions, and radiative sources and sinks of heat, is written as

$$\frac{T}{\theta}\frac{d\theta}{dt} = -\frac{\partial}{\partial z}\frac{H}{\rho c_p},$$ (3.1.2)

where θ and T are potential and absolute temperatures and c_p is the heat capacity of air at constant pressure.

If the left-hand side of Equation (3.1.2) has order 0.1×10^{-3} °C/s and the characteristic value of $|H|/\rho c_p$ is equal to 2.0×10^{-2} °C m/s then for a thickness of the layer within the limits of which the relative variation of the vertical heat flux H is less than 20% the estimate $h_2 > 40$ m is valid.

Similarly, if the variation of specific humidity q in time has order 1.5×10^{-7} g/kg s, and the vertical flux E of the water vapour is 0.5×10^1 g/m²s, then on the basis of the averaged budget equation for specific humidity

$$\frac{dq}{dt} = -\frac{\partial}{\partial z}\frac{E}{\rho},$$ (3.1.3)

and of the assumption that water vapour phase transitions are absent, the thickness h_3 of the layer where the vertical flux of water vapour varies only up to 20% is $h_3 > 50$ m.

Thus, conditions of approximate constancy with height for the vertical fluxes of momentum, heat and moisture are fulfilled simultaneously if the thickness h of the surface atmospheric layer is defined as $h = \min(h_1, h_2, h_3)$.

Let us now find the characteristic relaxation time t_a for the surface atmospheric layer. On the basis (3.1.1) the time has to be equal to $t_a \approx u_0 h/0.2(\tau/\rho)$, which, after substitution of the typical values u_0, h and $|\tau|/\rho$ yields $t_a \approx 2$ hours. The characteristic time t_w of the wind wave development has the same order. And, indeed, judging by the measurement data the limiting root mean square height of wind waves is achieved at the fetch X (a distance counted in the direction of the wave propagation), which is $X \approx 10^4 u_a^2/g$, where u_a is the mean wind velocity at the standard measurement level in the surface atmospheric layer, g is gravity. Assuming as a rough estimate that $X \approx c_0 t_w$, where c_0 is a phase velocity complying with the frequency of maximum in the spectrum of wind waves, we find that $t_a \approx 10^4 u_a^2/g c_0$. Next, assuming that the entire wind momentum $\mathcal{M}_a \approx \rho u_a h$ in the surface atmospheric layer is spent only on wave development, and for the wave momentum $\mathcal{M}_w$ the estimate $\mathcal{M}_w \approx \rho_w g a^2/c_0$ is valid (here a is the characteristic wave height; ρ_w and ρ are water and air densities), we have that $c_0 \approx (\rho_w/\rho)(ga^2/u_a h)$. Substitution of this expression into t_w yields $t_w \approx 10^4(\rho/\rho_w)(u_a^3 h/g^2 a^2)$. From this it follows at $(\rho/\rho_w) \approx 10^{-3}$, $u_a \approx 10$ m/s, $h \approx 100$ m, $a \approx 1$ m, $g \approx 10$ m/s² that $t_w \approx 3$ h. Thus, the statistical characteristics of the quasi-stationary surface atmospheric layer have to depend not only on height over the underlying surface and other determining parameters, but also on the stage of wind wave development.

3.2 Vertical distribution of the mean velocity over an immovable smooth surface; viscous sublayer; logarithmic boundary layer

We orient the x axis along the tangential wind stress. Then the condition of approximate constancy with height of the momentum flux takes the form

$$\tau/\rho = -\overline{u'w'} + v\frac{du}{dz} \approx const, \tag{3.2.1}$$

where u is a component of mean wind velocity along the axis x; u' and w' are turbulent fluctuations of velocity along axes x and z; v is the kinematic viscosity; the overbar signifies averaging.

Equation (3.2.1) does not permit determination of the unique vertical distribution of mean velocity because, apart from u, it contains one more unknown function: the Reynolds stress $-\overline{u'w'}$. But some conclusions about the possible form of the function $u(z)$ can be derived with the help of dimensional analysis. Indeed, the mean velocity u near a wall depends only on the friction stress τ, distance z from the underlying surface and also on the kinematic viscosity v and density ρ of the medium. In addition, τ and ρ can only be in the form of the combination τ/ρ not containing the dimension of mass. Usually, instead of τ/ρ it is customary to use the value $u_* = (\tau/\rho)^{1/2}$ with dimensions of velocity. This value, called *friction velocity* (or *dynamic velocity*), defines the velocity scale for the flow near a wall. On the basis of the π-theorem of dimensional analysis the dependence of u on u_*, z and v can be presented as

$$u = u_* f_u\left(\frac{zu_*}{v}\right), \tag{3.2.2}$$

where $f_u(zu_*/v)$ is a universal function of the argument zu_*/v.

Breaking the established order of discussion somewhat we recall the formulation and formal proof of the π-theorem which forms the central statement of dimensional analysis and which will be repeatedly applied subsequently.

The π-theorem. Any dependence among $n + 1$ dimensional variables of which the $k \leq n$ variables have independent dimensions can be represented in the form of a dependence among the $n + 1 - k$ non-dimensional combinations.

Let there be a functional dependence

$$a = f(a_1, \ldots, a_k, a_{k+1}, \ldots, a_n), \tag{3.2.3}$$

where the arguments $a_1, \ldots, a_k$ have independent dimensions (i.e. their dimensions cannot be expressed by means of the dimensions of the remaining variables), and the dimensions of the variable a being determined and the arguments $a_{k+1}, \ldots, a_n$ are expressed by means of dimensions of the first k arguments as

$$\left. \begin{aligned} [a] &= [a_1]^p \ldots [a_k]^r, \\ [a_{k+1}] &= [a_1]^{p_{k+1}} \cdots [a_k]^{r_{k+1}}, \\ [a_n] &= [a_1]^{p_n} \ldots [a_n]^{r_n} \end{aligned} \right\} \tag{3.2.4}$$

here, as is common practice, the symbol [ ] denotes a dimension.

Let us pass from one system of units of measurement to another inside a given class so that the first k arguments $a_1, \ldots, a_k$ change β_i $(i = 1, \ldots, k)$ times. Then,

$$a_1' = \beta_1 a_1, \ldots, a_k' = \beta_k a_k.$$

Accordingly, the variable a is determined and all of the remaining arguments $a_{k+1}, \ldots, a_n$ change too, and relationship (3.2.3) is rewritten as

$$a' = f(a_1', \ldots, a_k', a_{k+1}', \ldots, a_n'). \tag{3.2.5}$$

However, by virtue of (3.2.4),

$$a' = \beta_1^p \ldots \beta_k^r a,$$

$$a_{k+1}' = \beta_1^{p_{k+1}} \ldots \beta_k^{r_{k+1}} a_{k+1},$$

$$a_n' = \beta_1^{p_n} \ldots \beta_k^{r_n} a_n.$$

Hence, substitution of these into (3.2.5) yields

$$\beta_1^p \ldots \beta_k^r a = f(\beta_1 a_1, \ldots, \beta_k a_k, \beta_1^{p_{k+1}} \ldots \beta_k^{r_{k+1}} a_{k+1}, \ldots, \beta_1^{p_n} \ldots \beta_k^{r_n} a_n).$$

Since β_i is an arbitrary change in scales of the units of measurement, we can choose it such that $\beta_i a_i = 1$. Then

$$\frac{a}{a_1^p \ldots a_k^r} = f\left(1, \ldots, 1, \frac{a_{k+1}}{a_1^{p_{k+1}} \ldots a_k^{r_{k+1}}}, \ldots, \frac{a_n}{a_1^{p_n} \ldots a_k^{r_n}}\right)$$

or, which is equivalent,

$$\frac{a}{a_1^p \ldots a_k^r} = F\left(\frac{a_{k+1}}{a_1^{p_{k+1}} \ldots a_k^{r_{k+1}}}, \ldots, \frac{a_n}{a_1^{p_n} \ldots a_k^{r_n}}\right). \tag{3.2.6}$$

If we introduce the designations

$$\Pi = \frac{a}{a_1^p \ldots a_k^r}, \quad \Pi_1 = \frac{a_{k+1}}{a_1^{p_{k+1}} \ldots a_k^{r_{k+1}}}, \ldots, \Pi_{n-k} = \frac{a_n}{a_1^{p_n} \ldots a_k^{r_n}},$$

the dependence (3.2.6) is rewritten in the form

$$\Pi = F(\Pi_1, \ldots, \Pi_{n-k}). \tag{3.2.7}$$

Thus, the initial dependence (3.2.3) among the $n + 1$-dimensional variables is actually reduced to dependence (3.2.7) among the $n + 1 - k$ non-dimensional combinations, QED.

Now, we return to (3.2.2) and, following Monin and Yaglom (1965), set the form of the function $f_u(zu_*/v)$ for two limiting cases: large and small values of the argument zu_*/v. First we recollect that on the wall (at $z = 0$) the mean velocity u and velocity fluctuations u' and w' vanish because of the no-slip conditions. Hence, in the vicinity of the wall (at small zu_*/v), one can select a layer called the *viscous sublayer* within the limits of which the viscous stress $\rho v\, du/dz$ is much larger than the Reynolds stress $-\overline{\rho u'w'}$. Then, according to (3.2.1),

$$\tau/\rho = v\, du/dz = const,$$

hence,

$$u = \frac{u_*^2}{v} z, \qquad (3.2.8)$$

that is,

$$f_u\left(\frac{zu_*}{v}\right) = \frac{zu_*}{v}, \qquad (3.2.9)$$

that is, equivalent to the replacement of the function $f_u(zu_*/v)$ by the Taylor series expansion in powers of orders zu_*/v and ignoring all terms but the first.

The combination v/u_* appearing in (3.2.8) is the meaning of the scale of viscous sublayer thickness. The thickness δ_v of this sublayer is defined by the expression $\delta_v = \alpha_v v/u_*$, where α_v is a numerical factor depending on the selected method of definition of the upper boundary of the viscous sublayer as the level on which the Reynolds stress is only a fraction of the viscous stress. It is usually accepted that $\alpha_v = 5$.

At a considerable distance from the wall (at large zu_*/v) the relation between the viscous stress and the Reynolds stress becomes inverse: the former is many times less than the latter. In this case the term $v\, du/dz$ in (3.2.1) can be neglected. Thus it is assumed that the vertical velocity gradient (but not the velocity itself, whose absolute value depends on the law of its changing in the vicinity of the wall where viscosity becomes apparent) is determined only by u_* and z. Then, from dimensional considerations,

$$\frac{du}{dz} = A \frac{u_*}{z}, \qquad (3.2.10)$$

where A is a universal constant.

Integrating (3.2.10) over z we have

$$u = Au_* \ln z + A_1, \tag{3.2.11}$$

where A_1 is a constant depending on kinematic viscosity.

The layer of liquid where the relation (3.2.11) is fulfilled is called the *logarithmic boundary layer*. The height δ_l of its lower boundary, as well as δ_v, should be a function of u_* and v, so that from dimensional considerations $\delta_l = \alpha_l v / u_*$, where α_l is one more numerical constant which, according to experimental data, is equal to 30.

There is the so-called *buffer sublayer* between the viscous sublayer and the logarithmic layer where the viscous stress and Reynolds stress have the same order of magnitude, and the linear profile of velocity turns smoothly into the logarithmic profile.

Let us present A_1 as $A_1 = u_*(A \ln u_*/v + B)$, where B is a new universal constant, and rewrite Equation (3.2.11) as follows:

$$u = u_*(A \ln zu_*/v + B) \quad \text{at } zu_*/v > \alpha_l. \tag{3.2.12}$$

In this case the function $f_u(zu_*/v)$ in (3.2.2) takes the form $f_u(zu_*/v) = A \ln zu_*/v + B$. According to data from measurements in smooth pipes, in rectangular channels with smooth walls, and in the boundary layer over a smooth, flat surface, the values of the constant A (or of the von Karman constant $\kappa = 1/A$ that is uniquely connected to it) and of the constant B turn out to be equal to 2.5 and 5.1 respectively. These values provide a sufficiently precise reproduction of the profile of u/u_* for $zu_*/v < 5$ and $30 < zu_*/v < 500$. But in the buffer sublayer located between the viscous sublayer and the logarithmic boundary layer, that is, at $5 < zu_*/v < 30$ the values of u/u_*, calculated from Equations (3.2.8) and (3.2.12), and observed values are slightly different (see Figure 3.1). To avoid unnecessary complications and at the same time to ensure the right asymptotes the domain of definition of the solution (3.2.8) extends to $zu_*/v = 11.1$, and everywhere after that (for $zu_*/v > 11.1$) the vertical profile of the mean velocity is described by Equation (3.2.12). In other words, it is suggested that the buffer sublayer has zero thickness, and the thickness of the viscous sublayer is equal to $\delta_v = \delta_l = 11.1(v/u_*)$.

3.3 Vertical distribution of the mean velocity over an immovable rough surface; roughness parameter; hydrodynamic classification of underlying surfaces

We now consider the vertical distribution of the mean velocity over an immovable rough surface where the average height h_0 of undulations is not

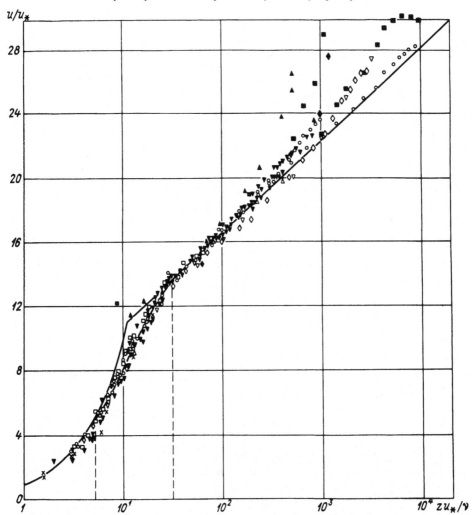

Figure 3.1 Vertical profile of mean velocity in the turbulent flow over a smooth surface. Dots are experimental data from different authors; solid lines are described by Equation (3.2.8) at $zu_*/v < 5$ and by Equation (3.2.12) at $zu_*/v > 30$.

small compared with the thickness depth scale v/u_* of the viscous sublayer. Because elements of roughness have a considerable effect on the distribution of the mean velocity in the vicinity of the underlying surface, then (3.2.2) is replaced by the more general expression

$$u = u_* f_u\left(\frac{zu_*}{v}, \frac{h_0 u_*}{v}, \dots, \right). \tag{3.3.1}$$

Here the ellipsis means that in (3.3.1) the non-dimensional parameters characterizing a form of roughness elements and their mutual arrangement are omitted.

Let us discuss again the cases of large and small values of z. For small z compared with h_0 the above with regard to the dependence of mean velocity on the determining parameters is valid. In this case to use dimensional considerations for revealing general regularities of the vertical distribution of the mean velocity is unpromising. Another problem is the other limiting case when the distance z from the underlying surface is much more than the average height h_0 of roughness elements and the thickness scale v/u_* of the viscous sublayer, that is, $z \gg h_0 > v/u_*$. In this case the local properties of the underlying surface do not have any marked effect on the vertical distribution of the mean velocity, the number of determining parameters is reduced, and there is a formula for u (compare with (3.2.11)):

$$u \approx A u_* \ln z + A_1 \quad \text{for } z \gg h_0, \tag{3.3.2}$$

where, however, the constant A_1 prescribed by the condition of continuity of velocity at the lower boundary of the domain of definition of solution (3.3.2) depends on the dimensions, form and mutual arrangement of roughness elements controlling the vertical structure of a flow in the near-wall layer and on viscosity.

As before, we represent the constant A_1 as $A_1 = u_*(A \ln u_*/v + B)$ or, what amounts to the same, $A_1 = u_*(A \ln(u_* h_0/v) - A \ln h_0 + B)$ and then substitute it into (3.3.2). As a result this expression is rewritten as follows:

$$u = u_*(A \ln(z/h_0) + B'), \tag{3.3.3}$$

where $B' = u_*(A \ln(u_* h_0/v) + B)$ is the new numerical constant which is a function of the *surface Reynolds number* $Re_0 = u_* h_0/v$ and of the form and mutual arrangement of the roughness elements.

We take into consideration that $A = 1/\kappa$ and introduce the definition of the *roughness parameter* $z_0 = h_0 \exp(-\kappa B')$. Substitution of this expression in (3.3.3) yields

$$u = \frac{u_*}{\kappa} \ln \frac{z}{z_0}, \tag{3.3.4}$$

from which it follows that the roughness parameter represents a level where the mean velocity would vanish if the logarithmic law for the profile of the mean velocity were valid up to this level (actually it is valid for $z \gg z_0$).

For the underlying surface covered by homogeneous densely packed roughness elements, the roughness parameter z_0 should depend only on the surface Reynolds number Re_0 so that $z_0 = h_0 f(Re_0)$, where $f(Re_0)$ is a dimensionless universal function. We find asymptotes of this function for small and large values of Re_0. Let $Re_0 \ll 1$ (roughness elements are completely submerged in the viscous sublayer; the effect of the underlying surface on an ambient velocity is caused by the molecular viscosity force). Then the height h_0 of the roughness elements has to be excluded from the set of determining parameters. Hence, $f(Re_0) \sim Re_0^{-1}$, and the expression for the roughness parameter will take the form

$$z_0 = m_0 v/u_*, \tag{3.3.5}$$

where m_0 is a numerical factor.

As may be seen in this case, neither the roughness parameter nor, consequently, the profile of the mean velocity depends on the height of the roughness elements. The underlying surface appropriate for this case can be considered as *hydrodynamically smooth*.

On the other hand, at $Re_0 \gg 1$ (roughness elements protrude beyond the limits of the viscous sublayer; the effect of the underlying surface on an ambient velocity is caused by the normal pressure force on the roughness elements), kinematic viscosity has to be excluded from the set of determining parameters. This is possible if the function $f(Re_0)$ approaches a certain constant value. Designating it as m_1 we have

$$z_0 = m_1 h_0. \tag{3.3.6}$$

The surface complying with this case is termed *hydrodynamically rough*.

Equations (3.3.5) and (3.3.6) form the basis of the existing classification of underlying surfaces by the surface Reynolds number. According to this classification an underling surface is considered to be hydrodynamically smooth if $Re_0 < 4$, $m_0 \approx 0.1$; it is considered to be hydrodynamically rough if $Re_0 > 60$, $m_1 \approx 0.03$; and it is considered to be intermediate (slightly rough) if $4 < Re_0 < 60$. In the general case where the roughness elements have different forms and are randomly located with reference to each other, the proportionality factors in (3.3.5) and (3.3.6), and the limiting values Re_0 for the different types of underlying surface will differ from those mentioned above. In such a situation the hydrodynamic properties of the underlying surface should be described not in terms of the average height of the roughness elements, but, rather, in terms of the so-called *equivalent sand roughness* defined as the height of irregularities of the underlying surface covered by sand roughness which, for identical u_*, complies with the same logarithmic profile of the mean velocity as does that over the surface in question. If we now consider h_0 as the height of the equivalent sand roughness, then Equations (3.3.5) and (3.3.6) will be universal in the sense that the numerical factors m_0 and m_1 appearing there will not depend on the form and mutual arrangement of roughness elements.

3.4 Hydrodynamic properties of the sea surface

It is clear from general considerations that in moving away from the sea surface any disturbances created by waves must diminish. Because of this in the range of heights from the upper boundary of the surface atmospheric layer to the upper boundary of the *wave sublayer* (the sublayer within the limits of which the effect of wave disturbances on the mean wind velocity becomes apparent), the vertical structure of the stationary, horizontally homogeneous, neutrally stratified atmospheric boundary layer over the wave disturbed sea surface should remain the same as over an immovable underlying surface. In short, in the above-mentioned height range the logarithmic law for the profile of the mean velocity must be valid.

Up to the present time many gradient measurements of mean wind velocity in the surface atmospheric layer have been gathered. The vast majority of these measurements concerning comparatively great heights (5–10 m) demonstrate satisfactory fulfilment of the logarithmic law

$$u = \frac{u_*}{\kappa} \ln \frac{z}{z_0} \tag{3.4.1}$$

under neutral stratification. This experimental fact provides a possibility of determining, using the data from gradient measurement, the friction velocity u_* and the roughness parameter z_0 of the sea surface connected to the *resistance coefficient* $C_u = u_*^2/u_a^2$ by the relationship

$$z_0 = z_a \exp(-\kappa/\sqrt{C_u}), \tag{3.4.2}$$

where u_a is the mean wind velocity at the conventional measurement height z_a.

At the same time the experimental data referring to heights commensurate with wave heights demonstrate obvious deviations from the logarithmic distribution, and therefore do not admit their interpretation in terms of the hydrodynamic characteristics of the immovable underlying surfaces. Let us discuss first the experimental data on the logarithmic boundary layer. Formally, nothing prevents intepretation of these in terms of the resistance coefficient C_u or the roughness parameter z_0 of the sea surface. But a direct analogy between wind waves and roughness elements of the immovable underlying surfaces turns out to be absolutely inappropriate. A dramatic testimony to this is the cloud of dots on the graph of dependence of the roughness parameter z_0 on the average height h_0 of waves (Figure 3.2). As can be seen, the roughness parameter of the sea surface has strong variability, amounting to six orders of magnitude (see Kitaigorodskii, 1970). The same feature is typical of the resistance coefficient of the sea surface but by virtue of

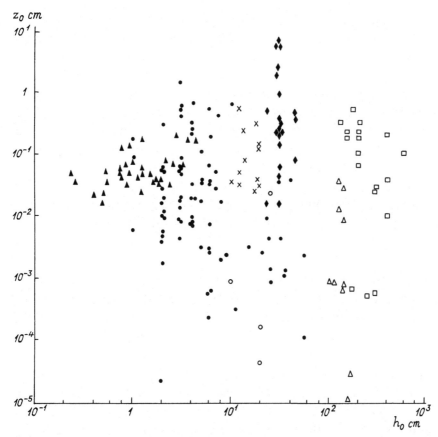

Figure 3.2 Dependence of the sea surface roughness z_0 on the average height h_0 of wind waves, according to Kitaigorodskii (1970). The various symbols indicate estimates by different authors.

the logarithmic character of dependence C_u on z_0 the dispersion of its values turns out to be less.

Let us return to the hydrodynamic classification of the underlying surfaces presented in the preceding section, in accordance with which the hydrodynamically smooth surface complies with

$$C_u = (\kappa^{-1} \ln z_a u_* / m_0 \nu)^{-2}, \qquad (3.4.3)$$

and the hydrodynamically rough surface with

$$C_u = (\kappa^{-1} \ln z_a / m_1 h_0)^{-2}. \qquad (3.4.4)$$

From this it follows that in the first case the resistance coefficient decreases with increasing wind velocity, and in the second it remains constant. Data from field and laboratory measurements over the water surface do not confirm this conclusion, the reason for which is usually associated with the dependence of C_u on the stage of wind wave development or, in other words, on the *wave age* c_0/u_*, where $c_0 = g/\omega_0$ is the characteristic phase velocity of energetically significant wind waves; ω_0 is the frequency of maximum in the spectrum of wind waves. However, we note that $C_u = (u_*/u_a)^2 = (\omega_0 u_*/g)^2(\omega_0 u_a/g)^{-2}$ and that in accordance with field measurement data $\omega_0 u_a/g \sim 1$, that is, $C_u \sim (\omega_0 u_*/g)^2 = (c_0/u_*)^{-2}$. In other words, the dependence of the resistance coefficient on c_0/u_* is predetermined by its expression, and in this sense does not contribute any additional information about the character of the changing hydrodynamic properties of the sea surface.

A clearer understanding of the question can be obtained from wave disturbances u_w of the mean wind velocity defined as

$$u_w = u - u_T, \tag{3.4.5}$$

where u is the actual wind velocity and u_T is the wind velocity over the underlying surface undisturbed by waves at the same friction velocity.

If there are no waves, then at the air–water interface the continuity condition for the momentum flux $\rho u_*^2 = \rho_w u_{*w}^2$ should be fulfilled, where subscript w means belonging to water. In this case the characteristic scale of turbulent fluctuations of velocity in the near-surface water layer is of the order of $u_{*w} = (\rho/\rho_w)^{1/2}u_*$. These velocity fluctuations will create vertical displacements of the air–water interface having the scale $\eta_T \sim g^{-1}u_{*w}^2 = g^{-1}(\rho/\rho_w)u_*^2$. But the thickness δ_v of the viscous sublayer is equal to $5\nu/u_*$; therefore at typical values of the friction velocity ($u_* \approx 0.3$ m/s) and coefficient of molecular viscosity ($\nu \approx 0.15 \times 10^{-6}$ m^2/s) the vertical displacements of the interface will be less than δ_v. Hence, in accordance with the hydrodynamic classification of underlying surfaces the surface in question can be considered to be smooth. Taking this into account we assume that the component u_T in (3.4.5) follows the logarithmic law for the mean velocity profile over the smooth surface, that is,

$$u_T = \frac{u_*}{\kappa} \ln \frac{z}{z_{0v}}, \tag{3.4.6}$$

where $z_{0v} = 0.1\nu/u_*$ is the roughness parameter of the smooth surface.

Combining (3.4.1), (3.4.5) and (3.4.6) we obtain

$$u_w = \frac{u_*}{\kappa} \ln \frac{z_{0v}}{z_0}, \tag{3.4.7}$$

from which it can be seen that z_0 describes not the mean wind velocity but, rather, its wave component, and because the latter is a function of c_0/u_*, then this, in fact, defines the dependence of the roughness parameter of the sea surface on the stage of development of wind waves. A. Yu. Benilov was the first to pay attention to this fact (personal communication). The results of gradient measurements of the mean wind velocity presented by him demonstrate that the dependence of the dimensionless (normalized by u_*) wave component u_w at height 10 m on c_0/u_* in the range of c_0/u_* values from 5 to 90 can be approximated by the polynomial of second degree:

$$u_w/u_* = -35 + 1.29(c_0/u_*)^{-3} - 6.9 \times 10^{-3}(c_0/u_w)^2.$$

Consequently, maximum negative values of u_w/u_* and, hence, maximum values of the roughness parameter z_0 conform to the initial stages of development of wind waves ($c_0/u_* < 30$). At $c_0/u_* \approx 30$–40, that is, for developed waves, the function u_w/u_* is close to zero and the sea surface resembles a smooth wall in its hydrodynamic properties ($z_0 \approx z_{0v}$). In the regime of attenuation of the swell when $c_0/u_w > 30$–40, values u_w/u_* become positive. From the viewpoint of the hydrodynamic classification of underlying surfaces, this is equivalent to the fact that in the regime of attenuation of the swell the sea surface turns out to be smoother than the immovable smooth wall.

Thus the process of development of wind waves is accompanied by significant reorganization of the regime of hydrodynamic resistance of the sea surface. In addition the most distinctive changes occur in the early stages of development of wind waves (at $c_0/u_* \leq 10$–20). This is the first important distinction of the marine surface atmospheric layer from the surface atmospheric layer over an immovable surface. Another distinction is the existence of an additional momentum flux near the air–water interface that is generated by wave disturbances of velocity and pressure. In the stationary case, as follows from the condition of constancy with the height of the friction stress, the appearance of the wave momentum flux has to be accompanied by vertical redistribution of the turbulent momentum flux. Accordingly, the vertical gradients of the mean wind velocity have to change as well: they decrease as compared with those predicted by the logarithmic law in the regime of wave generation, and increase in the regime of attenuation of the swell. But because the thickness of the wave sublayer at all possible values of c_0/u_* does not exceed 10 m (it equals approximately $0.1\lambda_0$, where λ_0 is the wavelength corresponding to a maximum in the wave spectrum), then at heights $z \gtrsim 10$ m the interpretation of gradient measurements of mean wind velocity in terms of u_* and z_0 remains justified.

There is an alternative to the method mentioned above describing the logarithmic profile of the mean wind velocity over a smooth surface. In the alternative version the mean velocity u_T of a smooth flow at a certain fixed height z coincides with the observed one at the same height, that is,

$$u_T = \frac{u_{*T}}{\kappa} \ln \frac{z_0}{z_{0v}}, \qquad u_T(z_a) = u(z_a), \qquad (3.4.8)$$

and the friction velocity does not coincide ($u_{*T} \neq u_*$). In taking into account that within the limits of the logarithmic boundary layer the square of the friction velocity is proportional to the momentum flux towards the underlying surface, such a description makes it possible to compare momentum fluxes towards the sea surface and the smooth wall.

Indeed, we define the resistance coefficients of the sea surface ($C_u = u_*^2/u_a^2$) and of the smooth wall ($C_{uT}^2 = u_{*T}^2/u_a^2$) and find an expression for their relative deviation $\delta C_u/C_u = (C_u - C_{uT})/C_u$. This expression connecting $\delta C_u/C_u$ with the relative difference of momentum fluxes at the sea surface and smooth wall has the form

$$\delta C_u/C_u = (1 - u_{*T}^2/u_*^2). \qquad (3.4.9)$$

It is evident now that in the presence of swell when $\delta C_u/C_u < 0$ and, hence, when $u_{*T}^2/u_*^2 > 1$, the wind obtains an additional momentum from waves. On the other hand, for developing waves when $\delta C_u/C_u > 0$ and $u_{*T}^2/u_*^2 < 1$ the momentum is transmitted from wind to waves (this determines the differences of the sea surface from a smooth wall). Finally, in the intermediate case, when the sea surface is close to the hydrodynamically smooth wall ($\delta C_u/C_u \approx 0$), waves can be considered as developed and, therefore, weakly interacting with the wind. In this case the complete momentum flux to the sea surface is spent not on developing waves but, rather, on forming a drift current by means of tangential stress and wave breakdown.

In conclusion, we list five of the most popular methods used to estimate the roughness parameter z_0 of the sea surface. The first is based on the Rossby assumption about the proportionality of z_0 and the average wave height. The latter, in the regime of developed waves, is a function of the friction velocity u_* and gravity g. Taking both these circumstances into account and using dimensional considerations we obtain the well-known Charnock formula $z_0 = mu_*^2/g$, where m is a numerical constant varying, according to different authors, from 0.011 to 0.08.

Such a significant spread in the values of m indicates that the roughness parameter should depend not only on the wave height but on some other wave characteristic, say the wavelength. In this case, as was postulated by Hsu (1974), the Charnock formula can be rewritten in the form $z_0 = m_1(H_0/L_0)(u_*^2/g)$, where H_0 and L_0 are the height and length of energetically significant waves with maximum frequency ω_0 in the wind wave spectrum; m_1 is another numerical constant close to 1.0. We note that the combination H_0/L_0 is nothing but the wave steepness, and that in deep water the phase velocity c_0 and wavelength L_0 are connected by the relation $c_0 \approx (gL_0/2\pi)^{1/2}$. Substitution of this relation into the expression mentioned above yields $z_0 = (m_1/2\pi)H_0(c_0/u_*)^{-2}$. Under conditions of developed waves when $c_0/u_* \approx 30$ the Hsu formula reduces to the form $z_0 \sim H_0$, implying that the roughness parameter is proportional to the height of energetically significant waves.

If, following Kitaigorodskii (1970), we replace H_0 by the root-mean-square wave height σ_η, defined by the equality $\sigma_\eta = (2 \int_0^\infty S(\omega)\, d\omega)^{1/2}$, where $S(\omega)$ is the spectral density (ω is the frequency of the waves), and take into account the mobility of the wave spectral components moving in deep water with phase velocities $c = g/\omega$, then the condition of proportionality of z_0 and σ_η takes the form

$$z_0 \sim \left(\int_0^\infty S(\omega) \exp(-2\kappa g/\omega u_*)\, d\omega \right)^{1/2}.$$

We approximate $S(\omega)$ in the form

$$S(\omega) = \begin{cases} \beta g^2 \omega^{-5} & \text{at } \omega \geq \omega_0, \\ 0 & \text{at } \omega < \omega_0, \end{cases} \tag{3.4.10}$$

and thus ignore the contribution of low-frequency waves with $\omega < \omega_0$ to σ_η and z_0. Here, β is a universal constant varying, according to different authors, from 0.4×10^{-2} to 1.0×10^{-2}. After substitution of (3.4.10) into the expression for z_0 and using the relation $\omega_0 = (\beta/2)^{1/4}(g/\sigma_\eta)^{1/2}$ connecting ω_0 and σ_η, we obtain

$$z_0 \sim \begin{cases} \sigma_\eta & \text{at } c_0/u_* \ll 1, \\ \sigma_\eta \exp(-\kappa c_0/u_*) & \text{at } c_0/u_* \approx O(1), \\ (\tfrac{3}{4}\beta)^{1/2}\kappa^{-2}u_*^2/g & \text{at } c_0/u_* \gg 1, \end{cases} \tag{3.4.11}$$

from which it can be seen that in the initial stage of wave development (at $c_0/u_* \ll 1$), when the flow above the waves is similar to the flow above the immovable roughness elements, the parameter z_0 is proportional to the root-mean-square wave height σ_η. In the second stage, complying with developing waves (at $c_0/u_* \approx O(1)$), z_0 turns out to depend both on root-mean-square height σ_η and on age c_0/u_*. Finally, at the stage of developed waves (at $c_0/u_* \gg 1$) the roughness parameter z_0 ceases to depend on wave age c_0/u_* and is determined only by u_* and g, as in the Charnock formula.

To confirm the ambiguous dependence of z_0 on c_0/u_* at the stage of developing waves we picture the dimensionless roughness parameter gz_0/u_*^2 in the form

$$gz_0/u_*^2 = f(c_0/u_*). \qquad (3.4.12)$$

We take into account the dispersion relation $c_0 = g/\omega_0$ and approximate the right-hand side of Equation (3.4.12) by the power dependence $f(c_0/u_*) \sim (\omega_0 u_*/g)^\gamma$. Then, from (3.4.12), at $\gamma = 0$ we have the Charnock formula $gz_0/u_*^2 = m$; at $b = -1$ we have the Toba and Koga formula $z_0\omega_0/u_* = n$, here $n = 0.025$ is a numerical constant. Verification of these formulae by data from field and laboratory measurements have shown (see Toba, 1991) that on the phase plane $(gz_0/u_*^2, \omega_0 u_*/g)$ experimental dots are grouped within the triangle formed by lines $gz_0/u_*^2 = m$, $gz_0/u_*^2 = n(\omega_0 u_*/g)^{-1}$ and $\omega_0 u_*/g = 0.05$. The latter conforms to the boundary value of the wave age $\omega_0 u_*/g$ separating the regime of mixed waves (wind waves and swell) from the regime of developing waves. In other words, experimental data demonstrate that at the stage of developing waves the dimensionless roughness parameter gz_0/u_*^2 is determined not only by the wave age $\omega_0 u_*/g$ but also by the fetch, for example. As a palliative aimed at practical use, Toba (1991) proposed the approximation formula $gz_0/u_*^2 = 0.020(\omega_0 u_*/g)^{-1/2}$ occupying, on the phase plane $(gz_0/u_*^2, \omega_0 u_*/g)$, an intermediate position between the Charnock formula and the Toba and Koga formula.

The fourth method basically reduces to a combination of the Hsu and Kitaigorodskii approaches. The expression for the roughness parameter has the form $z_0 \sim (c_0/u_*)^{-2}[\int_0^\infty S(\omega) \exp(-2\kappa g/\omega u_*) \, d\omega]^{1/2}$, where the proportionality factor equals 2.80, in accordance with Geernaert *et al.* (1986). As shown in the same work, the proportionality constant in (3.4.11) is equal to 0.028.

The fifth method, proposed by Donelan (1982), is conceptually close to that proposed by Kitaigorodskii. But the low-frequency components of wind waves are not excluded and are considered as new roughness elements characterized by their own roughness parameters. The latter for low-frequency and high-frequency spectral ranges is given, respectively, in the form $z_{0l} = m_2 \cdot (\int_0^{2\omega_0} S(\omega) \, d\omega)^{1/2}$, $z_{0s} = m_2 \cdot (\int_{2\omega_0}^\infty S(\omega_0) \, d\omega)^{1/2}$, where m_2 is a numerical constant equal to 0.0125. Thus it is suggested that all spectral components of wind waves are fixed. This restriction is weakened in the following manner. Resistance coefficients for low- and high-frequency spectral ranges are calculated with the help of known values of z_{0l} and z_{0s}. Then the first mentioned coefficient is multiplied by

$$|u_{10} - c_l/\cos \varphi|(u_{10} - c_l/\cos \varphi) \cdot (u_{10}^{-2} \cos \varphi),$$

the second one by $(u_{10} - c_l \lambda^2 u_{10}^{-2})$, that is, it is assumed that low-frequency waves are propagated at an angle φ to the wind, and high-frequency waves are propagated along the wind. Phase velocities of low-frequency (c_l) and high-frequency (c_s) waves are found by the empirical formulae $c_l = 0.83c_0$ and $c_s = 0.415c_0$. The wind velocity u_{10} at a height of 10 m is assumed to be preset. It is clear that a number of assumptions forming the basis of this and other methods of evaluation of the roughness parameter, and particularly the assumption of additive presentation of the roughness parameter that is intended to account for contributions of different spectral intervals into the formation of the resulting resistance, stand in need of an additional substantiation.

3.5 Wind–wave interaction

As is well-known, the turbulent velocity fluctuations in the near-surface layer of water are less than the wave disturbances, complying with the energy-carrying components of the wave spectrum. Because of this, it is considered that waves are potential and that their generation is due not to the tangential wind stress but, rather, to the force of normal pressure. On the other hand, motions in the near-surface air layer are turbulent in character and wave disturbances of velocity are quite obvious. These disturbances transform the vertical structure of the surface atmospheric layer, and, therefore, the momentum and energy transfer to waves. Thus, if the problem of simulating the evolution of the wind in the surface atmospheric layer or of waves in the upper layer of the ocean is posed, then its solution can be obtained only within the framework of a coupled wind–wave interaction model.

We consider one such model proposed by Benilov *et al.* (1978). Assume that wind and wave fields are horizontally homogeneous and non-stationary and that some more simplifying assumptions are fulfilled, namely: (i) the upper ocean layer disturbed by waves can be contracted into to a plane, that is, its effect can be taken into account by assigning proper boundary conditions at the air–water interface; (ii) breakdown of waves generating the conversion of momentum and energy of waves into the momentum and energy of the drift current, and into the energy of small-scale turbulence in water is absent; (iii) wind momentum and energy fluxes supporting the drift current are also absent; (iv) in the surface atmospheric layer one can select the logarithmic layer and an underlying layer between the roughness level and the free ocean surface, within the limits of which the mean wind velocity and viscous stress are equal to zero; (v) because of non-commensurability of vertical scales of the above-mentioned layers the contribution of the latter to the integral (over the entire surface atmospheric layer) turbulent energy dissipation is negligible.

Let us write down the evolution equations for the momentum:

$$\frac{\partial u}{\partial t} = \frac{\partial}{\partial z}\frac{\tau}{\rho},$$

(3.5.1)

for the kinetic energy of mean motion

$$\frac{\partial}{\partial t}\frac{u^2}{2} = u\frac{\partial}{\partial z}\frac{\tau}{\rho},$$

(3.5.2)

and for the turbulent energy

$$\frac{\partial b}{\partial t} = \frac{\partial}{\partial z}\frac{q}{\rho} + \frac{\tau}{\rho}\frac{\partial u}{\partial z} - \frac{\varepsilon}{\rho},$$

(3.5.3)

in the surface atmospheric layer. Here b is the kinetic energy of turbulent fluctuations of velocity normalized to the air density (ρ); ε and q are the dissipation rate and diffusion flux, respectively, of the turbulent energy; other symbols are the same.

Momentum and turbulent energy fluxes at the air–water interface which, according to the problem conditions, are spent only on wave generation, can be determined from the evolution equations for the momentum $\mathcal{M}_w$ and energy E_w of the wave field. On the basis of assumptions (i)–(iv) these equations take the form

$$\frac{\partial \mathcal{M}_w}{\partial t} = \tau; \qquad \frac{\partial E_w}{\partial t} = -q \quad \text{at } z = 0.$$

(3.5.4a,b)

These are used as the dynamic boundary conditions at the air–water interface. One more condition at this surface is the continuity of the mean velocity in the absence of drift current. It is written in the form

$$u = 0 \quad \text{at } z = 0.$$

(3.5.5)

The following conditions are given at the upper boundary h of the surface atmospheric layer:

$$u = u_h = const, \qquad \tau, q, b = 0 \quad \text{at } z = h.$$

(3.5.6)

The first means choosing a fixed mean wind velocity, the others refer to vanishing of turbulent fluxes of momentum and energy, and also of the turbulent energy itself. We note the approximate character of the last three conditions: they assume that interaction between the surface layer and the upper layers of the atmosphere is lacking.

We turn to the derivation of integral relations for the surface atmospheric layer, not specifying concrete initial conditions for the moment. For this purpose we integrate (3.5.1)–(3.5.3) over z from 0 through h and take advantage of conditions (3.5.4)–(3.5.6). As a result we obtain

$$\frac{\partial}{\partial t} \int_0^h u \, dz = u_h \frac{\partial h}{\partial t} - \frac{\partial \mathcal{M}_w / \rho}{\partial t}, \tag{3.5.7}$$

$$\frac{\partial}{\partial t} \int_0^h \frac{u^2}{2} \, dz = \frac{u_h^2}{2} \frac{\partial h}{\partial t} - \int_0^h \frac{\tau}{\rho} \frac{\partial u}{\partial z} \, dz, \tag{3.5.8}$$

$$\frac{\partial}{\partial t} \int_0^h b \, dz = \int_0^h \frac{\tau}{\rho} \frac{\partial u}{\partial z} \, dz - \frac{\partial E_w / \rho}{\partial t} - \int_0^h \frac{\varepsilon}{\rho} \, dz, \tag{3.5.9}$$

where the terms on the left-hand side describe the rate of change in the integral (within the limits of the surface atmospheric layer) momentum, kinetic energy of mean motion and kinetic energy of fluctuating motion; the first terms on the right-hand sides of (3.5.7) and (3.5.8) are changes in momentum and kinetic energy of mean motion due to variations in thickness of the surface layer; the second terms on the right-hand sides of (3.5.7) and (3.5.9) are changes in the momentum and kinetic energy of fluctuating motion due to their transport to waves; the second term on the right-hand side in (3.5.8) and the first term on the right-hand side in (3.5.9) are mutual conversions of the kinetic energy of mean and fluctuating motions; and, finally, the third term on the right-hand side in (3.5.9) is the integral (within the limits of the surface atmospheric layer) turbulent energy dissipation.

Let us now recall that the lifetime of turbulent formations in the surface atmospheric layer is much less than the relaxation time of the surface atmospheric layer (which allows us to ignore the term on the left-hand side of Equation (3.5.9)), and take into account the constancy of u_h in time. Then, based on (3.5.7)–(3.5.9) we obtain

$$\frac{d}{dt} \left[\int_0^h (u_h - u) \, dz - \frac{\mathcal{M}_w}{\rho} \right] = 0, \tag{3.5.10}$$

$$\frac{d}{dt} \left[\frac{1}{2} \int_0^h (u_h^2 - u^2) \, dz - \frac{E_w}{\rho} \right] = \int_0^h \frac{\varepsilon}{\rho} \, dz. \tag{3.5.11}$$

Next, according to assumptions (iv) and (v) the integral (within the limits of the surface atmospheric layer) turbulent energy dissipation may be replaced by the integral dissipation within the limits of the logarithmic layer. But in the logarithmic layer the local turbulent energy dissipation ε / ρ is determined only by the friction velocity u_* and the height z, so from dimensional considerations $\varepsilon/\rho \approx u_*^3/z$, and, therefore,

$$\int_0^h \frac{\varepsilon}{\rho}\, dz \sim \int_{z_0}^h \frac{u_*^3}{z}\, dz, \tag{3.5.12}$$

or, if we use the logarithmic distribution of the mean velocity in the layer (z_0, h),

$$\int_0^h \frac{\varepsilon}{\rho}\, dz \approx \gamma u_h u_*^2, \tag{3.5.13}$$

where γ is a non-dimensional numerical factor representing the product of the von Karman constant and the proportionality constant from (3.5.12).

We also take into account that $u_*^2 = \tau/\rho$ with $\tau/\rho = d(\mathscr{M}_w/\rho)/dt$ (see 3.5.4)). Because of this, after substitution of (3.5.13) into (3.5.11) we finally obtain

$$\frac{d}{dt}\left[\frac{1}{2}\int_0^h (u_h^2 - u^2)\, dz - \frac{E_w}{\rho} - \gamma u_h \frac{\mathscr{M}_w}{\rho} \right] = 0. \tag{3.5.14}$$

The system (3.5.10) and (3.5.14), together with initial conditions

$$u = u_n,\ \overline{\mathscr{W}}_w,\ E_w = 0 \quad \text{at } t = 0, \tag{3.5.15}$$

has the two first integrals: the law of conservation of momentum,

$$\rho \int_0^h (u_h - u)\, dz = \mathscr{M}_w \tag{3.5.16}$$

and the law of conservation of energy

$$\tfrac{1}{2}\rho \int_0^h (u_h^2 - u^2)\, dz = E_w + \gamma u_h \mathscr{M}_w. \tag{3.5.17}$$

The system (3.5.16) and (3.5.17) is not closed. To close it according to assumption (iv) we assume that

$$u = \begin{cases} \dfrac{u_*}{\kappa} \ln \dfrac{z}{z_0} & \text{at } z_0 \le z \le h,\, t > 0, \\[2mm] 0 & \text{at } 0 \le z \le z_0,\, t > 0, \end{cases} \tag{3.5.18}$$

and then we take advantage of the momentum and energy definitions

$$\mathscr{M}_w = \rho_w \int_0^\infty \omega S(\omega)\, d\omega, \tag{3.5.19}$$

$$E_w = \rho_w g \sigma_\eta^2 = \rho_w g \int_0^\infty S(\omega)\, d\omega, \tag{3.5.20}$$

and of the Phillips expression for the frequency spectrum $S(\omega)$ of wind waves

$$S(\omega) = \begin{cases} \beta g^2 \omega^{-5} & \text{at } \omega \geq \omega_0, \\ 0 & \text{at } \omega < \omega_0, \end{cases} \qquad (3.5.21)$$

where, as before, ω is the wave frequency; ω_0 is the frequency of the spectral maximum; σ_η is the root-mean-square wave height; ρ_w is density of sea water and β is a numerical universal constant.

Substitution of (3.5.21) into (3.5.19) and (3.5.20) yields

$$\mathcal{M}_w = \rho_w \frac{\beta}{3g} c_0^3, \qquad (3.5.22)$$

$$E_w = \rho_w \frac{\beta}{4g} c_0^4, \qquad (3.5.23)$$

where $c_0 = g/\omega_0$ is the phase velocity complying with the frequency of the maximum in the wind wave spectrum.

Now system (3.5.4a), (3.5.16)–(3.5.18), (3.5.22) and (3.5.23) is closed: it includes only one differential equation (3.5.4a) and five diagnostic relationships to determine six unknown functions: $\mathcal{M}_w$, E_w, u_*, z_0, c_0 and h. We find what can be obtained from the system without recourse to integration over time. Keeping this in mind we select any one of the six unknowns (say, u_*) as an independent variable and find the remaining ones as functions of that. But first we introduce the definition

$$C = \frac{1}{2} \int_0^h (u_h^2 - u^2) \, dz \Big/ \int_0^h (u_h - u) \, dz, \qquad (3.5.24)$$

describing the ratio between energy and momentum losses in the surface atmospheric layer, and rewrite (3.5.17) as

$$C\mathcal{M}_w = E_w + \gamma u_h \mathcal{M}_w,$$

from whence

$$E_w = u_h \mathcal{M}_w \left(\frac{C}{u_h} - \gamma \right).$$

In combination with (3.5.22) and (3.5.23) this equality gives

$$\frac{c_0}{u_h} = \frac{4}{3} \left(\frac{C}{u_h} - \gamma \right). \qquad (3.5.25)$$

Substituting (3.5.18) into (3.5.16) and taking into account (3.5.22) we obtain

$$\frac{gz_0}{u_h^2} = \frac{\rho_w}{\rho} \frac{\beta}{3} \left(\frac{c_0}{u_h} \right)^3 \frac{\xi e^{-\xi}}{1 - e^{-\xi}}, \qquad (3.5.26)$$

where $\xi = \kappa u_h / u_*$ is a dimensionless parameter connected to the resistance coefficient C_u by the relationship $\xi = \kappa C_u^{-1/2}$.

Considering that $gz_0 / u_h^2 = (gh/u_h^2)(z_0/h)$, and $(h/z_0) = e^\xi$ (see (3.5.18)) we obtain

$$\frac{gh}{u_h^2} = \frac{\rho_w}{\rho} \frac{\beta}{3} \left(\frac{c_0}{u_h}\right)^3 \frac{\xi}{1 - e^{-\xi}}. \tag{3.5.27}$$

Finally, in accordance with (3.5.20), (3.5.22) and (3.5.23),

$$\frac{g \mathcal{M}_w / \rho_w}{u_h^3} = \frac{\beta}{3} \left(\frac{c_0}{u_h}\right)^3, \tag{3.5.28}$$

$$\frac{g E_w / \rho_w}{u_h^4} = \frac{\beta}{4} \left(\frac{c_0}{u_h}\right)^4, \tag{3.5.29}$$

$$\frac{g \sigma_\eta}{u_h^2} = \frac{\beta^{1/2}}{3} \left(\frac{c_0}{u_h}\right)^2. \tag{3.5.30}$$

We only have to find the function C appearing in (3.5.25). To do this we substitute (3.5.18) into (3.5.24), as a result of which we obtain

$$\frac{C}{u_h} = (1 - e^{-\xi})^{-1} - \xi^{-1}. \tag{3.5.31}$$

It follows from (3.5.31) that the function C/u_h is limited at all values of its argument ξ in the range $[0, \infty]$ and obeys the inequality $0.5 \le C/u_h \le 1.0$, where the lower limit ($C/u_h = 0.5$) complies with $\xi = 0$; the upper limit ($C/u_h = 1.0$) complies with $\xi \Rightarrow 1$. From this and from conditions of positiveness of z_0, h and c_0 follows boundedness of γ: under variations of ξ in the range $[\xi_0, \infty]$ the constant γ obeys the inequality $0 \le \gamma \le (C/u_h)_{\xi = \xi_0}$, where $\xi_0 \ge 0$ describes an initial momentum flux to waves. The equality $\gamma = (C/u_h)_{\xi = \xi_0}$ meets physically realized situations, and in (3.5.25)–(3.5.28) it provides a continuous transfer to initial values of c_0, z_0, h and σ_η at $t \Rightarrow 0$. Thus, ξ_0 is a solution of the algebraic equation $(C/u_h)_{\xi_0} = \gamma$.

Formulae (3.5.25)–(3.5.31) allow us to find all the basic parameters of the surface atmospheric layer as functions of c_0/u_*. For real values of the resistance coefficient ($C_u \sim 10^{-3}$, $\xi \sim 10$) the component $e^{-\xi}$ in (3.5.26), (3.5.27) and (3.5.31) is much less than one. If it is ignored, then it follows from (3.5.25)–(3.5.31) that changes in c_0/u_h make up no more than 30% of maximum value, and that for developing waves ($\xi \Rightarrow \infty$) the relation c_0/u_h approaches its limiting value

$$\frac{c_0}{u_h} = \tfrac{4}{3}(1 - \gamma) = \begin{cases} 4/3 & \text{at } \gamma = 0, \\ 1 & \text{at } \gamma = 0.25, \\ 2/3 & \text{at } \gamma = 0.5, \end{cases}$$

inherent in maximum possible waves at fixed wind velocity. Based on (3.5.25) and (3.5.30) we have for such waves

$$\frac{g\sigma_\eta}{u_h^2} = \frac{8\sqrt{(\beta)}}{9}(1 - \gamma) = \begin{cases} 9 \times 10^{-2} & \text{at } \gamma = 0, \\ 5 \times 10^{-2} & \text{at } \gamma = 0.25, \\ 2 \times 10^{-2} & \text{at } \gamma = 0.5. \end{cases}$$

At the same time, according to observational data for the conditions of developed waves (see Benilov *et al.* (1978)), $c_0/u_h \approx 1.14$ and $g\sigma_\eta/u_h^2 \approx 5.2 \times 10^{-2}$. As can be seen, at $\gamma = 0.25$ the theoretical and experimental estimates are in agreement with each other. The same can be said about the roughness parameter of the sea surface (Figure 3.3).

Calculated, by Equation (3.5.27), values of gh/u_h^2 demonstrate that at $c_0/u_* \approx 10$, $u_h = 10\,\text{m/s}$ and $\gamma = 0.25$, the thickness of the surface atmospheric layer becomes roughly equal to 50 m, and at $c_0/u_* \approx 30$ (developed waves),

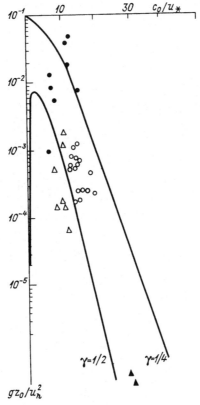

Figure 3.3 Dependence of the non-dimensional sea surface roughness gz_0/u_h^2 on c_0/u_*, according to Benilov *et al.* (1978). Various symbols show the estimates of different authors.

other things being equal, it is 200 m. Thus it turns out that in the process of wind wave development accompanied by momentum and energy transfer to waves the surface atmospheric layer is rebuilt up to heights of the order of hundreds of metres.

In conclusion, we adduce an asymptotic solution describing the evolution of wave parameters at large t. Previously, with the relationships (3.5.22), (3.5.25) and (3.5.31) and definitions τ and ξ, we reduce Equation (3.5.4a) to the form

$$\frac{d}{dt}\left(\frac{c_0}{u_h}\right) = \frac{\rho}{\rho_w}\frac{\kappa^2}{\beta}\frac{g}{u_h}\xi^{-2}\left(\frac{u_h}{c_0}\right)^2. \tag{3.5.32}$$

At large t (in other words, at large ξ) the expression for (c_0/u_h) takes the form

$$\frac{c_0}{u_h} = \frac{4}{3}\left(1 - \gamma - \frac{1}{\xi}\right). \tag{3.5.33}$$

Its substitution into (3.5.32) and elimination of the small terms on the right-hand side yield

$$\frac{4}{3}\frac{d\xi}{dt} = \frac{9}{16}\frac{\rho}{\rho_w}\frac{\kappa^2}{\beta}\frac{g}{u_h}(1 - \gamma)^{-2},$$

whence, after the introduction of non-dimensional time t' connected to t by the relation $t = (64/27)(\rho_w/\rho)(\beta/\kappa^2)(u_h/g)t'$, we obtain

$$\frac{d\xi}{dt'} = (1 - \gamma)^{-2}. \tag{3.5.34}$$

The solution of this equation at $t' \geq t'_1$ ($\xi > \xi_1$) has the form

$$\xi = \xi_1 + (1 - \gamma)^{-2}(t' - t'_1), \tag{3.5.35}$$

where $\xi_1 = \xi|_{t'=t'_1}$.

Knowing ξ it is easy to find the dependencies of wave characteristics on time. These dependencies, in accordance with (3.5.30) and (3.5.33), are written to within small terms as follows:

$$\frac{c_0}{u_h} = \tfrac{4}{3}(1 - \gamma)\left[1 - \frac{(1 - \gamma)}{(1 - \gamma)^2\xi_1 + (t' - t'_1)}\right], \tag{3.5.36}$$

$$\frac{g\sigma_\eta}{u_h^2} = \frac{8\sqrt{(\beta)}}{9}(1 - \gamma)^2\left[1 - \frac{2(1 - \gamma)}{(1 - \gamma)^2\xi_1 + (t' - t'_1)}\right], \tag{3.5.37}$$

$$\frac{u_*}{u_h} = \frac{\kappa(1 - \gamma)^2}{(1 - \gamma)^2\xi_1 + (t' - t'_1)}. \tag{3.5.38}$$

We pay attention to the decrease in u_*/u_h and the increase in c_0/u_h and $g\sigma_\eta/u_h^2$ over time. The last circumstance is obvious, but the first needs comment.

We have discussed above the limiting situation when the momentum and energy fluxes coming from the atmosphere are completely spent on wave generation (wave breakdown and drift current formation do not occur). In reality, everything is much more complicated. Momentum and energy of the wind are transferred simultaneously to waves and the drift current, and in doing so the momentum and energy transferred to waves is only partly spent on their development: at wave breakdown a part of the wave momentum is transformed into the momentum of the drift current, and a part of the wave energy into the energy of small-scale turbulence in the surface ocean layer. A solution of the wind–wave interaction problem, as applied to this more realistic situation, was obtained by Benilov (personal communication). As a result it was possible to clarify that establishing the dynamic equilibrium between waves and wind at the stage of wave breakdown occurs much more slowly than at the wave growth stage, and that at the intermediate stage when wave phase velocities are still increasing but wave breakdown already comes to manifest itself, the wave momentum and energy fluxes turn out to be much lower than the momentum and energy fluxes caused by wave breakdown. This conclusion is confirmed by the estimate presented by Kitaigorodskii (1970), according to which the wave momentum flux is of the order of 10% of the total momentum flux, while its remaining part is redistributed by some (still not established) means between two other components controlling the processes of wave breakdown and drift current maintenance.

3.6 Vertical distribution of the temperature and passive admixture over an immovable surface

In a similar way, as was demonstrated in Section 3.1, it can be shown that in a near-wall layer the condition of approximate constancy of the vertical flux F of any passive admixture has to be fulfilled, that is,

$$F(z) = \rho\overline{c'w'} - \rho\chi\frac{\mathrm{d}c}{\mathrm{d}z} \approx const, \tag{3.6.1}$$

where c is the mean concentration of the passive admixture, c' is the departure from its mean value, χ is the molecular diffusion coefficient; other symbols are the same.

From general considerations, it is clear that the profile of c should depend on the admixture transfer intensity for which the flux F serves as its measure, on statistical characteristics of the velocity field determined by parameters u_* and v, and on the molecular diffusion coefficient χ. Then, on the basis of the π-theorem of dimensional analysis

$$c - c_0 = \frac{F_0/\rho}{\kappa u_*} f_c\left(\frac{zu_*}{v}, Sc\right), \tag{3.6.2}$$

where $f_c(zu_*/v, Sc)$ is a new non-dimensional universal function satisfying the condition $f_c(0, Sc) \equiv 0$; $Sc = v/\chi$ is the molecular Schmidt number; subscript 0 indicates belonging to the underlying surface ($z = 0$); the von Karman constant κ is included in the multiplier before the function $f_c(zu_*/v, c_c)$ for convenience.

Let us consider the case of a smooth surface and $z \ll v/u_*$. At values of the number Sc not very different from one the condition $z \ll v/u_*$ provides the existence not only of the viscous sublayer but of a *molecular diffusion sublayer*. In the latter, on the basis of (3.6.1),

$$c - c_0 = \frac{F_0/\rho}{\chi} z \tag{3.6.3}$$

or (compare to (3.6.2) and (3.6.3)) $f_c(zu_*/v, Sc) = \kappa(zu_*/v)Sc$ and the thickness δ_c of the molecular diffusion sublayer should be connected with the determining parameters by the relation $\delta_c = \psi(Sc)v/u_*$.

In the other limiting case, where $z \gg v/u_*$ and, other things being equal, the vertical gradient of the admixture concentration should be a function of F_0/ρ, u_* and z only, then from dimensional considerations

$$\frac{dc}{dz} = \frac{F_0/\rho}{\alpha \kappa u_* z}, \tag{3.6.4}$$

and hence,

$$c - c_0 = \frac{F_0/\rho}{\alpha \kappa u_*} \ln z + C; \quad f_c\left(\frac{zu_*}{v}, Sc\right) = \frac{1}{\alpha} \ln \frac{zu_*}{v} + C_1, \tag{3.6.5}$$

where α is a numerical constant of order unity; C is the constant of integration defined by the condition of equal concentrations joining at the boundary with the underlying layer and thereby depending on the concentration change in the thin near-wall layer; $C_1 = (F_0/\rho\kappa u_*)^{-1}C - \alpha^{-1} \ln u_*/v$.

To estimate the constant C, several methods have been proposed. The simplest of these is to use the Reynolds hypothesis (or, as it is often called, the *Reynolds analogy*) about the identity of mechanisms of the momentum and admixture transfer in a layer of approximate constancy of fluxes. This is equivalent to assigning the relation

$$\frac{F_0/\rho}{(u_2 - u_1)(c_2 - c_1)} = \frac{\tau/\rho}{(u_2 - u_1)^2},$$

where u_1, u_2 and c_1, c_2 are values of the mean velocity and admixture concentration at two fixed heights.

Comparing (3.6.3) to (3.2.8), and (3.6.5)–(3.2.12), we see that the relation mentioned above is fulfilled only at $Sc = 1$, $\alpha = 1$ in the entire boundary layer including the viscous-buffer layer. In other words, its introduction is equivalent to discounting changes in admixture diffusion affected by the molecular Schmidt number. The latter amounts to 0.62 for water vapour.

Turning to discuss admixture diffusion near a rough wall, we note that the Reynolds analogy in this case is not fulfilled in general because the momentum exchange does not now depend on viscosity, and the diffusion continues to be controlled by molecular transfer. In such a situation there is nothing to be done but to present an expression for the profile of the admixture concentration in the form

$$c - c_0 = \delta c_0 + \frac{F_0/\rho}{\alpha_c \kappa u_*} \ln \frac{z}{z_0}, \tag{3.6.6}$$

where α_c is inversely proportional to the turbulent Schmidt number (see below), and then to use empirical data (say, data from laboratory experiments) to determine the parameter δc_0 meaning a near-surface concentration jump. Based on dimensional considerations it was shown by Yaglom and Kader (1974) that, for temperature, the numerical parameter $\delta\theta_0/T_*$ (here, $T_* = -H/\rho c_p \kappa u_*$ is the temperature scale) is a function of the surface Reynolds number Re_0 and is associated with it by the relation $\delta\theta_0/T_* = 0.18 Re_0^{0.5}$.

But because the discussion has turned to temperature, it is worth remembering that its identification with a passive admixture, strictly speaking, has no justification due to the appearance in heated liquids of additional buoyancy forces and the dependence of molecular viscosity and thermal conductivity on temperature.

3.7 Coefficients of resistance, heat exchange and evaporation for the sea surface

We have already mentioned the sea surface resistance coefficient several times. And now it is time to explain its meaning, and that of the heat exchange and evaporation coefficients related to it. We give the names *resistance coefficient*, *heat exchange coefficient* (*Stanton number*), and *evaporation coefficient*

(*Dalton number*) to non-dimensional integral characteristics of momentum, heat and moisture exchange defined, respectively, as

$$C_u = \frac{\tau/\rho}{u^2}, \qquad C_\theta = \frac{-H/\rho c_p}{u(\theta - \theta_0)}, \qquad C_q = \frac{-E/\rho}{u(q - q_0)}. \qquad (3.7.1)$$

Here θ_0 and q_0 are temperature and air specific humidity at the underlying surface; u, θ and q are the mean wind velocity, temperature and specific humidity at a fixed height z.

According to these definitions and Equations (3.4.1) and (3.6.5) describing the vertical distributions of the mean wind velocity and an arbitrary passive admixture in the logarithmic boundary layer, the expressions for C_u, C_θ and C_q can be presented in the form

$$\left. \begin{array}{l} C_u = \dfrac{\kappa^2}{\ln^2(z/z_0)}, \qquad C_\theta = \dfrac{\alpha_\theta^0 C_u}{(1 + C_u^{1/2}(\alpha_\theta^0/\kappa)(\delta\theta_0/T_*))}, \\[4mm] C_q = \dfrac{\alpha_q^0 C_u}{(1 + C_u^{1/2}(\alpha_q^0/\kappa)(\delta q_0/q_*))}, \end{array} \right\} \qquad (3.7.2)$$

where $q_* = -E/\rho\kappa u_*$ is the specific humidity scale; α_θ^0 and α_q^0 are inverse values of the turbulent Prandtl number (k_θ/k) and of the Schmidt number (k_q/k); k, k_θ and k_q are coefficients of turbulent viscosity, thermal conductivity and diffusion; other designations are the same.

We emphasize once more that Equations (3.7.2) are valid only in the logarithmic boundary layer within the limits of which the effect of molecular exchange on the vertical transfer of the momentum, heat and moisture can be neglected. Considering this and also the fact that for air the molecular Prandtl number (ratio between coefficients of molecular viscosity and thermal conductivity) and molecular Schmidt number (ratio between coefficients of molecular viscosity and diffusion) are close to 1, it may be assumed that the coefficients α_θ and α_q depend only on the stratification of the surface atmospheric layer. Experimental data systematized by Busch (1977) demonstrate an increase in the coefficient α_θ from 1 for neutral stratification, through 3.5 for unstable stratification. This fact is interesting in itself but for us it is important that under conditions of a weakly stratified liquid the coefficients α_θ and α_q do not differ very much from 1 and, hence, in the logarithmic boundary layer the coefficients of turbulent viscosity, thermal conductivity and diffusion are equal. But if this is the case then in all cases for hydrodynamically smooth surfaces $(\delta\theta_0 = \delta q_0 = 0)$ on the basis of (3.7.2), we obtain $C_u = C_\theta = C_q$. Equality of the coefficients of resistance, heat transfer and evaporation is a direct result of the Reynolds analogy.

In a preceding section we have already noted that extension of the Reynolds analogy to the case of a hydrodynamically rough surface has no justification because of the influence of molecular transfer in the immediate vicinity of the underlying surface. Therefore, until we have reliable information about the dependence of α_θ^0, α_q^0, $\delta\theta_0$ and δq_0 on the determining parameters, one has to be satisfied with different suggestions of the Reynolds analogy kind or with simply rejecting the use of Equations (3.7.2) for C_θ and C_q and restricting ourselves to experimental estimates of these coefficients. Each of the alternative variants has its own advantages and disadvantages, but the second, because of the absence of an exact analogy between the immovable rough wall and the wave-covered sea surface, is considered to be preferable. We use what can be obtained from analysis of experimental data.

Experimental estimates of the coefficients of resistance, heat exchange and evaporation are presented in Table 3.1. The analysis attests to the fact that available estimates differ quite perceptibly from each other. There are several causes, mainly the adoption of different (eddy correlation, dissipation and gradient, see below) methods of determining eddy fluxes of momentum, heat and moisture; different periods of averaging (sample lengths); and different frequency responses of measuring equipment. One more source of uncertainty is errors in measuring sea surface temperature, which is sometimes identified with the temperature of a certain finite water volume, or with the temperature of the upper mixed layer. And, certainly, one should not neglect the influence of waves and appropriate changes in the hydrodynamic properties of the ocean surface, and the effects of stratification of the surface atmospheric layer.

To illustrate the role of stratification let us direct our attention to the similarity theory, developed by Monin and Obukhov (1954), for a thermally stratified medium.

3.8 The Monin–Obukhov similarity theory

The vertical structure of a thermally stratified surface atmospheric layer is determined by the tangential wind stress τ, heat flux H, density ρ, molecular viscosity ν and thermal conductivity χ, air heat capacity c_p, buoyancy parameter $\beta = g/T_0$, and also by one or other parameter describing the geometrical properties of the underlying surface. The creation of the similarity theory is based on the following hypothesis (see Monin and Yaglom, 1965): *in the domain of developed turbulence for every interval of the spectrum, excluding the dissipation interval, the turbulent regime does not depend on molecular constants, and at heights much larger than the mean size of the underlying surface undulations, properties of the surface do not affect the*

Table 3.1. *Coefficients of resistance, heat exchange, and evaporation according to different authors*

Author	Range of change of wind velocity (m/s)	$C_u\,10^{-3}$	$C_\theta\,10^{-3}$	$C_q\,10^{-3}$
Sheppard (1958)	from 1 to 20	$0.80 + 0.114u_{10}$	—	—
Deacon and Webb (1962)	–	$1.10 + 0.04u_{10}$	—	—
Deacon and Webb (1962)	<14	$1.10 + 0.07u_{10}$	—	—
Zubkovskii and Kravchenko (1967)	–	$0.72 + 0.12u_{10}$	—	—
Weller and Burling (1967)	from 4 to 16	1.31 ± 0.36	—	—
Smith (1967)	from 4 to 16	1.03 ± 0.18	—	—
Makova (1968)	<15	$0.13u_{10}$	—	—
	≥15	$-13.1 + u_{10}$	—	—
Hasse (1968)	from 4 to 16	1.21 ± 0.24	—	—
Wu (1969)	≤1	$1.25u_{10}^{-0.2}$	—	—
	from 1 to 15	$0.5u_{10}^{0.5}$	—	—
	>15	2.6	—	—
Kurnetsov (1970)	≤10.4	$1.72 - 0.098u_{10}$	—	—
	>10.4	$1.00 + 1.26\,(u_{10} - 11)^{1/2}$	—	—
Smith (1970)	from 4 to 16	1.35 ± 0.34	—	—
Miyake et al. (1970)	from 4 to 16	1.09 ± 0.17	—	—
Hicks and Dyer (1970)	from 2 to 10	1.10 ± 0.10	—	1.40 ± 0.30
Pond et al. (1971)	from 4 to 7.5	1.44 ± 0.40	1.20 ± 0.25	—
Sheppard et al. (1972)	from 3 to 16	$0.36 + 0.10u_{10}$	—	—
Denman and Miyake (1973)	from 4 to 18	$1.29 + 0.03u_{10}$	—	—

		$1.2Re_0^{0.15}$	$Re_0^{0.11}$	$Re_0^{0.11}$
Kitaigorodskii et al. (1973)	from 1 to 22		1.36 ± 0.40	–
Pond et al. (1974)	from 2 to 8	1.49 ± 0.28	1.41 ± 0.18	1.47 ± 0.64
	from 2.5 to 8	1.48 ± 0.21	–	–
Wieringa (1974)	from 5 to 15	$0.86 + 0.058u_{10}$	1.28	1.46
Dunckel et al. (1974)	from 4.5 to 11	1.56		
Smith (1974)	from 3 to 10	1.20	1.20 ± 0.30	1.30 ± 0.50
Smith and Banke (1975)	from 7 to 21	$0.63 + 0.066u_{10}$	–	–
Kondo (1975)	from 2 to 8	$0.87 + 0.067u_{10}$	1.08 + 0.03	$1.08 + 0.03\,(u_{10} - 2)$
	from 8 to 25	$1.2\ + 0.025u_{10}$	–	–
	>25	$0.073u_{10}$		
Tsukamoto et al. (1975)	from 3 to 13	1.32	1.28	1.40
Garratt and Hyson (1975)	from 4 to 16	–	1.2	1.6
Emmanuel (1975)	from 2.7 to 8	1.15 ± 0.20	1.34 ± 0.30	1.10 ± 0.30
Krügermeyer (1976)	from 2.5 to 11	$1.34(1 - 0.331S)$	$1.42(1 - 0.455S)$	$1.20(1 - 0.394S)$
Friehe and Schmitt (1976)	from 2 to 21	–	1.32 ± 0.007	1.32 ± 0.07
Krügermeyer et al. (1978)	from 3 to 8	1.30		
Hasse et al. (1978)	from 1 to 12	1.25	1.34	1.15
Wu (1980)	from 1 to 21	0.80 + 0.063		
Smith (1980)	from 6 to 22	$0.61 + 0.063u_{10}$		
Donelan (1982)	from 4 to 17	$0.37 + 0.137u_{10}$		
Launiainen (1983)	from 1 to 21	$0.80 + 0.065u_{10}$		$0.82 + 0.041u_{10}$
Geernaert et al. (1986)	from 5 to 21	$0.43 + 0.097u_{10}$		

Note: dashes point to unavailable information; $Re = u_* z/\nu$ is the roughness Reynolds number; $S = 3.55(T_{10} - T_0)/u_{10}^2$ is the dimensionless stratification parameter (an analogue of the Richardson number); T_{10} and u_{10} are the temperature and wind velocity (in m/s) at a height of 10 m; T_0 is the sea surface temperature.

137

vertical changes in statistical characteristics of the hydrodynamic fields directly. The number of determining parameters can be reduced by two when it is considered that the mean velocity and temperature do not include mass dimensions and, hence, ρ and c_p have to be incorporated only into combinations τ/ρ and $H/\rho c_p$. Thus, the mean velocity and temperature vertical distributions should be determined by the values of parameters τ/ρ, β and $H/\rho c_p$, from which the following unique (to within a numerical factor) combination is formed:

$$L = \frac{u_*^3}{\kappa \beta(-H/\rho c_p)},$$

which is called *the Monin–Obukhov length scale*.

According to the above-mentioned similarity principle, the dimensionless vertical gradients of the mean velocity u and potential temperature θ should serve as universal functions of the dimensionless relation z/L, that is,

$$\frac{\partial u}{\partial z} = \frac{u_*}{\kappa z} \Phi_u\left(\frac{z}{L}\right), \qquad \frac{\partial \theta}{\partial z} = \frac{T_*}{z} \Phi_\theta\left(\frac{z}{L}\right), \tag{3.8.1}$$

where $u_* = (\tau/\rho)^{1/2}$ and $T_* = (-H/\rho c_p)/\kappa u_*$ are the friction velocity and temperature scale; $\Phi_u(z/L)$ and $\Phi_\theta(z/L)$ are the universal functions of the argument z/L. The latter represents a criterion of the hydrostatic stability: at $z/L < 0$ the stratification is unstable; at $z/L > 0$ it is stable; and at $z/L = 0$ it is neutral. We note that an increase in the length scale L in z/L is equivalent to a decrease in the height z, and vice versa. Therefore, at sufficiently small heights the effect of the stratification does not manifest itself.

Integrating (3.8.1) we arrive at the expressions

$$\left. \begin{array}{l} u = \dfrac{u_*}{\kappa}\left[\ln\dfrac{z}{z_0} - \psi_u\left(\dfrac{z}{L}\right)\right], \\[3mm] \theta - \theta_0 = T_*\left[\ln\dfrac{z}{z_0} - \psi_\theta\left(\dfrac{z}{L}\right)\right], \end{array} \right\} \tag{3.8.2}$$

where

$$\psi_{u,\theta} = \int_0^{z/L} \frac{1 - \Phi_{u,\theta}(\xi)}{\xi}\, d\xi$$

and, also, it is assumed that $z/z_0 \ll 1$, $\Phi_{u,\theta}(0) = 1$.

Let us discuss some asymptotic regimes.

Neutral stratification. In this case (complying with $z/L \Rightarrow 0$) the buoyancy force does not participate in the formation of the vertical distribution of mean velocity and potential temperature, and the buoyancy parameter β is excluded from the number of determining parameters. It is impossible to form a non-dimensional combination from the remaining parameters u_*, $H/\rho c_p$ and z. As applied to (3.8.1) it means that at $z/L \Rightarrow 0$ the functions $\Phi_u(z/L)$ and $\Phi_\theta(z/L)$ approach 1 asymptotically. Then

$$\frac{\partial u}{\partial z} = \frac{u_*}{\kappa z}, \qquad \frac{\partial \theta}{\partial z} = \frac{T_*}{z}, \tag{3.8.3}$$

from which, after integration, we obtain usual formulae of the theory of the logarithmic boundary layer:

$$u = \frac{u_*}{\kappa} \ln \frac{z}{z_0}, \qquad \theta - \theta_0 = T_* \ln \frac{z}{z_0}. \tag{3.8.4}$$

Stratification close to neutral (small values of z/L). We represent the functions $\Phi_u(z/L)$, $\Phi_\theta(z/L)$ as power series in the vicinity of the point $z/L = 0$ and restrict ourselves to the first two terms of the expansion. As a result we obtain

$$\Phi_{u,\theta}(z/L) = 1 + \beta_{u,\theta} z/L. \tag{3.8.5}$$

Substitution of (3.8.5) into Equations (3.8.1) and their subsequent integration yields the so-called log-linear law for profiles of the mean velocity and potential temperature

$$\left. \begin{aligned} u &= \frac{u_*}{\kappa}\left(\ln \frac{z}{z_0} + \beta_u \frac{z}{L} \right), \\ \theta - \theta_0 &= T_*\left(\ln \frac{z}{z_0} + \beta_\theta \frac{z}{L} \right), \end{aligned} \right\} \tag{3.8.6}$$

where it is assumed that $z_0/z \ll 1$.

The numerical constants β_u, β_θ appearing in Equations (3.8.5) and (3.8.6) must be positive. Indeed, an enhancement of instability should be accompanied by an intensification of the turbulent exchange that, in turn, should lead to a decrease in the vertical gradients of the mean velocity and potential temperature. Hence, corrections to the logarithmic terms on the right-hand sides of Equations (3.8.6) have to be negative for unstable stratification ($z/L < 0$), and positive for stable stratification ($z/L > 0$). This is possible when β_u, $\beta_\theta > 0$.

Strongly unstable stratification (large negative values of z/L). We take advantage of the fact that the condition $z/L \Rightarrow -\infty$ conforms to either unlimited growth of $-H/\rho c_p$ at fixed u_* or decrease in u_* at fixed $-H/\rho c_p$.

It is natural to assume that in this case the friction velocity should be excluded from the number of parameters determining the vertical structure of the temperature field, but not the velocity field. This condition is fulfilled if at $z/L \Rightarrow -\infty$ the function $\Phi_\theta(z/L)$ in (3.8.1) approaches $(z/L)^{-1/3}$ asymptotically. If we assume, in addition, that at $z/L \Rightarrow -\infty$ the relation $\Phi_u(z/L)/\Phi_\theta(z/L)$ approaches some certain finite limit, then on the basis of (3.8.1) we have

$$
\left.
\begin{aligned}
u &= \frac{a_u}{\kappa} u_* L^{1/3}(z^{-1/3} - z_0^{-1/3}), \\[2mm]
\theta - \theta_0 &= a_\theta T_\theta L^{1/3}(z^{-1/3} - z_0^{-1/3}),
\end{aligned}
\right\}
\tag{3.8.7}
$$

where a_u and a_θ are numerical constants.

Strongly stable stratification (large positive values of z/L). We again take advantage of the fact that large positive values of z/L are equivalent either to large z at fixed positive L, or to fixed z at small positive L. But the large positive values of $-H/\rho c_p$ (an abrupt temperature inversion) comply with small positive L, and under such conditions large eddies degenerate because of expenditure of energy on work against buoyancy forces. This leads to a reduction in the exchange between layers, which causes the turbulence to be local in character. But then the characteristics of the turbulent exchange and the vertical gradients of the mean velocity and potential temperature related to them have to cease depending on the vertical coordinate z. This last condition is equivalent to $\Phi_u(z/L) \Rightarrow (z/L)$, $\Phi_\theta(z/L) \Rightarrow (z/L)$ at $z/L \Rightarrow \infty$. In this case from (3.8.1) it follows that

$$
\left.
\begin{aligned}
u &= \frac{a_u'}{\kappa} \frac{u_*}{L} z, \\[2mm]
\theta - \theta_0 &= a_\theta' \frac{T_*}{L} z,
\end{aligned}
\right\}
\tag{3.8.8}
$$

where it is assumed as before that $z_0/z \ll 1$.

In the general case, in order to estimate the mean velocity and potential temperature according to Equations (3.8.2) it is necessary to know the universal functions $\psi_u(z/L)$ and $\psi_\theta(z/L)$. The first information about them was obtained by Monin and Obukhov (1954) from measurements of the mean wind velocity and temperature in the surface atmospheric layer over land. The results of data handling confirmed the presence of the asymptotic properties of the function $\Phi_u(z/L)$ predicted by the similarity theory. Later it was found that these results were in good agreement with estimates of the function $\Phi_u(z/L)$ from data of covariance measurements.

Conclusions obtained relating to the thermally stratified medium are easily generalized in the case of an allowance for humidity stratification. Such consideration reduces to replacement of the expression for the Monin–Obukhov length scale with

$$L = u_*^2/[\kappa^2(\beta T_* + 0.61gq_*)],$$ (3.8.9)

where $q_* = (-E/\rho)/\kappa u_*$ is the specific humidity scale, and to the addition to (3.8.2) of a similar formula for the vertical profile of the specific humidity q:

$$q - q_0 = q_*\left[\ln\frac{z}{z_0} - \psi_q\left(\frac{z}{L}\right)\right],$$ (3.8.10)

where $\psi_q(z/L)$ is a universal function having the same asymptotic properties as $\psi_\theta(z/L)$.

We demonstrate that the replacement of the Monin–Obukhov length scale by its modified expression (3.8.9) is equivalent to consideration of humidity stratification. Indeed, the equation of dry air state has the form $p = R\rho T$, where p, ρ and T are pressure, density and temperature; R is the gas constant. In the presence of water vapour when the air transforms into a gas mixture the gas constant $R = R_*/\mu$ (here R_* is the universal gas constant; μ is the relative molecular weight) becomes a variable: the relative molecular weights of dry air (μ_d) and water vapour (μ_w) are equal to 28.9 and 18 respectively. When considering water vapour as the ideal gas the equation of the moist air state can be presented in the form

$$p = (R_*/\mu)\rho T = (R_*/\mu_d)\rho_d T + (R_*/\mu_w)\rho_w T$$

from which, with $R = R_*/\mu_d$, it follows that

$$p = R\rho T[1 + (\mu_d/\mu_w - 1)q] = R\rho T(1 + 0.61q)$$

where $q = \rho/(\rho_w + \rho_d)$ is the specific humidity; ρ_d and ρ_w are the densities of dry air and water vapour. We take the logarithm of this equation and vary it, taking into account the fact that the density change depends only on changes in temperature and humidity but not on pressure (the *Boussinesq approximation*). As a result we have $\rho'/\rho = -T'/T - 0.61q'/(1 + 0.61q) \approx -T'/T - 0.61q'$, where ρ', T' and q' are fluctuations of density, temperature and specific humidity. Hence, the expression for the buoyancy flux $(g/\rho)\overline{\rho'w'}$ takes the form $(g/\rho)\overline{\rho'w'} = -\beta\overline{T'w'} - 0.61g\overline{q'w'} = \beta(-H/\rho c_p) + 0.61g(-E/\rho)$. Its substitution into L instead of $\beta(-H/\rho c_p)$ and using the definitions of T_* and q_* yield (3.8.9).

Turning to (3.8.1) we note (see Busch, 1977) that the overwhelming majority of experimental data on the vertical distribution of mean wind velocity, potential temperature and specific humidity are described very well by the expressions

$$
\Phi_u\left(\frac{z}{L}\right) = \begin{cases} (1 + \beta_u z/L) & \text{at } z/L \geq 0, \\ (1 - \gamma_u z/L)^{-1/4} & \text{at } z/L < 0, \end{cases}
$$

$$
\frac{\Phi_\theta(z/L)}{\Phi_\theta(0)} = \frac{\Phi_q(z/L)}{\Phi_q(0)} = \begin{cases} (1 - \beta_\theta z/L) & \text{at } z/L \geq 0, \\ (1 - \gamma_\theta z/L)^{-1/2} & \text{at } z/L < 0, \end{cases}
$$

(3.8.11)

at $\beta_u = 5$, $\gamma_u = 15$, $\beta_\theta = 6$, $\gamma_\theta = 9$.

We also note that the linear dependence of functions Φ_u, Φ_θ and Φ_q for the stable atmospheric stratification ($z/L > 0$) is universally accepted. As for the form of the function Φ_u for the unstable stratification ($z/L < 0$), there is no common opinion on this score. More often, the KEYPS equation (the abbreviation of the names of five authors – Kazanskii, Ellison, Yamamoto, Panofsky and Sellers) is used. It has the form $(\Phi_u^4 - \gamma(z/L)\Phi_u^3) = 1$ and follows from the interpolation formula $k_M = \kappa u_* z(1 - \gamma Rf)^{-1/4}$, after substitution of the relations

$$
k_M \ (\equiv u_*^2/(\partial u/\partial z) = \kappa u_* z/\Phi_u(z/L),
$$

$$
Rf \equiv \beta(-H/\rho c_p)/u_*^2(\partial u/\partial z) = (z/L)/\Phi_u(z/L)
$$

connecting the coefficient of vertical turbulent viscosity k_M, and the flux Richardson number Rf with $\Phi_u(z/L)$. The indicated interpolation formula yields correct results at large (positive and negative) and close to zero values of z/L. Indeed, at $z/L \Rightarrow 0$ (that is, under conditions of neutral stratification) $Rf \Rightarrow 0$ and $k_M \approx \kappa u_* z$; at $|z/L| \gg 1$ (that is, under conditions of strongly unstable stratification) $-Rf \sim |z/L|^{4/3}$ and $k_M \sim \kappa u_* z|z/L|^{1/3}$; finally, at $z/L \Rightarrow \infty$ (that is, under conditions of strongly stable stratification) $Rf \Rightarrow 1/\gamma$ and $k_M/z \Rightarrow 0$. Here and above, γ is an empirical constant varying within the limits from 9 to 18 according to different authors.

Values $\Phi_\theta(0)$ and $\Phi_q(0)$ also serve as the subject of discussion: generally, they are specified as identical, that is, equivalent to the assumption about the similarity of heat and moisture transfer. There is also another point of view according to which the similarity condition is fulfilled not for heat and moisture transfers, but, rather, for momentum and moisture transfers, that is, Φ_q is equal to Φ_u rather than to Φ_θ.

Now with data on the universal functions Φ_u, Φ_θ and Φ_q it is possible to obtain expressions specifying the relationships between the coefficients of resistance, heat exchange and evaporation, on the one hand, and the parameter z/L describing atmospheric stratification, on the other hand. These expressions take the form

$$
\left.\begin{aligned}
C_{\mathrm{u}} &= \frac{\kappa^2}{[\ln(z/z_0) - \psi_{\mathrm{u}}(z/L)]^2}, \\
C_{\theta,q} &= \alpha_{\theta,q}^0 C_{\mathrm{u}} \frac{[\ln(z/z_0) - \psi_{\mathrm{u}}(z/L)]}{[\ln(z/z_0) - \psi_{\theta,q}(z/L)]}.
\end{aligned}\right\}
\tag{3.8.12}
$$

The first equation bears a direct relation to the explanation of the following interesting circumstance. When averaging estimates of C_{u} by different authors, it turns out that the resistance coefficient can be considered constant within the wind velocity range from 4 to 15 m/s and equal to $(1.3 \pm 0.3) \times 10^{-3}$. But when grouping preliminary estimates according to stratification conditions and subsequently averaging them it turns out (see Bjutner, 1978) that for neutral stratification, or close to neutral stratification, the resistance coefficient increases from 1.1×10^{-3}, at a wind velocity of 4 m/s, to 1.8×10^{-3} at a wind velocity of 14 m/s. As is well known, the marine surface atmospheric layer is characterized by unstable stratification, which leads to an increase in the resistance coefficient; in this case the weaker the wind, the more does the influence manifest itself. Because of this, neglect of the stratification effect leads to a decrease in the difference between estimates of C_{u} at low and high wind velocities, thereby masking changes in the resistance coefficient.

3.9 Transformation of the thermal regime of the surface atmospheric layer in the presence of wind–wave interaction[1]

The problem in the title of this section is interesting in itself. But there is a further motive for its solution, i.e. the non-trivial issue, which has not been properly elucidated as yet, about the variability of heat exchange between the ocean and the atmosphere during the process of wind wave development.

So, let us consider the problem of thermal regime transformation of the surface atmospheric layer in the presence of wind–wave interaction. As a basis we use the model of wind–wave interaction presented in Section 3.5. We assume at the same time that the air temperature beyond the surface atmospheric layer (T_a) and the sea surface temperature (T_0) remain constant in time, and the eddy heat flux at the upper boundary of the surface atmospheric layer is equal to zero. Then in the absence of radiative sources and sinks of heat and latent heating due to water vapour phase transitions,

---

[1] This section accords with the results of Benilov's research (personal communication).

the equation for the integral, within the surface layer, temperature defect $(T - T_0)$ is written in the form

$$\frac{\mathrm{d}}{\mathrm{d}t} \int_0^h (T - T_a)\, \mathrm{d}z = H_0/\rho c_p, \tag{3.9.1}$$

where, as before, $H_0/\rho c_p$ is the heat flux at the water–air interface normalized to the volume heat capacity of air.

We will assume the temperature difference $(T_0 - T_a)$ to be small, that is, the stratification of the surface atmospheric layer is close to the neutral one. In this case to approximate the vertical temperature profile within the height range from the roughness level z_0 to the upper boundary h of the surface atmospheric layer, one can use the logarithmic law

$$T - T_0 = \delta T_0 + T_* \ln z/z_0, \qquad z \le z \le h, \tag{3.9.2}$$

where δT_0 is the near surface temperature jump, $T_* = -(H_0/\rho c_p)/\kappa u_*$ is the temperature scale.

To avoid unnecessary complications we assume as a first approximation that the vertical distribution of temperature within the height range from the water–air interface $(z = 0)$ to the roughness level $(z = z_0)$ obeys the linear relationship

$$T - T_0 = \frac{z}{z_0}\, \delta T_0, \qquad 0 \le z \le z_0. \tag{3.9.3}$$

Substitution of (3.9.2) and (3.9.3) into (3.9.1) yields

$$\frac{\mathrm{d}}{\mathrm{d}t}\left[\frac{z_0}{2}\, \delta T_0 + T_* h\left(1 - \frac{z_0}{h}\right)\right] = -\frac{H_0}{\rho c_p}. \tag{3.9.4}$$

We take into account the fact that $-H_0/\rho c_p = \kappa u_* T_*$ and that in accordance with (3.9.2) for T_* the equality $T_* = (T_a - T_0 - \delta T_0)/\xi$ is valid, where $\xi = \ln h/z_0 = \kappa u_h/u_*$ (see Section 3.5). After substitution of these expressions into (3.9.4) we have

$$\frac{\mathrm{d}}{\mathrm{d}t}\, h\left[\theta\frac{1 - \mathrm{e}^{-\xi}}{\xi} + \tfrac{1}{2}(1 - \theta)\mathrm{e}^{-\xi}\right] = \kappa u_* \frac{\theta}{\xi}, \tag{3.9.5}$$

where $\theta = 1 + \delta T_0/(T_0 - T_a)$ is a new dependent variable uniquely connected to the non-dimensional (normalized to $(T_0 - T_a)$) surface temperature jump.

We define the non-dimensional thickness D of the surface atmospheric layer, non-dimensional time τ and non-dimensional heat content defect Q by the formulae

$$D = h \bigg/ \left(\beta \frac{\rho_w}{\rho} \frac{u_h^2}{g} \right), \qquad \tau = t \bigg/ \left(\frac{\beta}{\kappa^2} \frac{\rho_w}{\rho} \frac{u_h}{g} \right),$$

$$Q = D \left[\theta \frac{1 - e^{-\xi}}{\xi} + \tfrac{1}{2}(1 - \theta)e^{-\xi} \right],$$

(3.9.6)

where the symbols have the same meaning as in Section 3.5.

According to the last of the displayed formulae,

$$\frac{\Theta}{\xi^2} = \frac{2}{2 - (2 + \xi)e^{-\xi}} \left(\frac{Q}{D} - \tfrac{1}{2}e^{-\xi} \right) \frac{1}{\xi}.$$

(3.9.7)

Therefore incorporating an allowance for (3.9.6) Equation (3.9.5) reduces to the form

$$\frac{dQ}{d\tau} = \frac{2}{2 - (2 + \xi)e^{-\xi}} \left(\frac{Q}{D} - \tfrac{1}{2}e^{-\xi} \right) \frac{1}{\xi}.$$

(3.9.8)

As in Section 3.5, we consider the developed wave stage. At this stage for $\xi \gg 1$ we have

$$\frac{dQ}{d\tau} \approx \frac{Q}{D\xi},$$

(3.9.9)

$$D \approx \frac{1}{3} \left(\frac{c_0}{u_h} \right)^3 \xi,$$

(3.9.10)

$$\frac{c_0}{u_h} \approx \tfrac{4}{3}(1 - \gamma),$$

(3.9.11)

or, after substitution of (3.9.10) and (3.9.11) into (3.9.9),

$$\frac{dQ}{d\tau} \approx \tfrac{81}{64}(1 - \gamma)^{-3} \frac{Q}{\xi^2}.$$

(3.9.12)

We take advantage of the fact that $dQ/d\tau \equiv (dQ/d\xi)(d\xi/d\tau)$, and in accordance with Equation (3.5.34), rewritten in terms of non-dimensional time τ,

$$\frac{d\xi}{d\tau} \approx \tfrac{27}{64}(1 - \gamma)^{-2},$$

(3.9.13)

$$\frac{dQ}{d\xi} \approx \frac{3}{(1 - \gamma)} \frac{Q}{\xi^2}.$$

(3.9.14)

Integrating (3.9.14) and using the initial condition $Q = Q_0$ at $\xi = \xi_0$

we have

$$Q = Q_m \exp\left(-\frac{3\xi^{-1}}{1-\gamma}\right). \tag{3.9.15}$$

From this it follows that at $\xi \Rightarrow \infty$ (or at $\tau \Rightarrow \infty$, see (3.9.13)) the heat content defect in the surface atmospheric layer approaches its limiting value $Q_m = Q_0' \exp(3\xi_0^{-1}/(1-\gamma)$, and the rate of its change and the heat flux at the water–air interface approach zero. The last circumstance indicates that at the end of the wave growth stage the heat exchange between the ocean and the atmosphere almost stops, and the evolution of the thermal regime of the surface atmospheric layer (the decrease in the temperature defect) occurs mainly due to heat redistribution between the surface atmospheric layer and the air entrained from the upper atmospheric layers.

Let us return to Equation (3.9.7) and substitute (3.9.10) and (3.9.15) into it. Then omitting terms with the factor $e^{-\xi}$ we obtain

$$\theta \approx \xi \frac{Q}{D} = \frac{81}{64} \frac{Q_m}{(1-\gamma)^3} \exp\left(-\frac{3\xi^{-1}}{1-\gamma}\right). \tag{3.9.16}$$

We use (3.9.16) to estimate the heat exchange coefficient C_T. First we note that

$$\frac{T_*}{T_0 - T_a} = \frac{\theta}{\xi} \tag{3.9.17}$$

and

$$T_* = \frac{-H/\rho c_p}{\kappa u_*} = C_T \frac{u_h}{\kappa u_*}(T_a - T_0) = -C_T \frac{\xi}{\kappa^2}(T_0 - T_a). \tag{3.9.18}$$

Substituting (3.9.17) and (3.9.18) into (3.9.16) and solving the expression obtained with respect to C_T we find

$$C_T \approx \tfrac{81}{64}(1-\gamma)^{-3} \frac{\kappa^2}{\xi^2} Q_m \exp\left(-\frac{3\xi^{-1}}{1-\gamma}\right),$$

or taking into account the fact that $\xi^2 = \kappa^2 C_u^{-1}$ (see the definition of ξ in Section 3.5),

$$C_T \approx \frac{81}{64} \frac{C_u}{(1-\gamma)^3} Q_m \exp(-3\sqrt{(C_u)}/\kappa(1-\gamma)). \tag{3.9.19}$$

At $\gamma = 0.25$ and observed values of the resistance coefficient $C_u \approx 10^{-3}$ the index of the exponent in (3.9.19) is equal to 0.4. Therefore, presenting the exponent as a series and discarding all the terms of the expansion, with the exception of the first two, we obtain the expression

$$C_T \approx \frac{81}{64} \frac{Q_m}{(1-\gamma)^3} \left(1 - \frac{3\sqrt{C_u}}{\kappa(1-\gamma)}\right) C_u, \qquad (3.9.20)$$

which conforms qualitatively to the expression obtained on the basis of experimental data.

3.10 Methods for estimating surface fluxes of momentum, heat and humidity

At the present time, four methods of estimating momentum, heat and humidity surface fluxes have gained wide acceptance; they are eddy correlation, dissipation, gradient and bulk methods. The first reduces to calculation of the covariances $u'w'$, $w'T'$ and $w'q'$ from measured values of fluctuations of the wind velocity u', w', temperature T' and specific humidity q', and to subsequent averaging of these covariances in time, as we have already discussed in Section 1.4. We now consider the other three methods.

Dissipation method. We start with the derivation of an equation for velocity, temperature and specific humidity dispersions. For this purpose we represent the pressure p and density ρ as the sum of their values complying with the hydrostatic equation and depending only on height, and on departures from them. We ascribe the subscript S for the former, and subscript D for the latter. Then taking into account that $\rho_D \ll \rho_S$ to within the terms of the second order in (ρ_D/ρ_S) we have $\nabla p/\rho \approx \nabla p_S/\rho_S + \nabla p_D/\rho_S - (\rho_D/\rho_S)\nabla p_S/\rho_S$, where ∇ is the gradient operator. For ease of subsequent presentation we turn to tensor designations, setting $x_1 = x$, $x_2 = y$, $x_3 = z$, and assume as usual that all subscripts which are encountered hereinafter take on values 1, 2 and 3, and an subscript which is repeated twice in any term is the equivalent of summation over all values of this subscript. In tensor designations the expression for $\nabla p/\rho$ with an allowance for the hydrostatic equation is $\partial p/\rho\, \partial x_i = -g\delta_{3i}$, and the equation of state $\rho_D/\rho_S \approx -T_D/T_S - 0.61q_D$ (see Section 3.8) transforms into the form $\partial p/\rho\, \partial x_i \approx \partial p_S/\rho_S\, \partial x_i + \partial p_D/\rho_S\, \partial x_i - g(T_D/T_S + 0.61q_D)\delta_{3i}$, where δ_{ij} is the Kronecker symbol equal to zero at $i \neq j$, and to 1 at $i = j$. Substitution of the expression for $\partial p/\rho\, \partial x_i$ into the Navier–Stokes equation, previously rewritten in divergence form after rejection of small (within the surface atmospheric layer) terms describing the Coriolis acceleration and changing ρ_S and T_S with height yields

$$\partial u_i/\partial t + \partial u_i u_k/\partial x_k = -\partial p_D/\rho_0\, \partial x_i + (\beta T_D + 0.61 g q_D)\delta_{3i} + v\, \partial^2 u_i/\partial x_k^2, \quad (3.10.1)$$

where u_i and u_k are instantaneous values of the ith and kth velocity components; v is the coefficient of kinematic viscosity; $\beta = g/T_0$ is the buoyancy parameter; ρ_0 and T_0 are reference values of density and absolute temperature.

We decompose u_i, u_k, p_D, T_D and q_D into mean and fluctuating components. The first will be designated by overbars, the second by primes. Then for $\bar{u}_i$ we come to the Reynolds equation:

$$\partial \bar{u}_i/\partial t + \partial(\bar{u}_i \bar{u}_k + \overline{u_i' u_k'})/\partial x_k = -\partial \bar{p}/\rho_0 \, \partial x_i + (\beta \bar{T} + 0.61 g\bar{q})\delta_{3i} + v \, \partial^2 \bar{u}_i/\partial x_k^2,$$

(3.10.2)

and, for u_i', to the equation

$$\partial u_i'/\partial t + \partial(u_i' \bar{u}_k + u_i u_k' + u_i' u_k' - \overline{u_i' u_k'})/\partial x_k$$

$$= -\partial p'/\rho_0 \, \partial x_i + (\beta T' + 0.61 g q')\delta_{3i} + v \, \partial^2 u_i'/\partial x_k^2, \quad (3.10.3)$$

is obtained by the term-wise subtraction of (3.10.2) from (3.10.3). Hereinafter the subscript D is omitted.

We multiply Equation (3.10.3) by u_j', and the same equation written for u_j' by u_i'; then we sum them, take advantage of the identities

$$(u_j' \, \partial p'/\partial x_i - u_i' \, \partial p'/\partial x_j) = (\partial u_j' p'/\partial x_i + \partial u_i' p'/\partial x_j) - p'(\partial u_j'/\partial x_i + \partial u_i'/\partial x_j),$$

$$(u_j' \, \partial^2 u_i'/\partial x_k^2 + u_i' \, \partial^2 u_j'/\partial x_k^2) = [\partial^2 u_i' u_j'/\partial x_k^2 - 2(\partial u_i'/\partial x_k)(\partial u_j'/\partial x_k)]$$

and the continuity equation (3.10.3) for the mean ($\partial \bar{u}_k/\partial x_k = 0$) and fluctuating ($\partial u_k'/\partial x_k = 0$) motions, and then we average the expressions obtained. As a result we have

$$\frac{\partial}{\partial t}\overline{u_i' u_j'} + \bar{u}_k \frac{\partial}{\partial x_k}\overline{u_i' u_j'} + \left(\overline{u_j' u_k'}\frac{\partial \bar{u}_i}{\partial x_k} + \overline{u_i' u_k'}\frac{\partial \bar{u}_j}{\partial x_k}\right) + \frac{\partial}{\partial x_k}\overline{u_i' u_j' u_k'}$$

$$= -\frac{1}{\rho_0}\left(\frac{\partial}{\partial x_i}\overline{u_j' p'} + \frac{\partial}{\partial x_j}\overline{u_i' p'}\right) + \frac{\overline{p'}}{\rho_0}\left(\frac{\partial u_j'}{\partial x_i} + \frac{\partial u_i'}{\partial x_j}\right) + (\beta \overline{u_i' T'} + 0.61 g\overline{u_i' q'})\delta_{3i}$$

$$+ (\beta \overline{u_j' T'} + 0.61 g\overline{u_j' q'})\delta_{3j} + v\frac{\partial^2}{\partial x_k^2}\overline{u_i' u_j'} - 2v\overline{\frac{\partial u_i'}{\partial x_k}\frac{\partial u_j'}{\partial x_k}}. \quad (3.10.4)$$

Here the first term on the left-hand side describes time change in the covariance $\overline{u_i' u_j'}$, the second term describes its transfer by mean motion, the third term describes the rate of generation due to the interaction of mean and fluctuating motions, the fourth term describes the divergence of the transfer by velocity fluctuations. The first four terms on the right-hand side describe the changes created by the interaction of pressure with the velocity field, and by buoyancy forces. And, finally, the last two terms describe the intensity of molecular diffusion and the rate of dissipation.

From (3.10.4) with $i = j$ the budget equation for the velocity dispersion $\overline{u'^2}$ follows:

$$\frac{\partial}{\partial t}\,\overline{u_i'^2} + \bar{u}_k\frac{\partial}{\partial x_k}\,\overline{u_i'^2} + 2\overline{u_i'u_k'}\frac{\partial \bar{u}_i}{\partial x_k} + \frac{\partial}{\partial x_k}\,\overline{u_k'u_i'^2}$$

$$= -\frac{2}{\rho_0}\frac{\partial}{\partial x_i}\,\overline{u_i'p'} + 2(\beta\overline{u_i'T'} + 0.61g\overline{u_i'q'})\delta_{3i} + \nu\frac{\partial^2\overline{u_i'^2}}{\partial x_k^2} - 2\nu\overline{\left(\frac{\partial u_i'}{\partial x_k}\right)^2} \quad (3.10.5)$$

and from this, after division of all terms by 2, the budget equation for turbulent kinetic energy $b = \overline{u'^2}/2$ follows:

$$\frac{\partial b}{\partial t} + \bar{u}_k\frac{\partial b}{\partial x_k} + \overline{u_i'u_k'}\frac{\partial \bar{u}_i}{\partial x_k} + \frac{\partial}{\partial x_k}\,\overline{u_k'(b + p'/\rho_0)}$$

$$= (\beta\overline{u_i'T'} + 0.61g\overline{u_i'q'})\delta_{3i} + \nu\frac{\partial^2 b}{\partial x_k^2} - \nu\overline{\left(\frac{\partial u_i'}{\partial x_k}\right)^2}, \quad (3.10.6)$$

where the last term on the right-hand side describing the rate of turbulent energy dissipation is usually designated as ε.

The budget equations for temperature $(\overline{T'^2})$ and specific humidity $(\overline{q'^2})$ dispersions are derived similarly. In this case the heat and humidity budget equations are initial equations. These equations, after being transformed to divergence form and rejecting small terms describing the radiative heat influx and water vapour phase conversions, take the form

$$\partial T/\partial t + \partial u_k T/\partial x_k = \chi_T\,\partial^2 T/\partial x_k^2, \quad (3.10.7)$$

$$\partial q/\partial t + \partial u_k q/\partial x_k = \chi_q\,\partial^2 q/\partial x_k^2, \quad (3.10.8)$$

where χ_T and χ_q are coefficients of the heat and water vapour molecular diffusion.

Decomposing u_k, T and q again into mean and fluctuating components, and averaging the equations obtained we find

$$\partial \bar{T}/\partial t + \partial(\bar{u}_k\bar{T} + \overline{u_k'T'})/\partial x_k = \chi_T\,\partial^2\bar{T}/\partial x_k^2; \quad (3.10.9)$$

$$\partial \bar{q}/\partial t + \partial(\bar{u}_k\bar{q} + \overline{u_k'q'})/\partial x_k = \chi_q\,\partial^2\bar{q}/\partial x_k^2. \quad (3.10.10)$$

Next, subtracting (3.10.9) and (3.10.10) from (3.10.7) and (3.10.8) and multiplying the resulting equations,

$$\partial T'/\partial t + \partial(u_k'\bar{T} + \bar{u}_k T' + u_k'T' - \overline{u_k'T'})/\partial x_k = \chi_T\,\partial^2 T'/\partial x_k^2; \quad (3.10.11)$$

$$\partial q'/\partial t + \partial(u_k'\bar{q} + \bar{u}_k q' + u_k'q' - \overline{u_k'q'})/\partial x_k = \chi_q\,\partial^2 q'/\partial x_k^2, \quad (3.10.12)$$

by T' and q', respectively, we have after averaging

$$\frac{\partial}{\partial t}\frac{\overline{T'^2}}{2} + \bar{u}_k \frac{\partial}{\partial x_k}\frac{\overline{T'^2}}{2} + \overline{u_k'T'}\frac{\partial \bar{T}}{\partial x_k} + \frac{1}{2}\frac{\partial}{\partial x_k}\overline{u_k'T'^2} = \chi_T \frac{\partial^2}{\partial x_k^2}\frac{\overline{T'^2}}{2} - \chi_T\overline{\left(\frac{\partial T'}{\partial x_k}\right)^2};$$

(3.10.13)

$$\frac{\partial}{\partial t}\frac{\overline{q'^2}}{2} + \bar{u}_k \frac{\partial}{\partial x_k}\frac{\overline{q'^2}}{2} + \overline{u_k'q'}\frac{\partial \bar{q}}{\partial x_k} + \frac{1}{2}\frac{\partial}{\partial x_k}\overline{u_k'q'^2} = \chi_q \frac{\partial^2}{\partial x_k^2}\frac{\overline{q'^2}}{2} - \chi_q\overline{\left(\frac{\partial q'}{\partial x_k}\right)^2}.$$

(3.10.14)

In Equations (3.10.13) and (3.10.14) the first terms on the left-hand side describe the change in dispersion over time; the second terms describe the ordered transfer; the third terms describe the generation due to interaction of the mean and fluctuating fields; the fourth terms describe the divergence of transfer by velocity fluctuations; the terms on the right-hand side describe, respectively, the molecular diffusion and the rate of equalization (degeneration) of temperature and specific humidity inhomogeneities.

If we adopt conditions of quasi-stationarity and horizontal homogeneity, ignore the divergence of the transfer by fluctuating motion and molecular diffusion in the vertical direction, and direct the x_1 axis along the tangential wind stress, then Equations (3.10.6), (3.10.13) and (3.10.14) are rewritten in the form

$$\left.\begin{aligned} -\overline{u_1'u_3'} &\equiv u_*^2 = [\varepsilon - (\beta\overline{u_3'T'} + 0.61g\overline{u_3'q'})](\partial\bar{u}_1/\partial x_3)^{-1}, \\ \overline{u_3'T'} &\equiv H/\rho c_p = -N_T(\partial\bar{T}/\partial x_3)^{-1}, \\ \overline{u_3'q'} &\equiv E/\rho = -N_q(\partial\bar{q}/\partial x_3)^{-1}. \end{aligned}\right\}$$

(3.10.15)

These relations can be used to estimate the eddy fluxes of momentum u_*^2, heat $H/\rho c_p$ and humidity E/ρ if the vertical gradients of the mean velocity, temperature and specific humidity, and also the rate $\varepsilon = \nu\overline{(\partial u'/\partial x_k)^2}$ of turbulent energy dissipation and the rates $N_T = \chi_T\overline{(\partial T'/\partial x_k)^2}$, $N_q = \chi_q\overline{(\partial q'/\partial x_k)^2}$ of equalization of temperature and specific humidity inhomogeneities are known. To determine the latter, information about fifteen spatial derivations is required, nine of them appearing in ε, and six in N_T and N_q. But if turbulence is considered to be isotropic and Taylor's hypothesis on frozen turbulence is applied, that is, $\partial/\partial x_i = \bar{u}^{-1}\,\partial/\partial t$ is assumed to be valid, then the expressions for ε, N_T, and N_q will take the form

$$\varepsilon = \frac{15\nu}{\bar{u}_i^2}\overline{\left(\frac{\partial u_1'}{\partial t}\right)^2}, \qquad N_T = \frac{3\chi_T}{\bar{u}_1^2}\overline{\left(\frac{\partial T'}{\partial t}\right)^2}, \qquad N_q = \frac{3\chi_q}{\bar{u}_1^2}\overline{\left(\frac{\partial q'}{\partial t}\right)^2}, \qquad (3.10.16)$$

from which it follows that the required estimates of ε, N_T and N_q can be obtained from measurements of the temporal derivatives of wind velocity, temperature and specific humidity fluctuations.

In practice, a frequent assumption is the existence of an inertial interval where the energy supply and dissipation are in fact absent, and there is only the cascade energy transfer from large eddies to smaller eddies. In this interval the spatial velocity spectrum is isotropic, it does not depend on viscosity and is defined only by the energy flux over the spectrum equal to the rate of viscous dissipation. Then, based on dimensional considerations, the spatial spectrum $F_{11}(k)$ of any velocity component has to obey the 'five-thirds power' law $F_{11}(k) = \alpha_1 \varepsilon^{2/3} k^{-5/3}$, where k is the wave number assigned in radians per unit length. Similarly, in the domain of overlapping of inertial intervals for velocity and temperature (humidity) where the viscous dissipation and degeneration of temperature (humidity) inhomogeneities become unimportant, the spatial temperature $F_T(k)$ and humidity $F_q(k)$ spectra are determined by two parameters: the viscous dissipation and the respective rate of equalization of inhomogeneities. In this case from dimensional considerations $F_T(k) = \beta_T N_T \varepsilon^{-1/3} k^{-5/3}$, $F_q(k) = \beta_q N_q \varepsilon^{-1/3} k^{-5/3}$, where $\alpha_1 = 0.55 \pm 0.11$, $\beta_T = 0.80 \pm 0.16$, and $\beta_q = 0.58 \pm 0.20$ are universal constants (Kolmogorov's constants).

So, if the spatial spectra $F_{11}(k)$, $F_T(k)$, $F_q(k)$, or the corresponding frequency spectra $F_{11}(f)$, $F_T(f)$, $F_q(f)$ defined in accordance with Taylor's hypothesis by transition from k to $2\pi f/\bar{u}_1$ (here f is a frequency) are known in the inertial interval then estimation of ε, N_T and N_q reduces to simple calculation by formulae.

Let us return to (3.10.15) and, just as in Section 3.8, introduce non-dimensional vertical gradients of the mean velocity $\Phi_u = (\kappa x_3/u_*) \partial \bar{u}_1/\partial x_3$ temperature $\Phi_T = (x_3/T_*) \partial \bar{T}/\partial x_3$ and specific humidity $\Phi_q = (x_3/q_*) \partial \bar{q}/\partial x_3$, where $T_* = (-H/\rho c_p)/\kappa u_*$ and $q_* = (-E/\rho)/\kappa u_*$ are temperature and specific humidity scales. Then instead of (3.10.15) we have

$$
\left.
\begin{aligned}
u_*^2 &= \left(\frac{\kappa x_3 \varepsilon}{\Phi_u - x_3/L}\right)^{3/2}, \\
H/\rho c_p &= (\kappa x_3 u_* N_T/\Phi_T)^{1/2}, \\
E/\rho &= (\kappa x_3 u_* N_q/\Phi_q)^{1/2},
\end{aligned}
\right\}
\tag{3.10.17}
$$

where L is the modified Monin–Obukhov length scale defined by Equation (3.8.9).

As may be seen, under known values ε, N_T and N_q, there is a single-valued correspondence between eddy fluxes of momentum, heat and humidity, on the one hand, and the universal functions Φ_u, Φ_T and Φ_q, on the other. To specify functions Φ_u, Φ_T and Φ_q, one can apply Equations (3.8.11) with $\Phi_T(0) = \Phi_q(0) = 1$ or (at large negative values of x_3/L) the KEYPS equation providing proper asymptotics in the regime of strongly unstable stratification. In any case, an application of approximating formulae for functions Φ_u, Φ_T and Φ_q allows us dispense with without measurement data on vertical profiles of the mean velocity, temperature and specific humidity, but in return it entails the realization of an iteration procedure the necessity of which is caused by the presence of fluxes sought in the expression for the Monin–Obukhov length scale.

The dissipation method of estimation of momentum, heat and humidity eddy fluxes has certain advantages as compared with the eddy correlation method: it does not need sensors to be fixed in a strictly prescribed direction; it excludes the necessity of measuring vertical velocity fluctuations, and thereby allows the use of moveable platforms. But it should be remembered that assumptions about local isotropy, the existence of an inertial interval, and smallness of the divergence of the vertical transfer by velocity fluctuations are the basis of this method. The last assumption may not be valid at large distances from the ocean surface. According to Large and Pond (1981, 1982) and Smith and Anderson (1984), discrepancies between estimates of fluxes resulting from the eddy correlation and dissipation methods do not usually exceed 40%.

Gradient method. The essence of this method is the determination of eddy fluxes of momentum, heat and humidity from measurements of vertical profiles of the mean wind velocity, temperature and humidity. The latter, in accordance with the Monin–Obukhov similarity theory, are described by Equations (3.8.2) and (3.8.10) involving the parameters u_*, T_*, q_*, z_0, T_0 and q_0 and the universal functions $\psi_u(z/L)$, $\psi_T(z/L)$ and $\psi_q(z/L)$. If these functions are known, they obey say, (3.8.11) or any other expressions with proper asymptotes at small and large positive and negative values z/L, see Section 3.8. Then to compile a closed system of equations providing a single-valued determination of the parameters sought, it is basically enough to take measurements of wind velocity, temperature and humidity at only two heights. The number of measurements required can be reduced by discarding the determination of z_0, T_0 and q_0, that is, by passing from u, T and q at a fixed height z to differences $u(z_2) - u(z_1)$, $T(z_2) - T(z_1)$ and $q(z_2) - q(z_1)$. But since these differences are small, particularly if the heights z_2 and z_1 are close to each other, then even small errors in the measurement of differences can entail marked errors in estimating fluxes. Because of this, measurements at three or more heights are preferred as this guarantees more reliable data on the vertical gradients of the meteorological elements.

It is obvious that in this case we do not eliminate errors caused by the still little known forms of the universal functions $\psi_u(z/L)$, $\psi_T(z/L)$ and $\psi_q(z/L)$, especially for inversion conditions, and by dependence of the Monin–Obukhov length scale on the fluxes sought, converting Equations (3.8.2) and (3.8.10) into a system of non-linear equations. The linear character of the formulae can be preserved if we turn from z/L to another stratification parameter (for example, the gradient Richardson number)

$$Ri = (\beta \, \partial T/\partial z + 0.61g \, \partial q/\partial z)(\partial u/\partial z)^{-2},$$

easily calculated from gradient measurements. It will be recalled that, according to the definition of the modified Monin–Obukhov length scale $z/L = (\kappa^2 z/u_*^2)(\beta T_* + 0.61gq_*)$, and thus $T_* = z\Phi_T^{-1}(z/L) \, \partial T/\partial z$, $q_* = z\Phi_q^{-1}(z/L) \, \partial q/\partial z$, $u_*^2/\kappa^2 = z\Phi_u^{-2}(z/L)(\partial u/\partial z)^2$. Hence, at $\Phi_T(z/L) = \Phi_q(z/L)$ the equality $z/L = \Phi_T^{-1}(z/L)\Phi_u^2(z/L)Ri$ occurs. From here, after expanding $\Phi_T(z/L)$, $\Phi_u(z/L)$ into series in powers of z/L and linearization, we finally have $z/L \approx Ri$.

But still the most serious source of errors in determining fluxes by the gradient method is errors of measurement themselves performed on buoys, ships and other moveable and/or bulky platforms, and accordingly they are subjected to the influence of inevitable distortions of the air flow. According to Blanc (1983) typical errors of such origin amount to from 15 to 35% for the eddy heat flux, from 15 to 105% for the eddy humidity flux, and from 10 to 40% for the momentum flux.

Bulk method. However paradoxical it may seem, the main disadvantage of this method is an excess of available, and often contradictory, possibilities. Indeed, this method is based on the presentation of momentum, heat and humidity fluxes in terms of characteristics included in the complex of standard ship measurements and on the use of the so-called bulk formulae

$$u_*^2 = C_u u^2, \quad -H/\rho c_p = C_T u(T - T_0), \quad -E/\rho = C_q u(q - q_0). \tag{3.10.18}$$

The bulk method would be quite attractive in itself if the coefficients of resistance (C_u), heat exchange (C_T) and evaporation (C_q) appearing in (3.10.18) were considered as numerical constants or, otherwise, if their dependence on the determining parameters were reliably established. We have neither (see Table 3.1), so the accuracy of fluxes obtained by this method leaves much to be desired. It is important to determine to which consequences the choice of one or other exchange coefficients will lead. An answer to this question was given by Blanc (1983). Using one and the same annual series of standard ship measurements on the Ocean Weather Ship C in the North Atlantic he made a comparison of fluxes calculated with the help of ten different schemes of assignment of resistance, heat exchange and evaporation coefficients. It turned out that mean deviations amount to ± 0.1 N/m^2 (or, in relative units, $\pm 45\%$) for the momentum flux; ± 17.5 W/m^2 ($\pm 70\%$) for the sensible heat flux; and ± 18.0 W/m^2 ($\pm 45\%$) for the latent heat flux, while maximum relative deviations reach 80%, 100% and 120% respectively.

Such discrepancies are certainly distressing. But even this is not the main consideration. The worst of this is that we do not know which one of the available schemes of assignment of resistance, heat exchange and evaporation coefficients is valid (there are as many opinions here as authors of schemes). It is clear that while estimating results obtained within the framework of the bulk method, one should be very careful and reasonable, and it will take a considerable amount of effort before these results can be qualified not only as promising but as sufficient.

3.11 Methods for estimating CO_2 flux at the ocean–atmosphere interface

In the state of thermodynamic equilibrium of the ocean and atmosphere the *solubility* (equilibrium concentration) of CO_2 in water c_s^O, as in fact of any other gas, is determined by its equilibrium concentration in air c_s^A, as well as by temperature T^O, and salinity S^O of sea water. In other words, CO_2 follows Henry's law, according to which $c_s^A = c_s^O G(T^O, S^O)$, where $G(T^O, S^O)$ is a solubility coefficient equal to the relation between equilibrium gas concentrations in air and water.

While passing from one equilibrium state to another the change δc^O in the total concentration of inorganic carbon in water, accompanied by a simultaneous change $\delta p_{CO_2}^A$ of partial pressure, is described by the formula $\delta c^O / c^O = B \delta p_{CO_2}^A / p_{CO_2}^A$ where B is the buffer factor or *Revelle factor* describing the so-called *buffer effect* – the dependence of the absorbing capacity of the ocean mainly on the content of the final product of the reaction of CO_2 with water (carbonate ions). Since carbon content in the form of carbonate ions in the surface ocean layer and carbon content in the form of CO_2 in the atmosphere are in ratio to each other as 1 to 10 then an increase in the atmospheric CO_2 concentration, say, by 10%, will result in an increase in the total carbon concentration in the upper ocean layer by only 1%. Available data demonstrate that for present-day conditions the Revelle factor varies within the range 0.090 at temperature 17.7 °C through 0.075 at 3.5 °C.

In actuality the atmosphere and the ocean are not in a state of thermodynamic equilibrium and, hence, at any point on the ocean surface at any moment in time, transfer of CO$_2$ molecules from one environment to the other takes place. Let us consider the CO$_2$ budget at the water–air interface. If the carbon dioxide flux Q_{CO_2} in the air layer closest to the surface is represented as $k_g^A(c_{CO_2}^A - c_s^A)$, and in the water layer closest to the surface as $k_g^O(c_s^O - c_{CO_2}^O)$, where k_g^O and k_g^A are the rates of gas exchange in water and air, and $c_{CO_2}^O$ and $c_{CO_2}^A$ are the volume concentrations of CO$_2$ in these environments, then the budget condition for CO$_2$ at the water–air interface can be written in the form $Q_{CO_2} = k_g^A(c_{CO_2}^A - c_s^A) = k_g^O(c_s^O - c_{CO_2}^O)$, or taking into account the definitions of the equilibrium concentration

$$Q_{CO_2} = K_g^O(c_{CO_2}^A/G - c_{CO_2}^O), \qquad (3.11.1)$$

where $K_g^O = (1/k_g^O + 1/Gk_g^A)^{-1}$ represents the *total resistance* and $1/k_g^O$ and $1/Gk_g^A$ represent *individual resistances* of the water–air interface to the gas exchange between them (Figure 3.4).

It is commonly considered that the gas exchange between two media is performed by molecular diffusion. If one adopts this viewpoint and assumes that the rate of gas transfer is equal to the coefficient of molecular diffusion divided by the thickness of the diffusive sublayer then, for example, for the oxygen whose coefficient of molecular diffusion in air is approximately 10^3 times more than in water, the total resistance will be stipulated only by the resistance in the liquid phase. This conclusion remains valid for carbon dioxide too, although it dissolves much better in water than does oxygen (the oxygen solubility coefficient G is equal to 30, and that of carbon dioxide is equal to 1). This is the result of the diversity of existent forms of CO$_2$ in sea water (see Section 2.7). Thus the intensity of CO$_2$ exchange between water and air is controlled by processes in the diffusive ocean sublayer closest to the surface.

But since we started by discussing the diffusive ocean sublayer it is worth noting that at small wind velocities when the ocean surface is hydrodynamically smooth its thickness δ_D^O is a function of the friction velocity u_{*w} in water and the coefficients of kinematic molecular viscosity v and diffusion χ. Then from dimensionality considerations we have $\delta_D^O = \chi u_{*w}^{-1} f(Sc)$, where $Sc = v/\chi$ is Schmidt's molecular number. We determine the form of the function $f(Sc)$. For this purpose we take advantage of the approximation $f(Sc) = 12.2 Sc^{2/3}$ proposed by Deacon (1977) for Sc ranging from 200 to 5000. Its substitution yields $\delta_D^O = 12.2\chi u_{*w}^{-1} Sc^{2/3}$ or, taking into account the definition of the thickness $\delta_v^O = 5v/u_{*w}$ of the viscous sublayer in water, we obtain $\delta_D^O = 2.4 Sc^{-1/3}\delta_v^O$. As can be seen, at $Sc = 10^3$ the thickness of the diffusion sublayer is a fraction of the thickness of the viscous sublayer. Next, according to the definition, $k_g^O = \chi/\delta_D^O$ so that $k_g^O = 0.082 u_{*w} Sc^{-2/3}$,

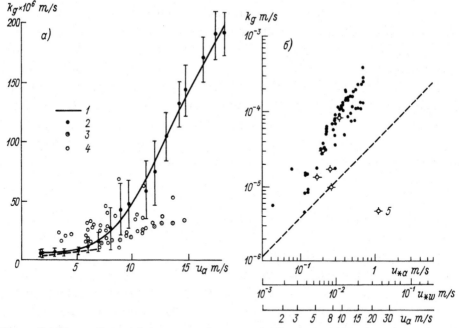

Figure 3.4 Dependence of the gas exchange rate k_g^O on wind velocity at a height of 10 m, according to Bjutner (1986) (*a*) and Coantic (1986) (*b*): (1) the curve drawn over mean weighted values (the latter were obtained by recalculation of laboratory measurements carried out before 1978); (2) mean values complying with data from these laboratory measurements in the range of wind velocity change ± 1 m/s (vertical lines are the root-mean-square deviations); (3) data from laboratory measurements in the presence of a heavy alcohol film on the surface of the water; (4) and (5) data of field measurements by the method of radon deficit; the remaining dots in part (*b*) are data from laboratory measurements carried out in 1978–1985; the dotted line is the result of a calculation by Equation (3.11.2).

whence, on the basis of the continuity condition $u_{*w} = (\rho_a/\rho_w)^{1/2}u_*$ for the momentum flux at the water–air interface, where u_* is the friction wind velocity and ρ_a and ρ_w are air and water densities, it follows that

$$k_g^O = 0.082(\rho_a/\rho_w)^{1/2}Sc^{-2/3}u_*. \qquad (3.11.2)$$

This expression results in understated (compared with experimental data) values for k_g^O at $u_* \geq 0.2$ m/s or at wind velocities over 6–7 m/s. Several explanations for this marked disagreement have been proposed. The most likely one is the rejection of an allowance for the amplification of gas exchange in the process of wind-wave breaking. This can be confirmed by the results of gas exchange measurements in the presence of the surface-active substances (for example, heavy alcohol) causing weakening or even complete cessation of wind-wave breaking and, accordingly, a decrease in k_g^O.

As is well known, wind wave breaking is accompanied by sporadic injections of turbulent energy and by the formation of intermittent turbulent spots in the ocean layer closest to the surface. This in turn, leads to the complete reorganization of its vertical structure. If this feature is taken into account but the ocean surface is still considered to be hydrodynamically smooth, then within the ocean layer closest to the surface one can detect the diffusion sublayer immediately adjoining the water–air interface where the coefficients of the turbulent v_T and diffusion χ_T are much less than v and χ, as well as the outer part of the viscous sublayer with $v_T \ll v$, $\chi_T \gg \chi$, and its underlying turbulent layer with $v_T \gg v$, $\chi_T \gg \chi$. As this takes place, the coefficients of eddy viscosity and diffusion in the viscous sublayer, according to Kitaigorodskii (1984), can justifiably be prescribed as squared, and in the turbulent layer as power (with an index of 4/3) functions of the vertical coordinate calculated from the water–air interface. In this case the thickness δ_D^O of the diffusion sublayer and the rate k_g^O of gas exchange take the form (see Kitaigorodskii, 1984)

$$\left.\begin{aligned}
\delta_D^O &= 2\left(\frac{\alpha}{\alpha_b}\right)^{1/2}\left(\frac{v^3}{\varepsilon(0)}\right)^{1/4} Sc^{-1/2}, \\
k_g^O &= \frac{1}{2}\left(\frac{\alpha}{\alpha_b}\right)^{-1/2}(v\varepsilon(0))^{1/4} Sc^{-1/2},
\end{aligned}\right\} \tag{3.11.3}$$

where $\varepsilon(0)$ is the rate of turbulent energy dissipation in the vicinity of the water–air interface; $\alpha \approx 1$ is the turbulent Prandtl number (the ratio of eddy viscosity and diffusivity); $\alpha_b \approx 0.1$ is the non-dimensional numerical constant in the expression for the eddy viscosity coefficient.

In (3.11.3) the only parameter depending on wind velocity is $\varepsilon(0)$. The simplest way to determine it is to use the condition of local balance between the turbulent energy production $n\varepsilon_0$ due to wind wave breaking and dissipation $\varepsilon(0)d$ in the layer with thickness d (the latter is defined as the thickness of the layer within the limits of which the influence of surface tension becomes apparent), that is, $n\varepsilon_0 = \varepsilon(0)d$, where $\varepsilon_0 = (\rho_a/\rho_w)mu_*^3$; $d = \gamma^{3/5}\varepsilon^{-2/5}(0)$; $\gamma = \sigma/\rho_w$; σ is the surface tension coefficient; $n \approx 1$ is the numerical constant. Substitution of $\varepsilon(0)$ into (3.11.3) gives, finally,

$$\left.\begin{aligned}
\delta_D^O &= 2\left(\frac{\alpha}{\alpha_b}\right)^{1/2}(v^3\gamma)^{1/4} Sc^{-1/2}\left(\frac{\rho_a/\rho_w}{mn}\right)^{5/12} u_a^{-5/4}, \\
k_g^O &= \frac{1}{2}\left(\frac{\alpha}{\alpha_b}\right)^{-1/2}(v/\gamma)^{1/4} Sc^{-1/2}\left(\frac{\rho_w/\rho_a}{mn}\right)^{5/12} u_a^{5/4}.
\end{aligned}\right\} \tag{3.11.4}$$

Comparison of (3.11.4) with (3.11.2) shows that the wind wave breaking provides stronger dependence of the gas exchange rate on wind velocity than is the case when this effect is being ignored.

Further stronger dependence of the gas exchange rate on wind velocity can be obtained by assuming that the ocean surface is hydrodynamically rough. Following Kerman (1984) we define the scale C_* of gas concentration by the equality $Q_{CO_2} = u_{*w} C_* s_B$, where $s_B = \alpha_B (u_*/u_{*C})^3$ is the relative area of breaking waves; u_{*C} is the critical friction wind velocity when the breaking of wind waves begins; $\alpha_B = 0.03$ is a numerical constant. On the other hand, as we know, $Q_{CO_2} = k_g^O \Delta c$, where $\Delta c = (c_{CO_2}^A/G - c_{CO_2}^O)$. Equating both equations we have $k_g^O = u_{*w}(\Delta c/C_*)^{-1} s_B$. We take advantage of the universal dependence for the mass transfer in the vicinity of the immoveable rough surface according to which $\Delta c/C_* = \alpha Re_0^{1/2} Sc^{2/3}$, where $Re_0 = u_{*w} h_0/v$ is the surface Reynolds number; $h_0 = a u_*^2/g$ is the root-mean-square height of breaking waves; $\alpha = 0.55$ and $a = 0.68$ are numerical constants. After substitution of these expressions into k_g^O and application of the continuity condition $u_{*w} = (\rho_a/\rho_w)^{1/2} u_*$ for the momentum flux at the water–air interface we obtain the following relation for the non-dimensional gas exchange rate $k_g^O/(vgu_{*C}^{-1})$:

$$k_g^O/(vgu_{*C}^{-1}) = mSc^{-2/3}(u_*/u_{*C})^{5/2}, \tag{3.11.5}$$

where $m = (\alpha_B/\alpha a^{1/2})(\rho_a/\rho_w)^{1/4}$ is a new numerical constant.

Comparison of (3.11.5) with (3.11.2) and (3.11.4) shows that a change in the hydrodynamic properties of the ocean surface is accompanied by an abrupt enhancement of the gas exchange rate, a conclusion confirmed in general by field and laboratory measurement data. In view of this we note that estimates of k_g^O are usually performed by indirect methods. The most precise of these under field conditions is the so-called radon deficit method, which is based on the assumption that a flux of radioactive gas ^{222}Rn from the ocean into the atmosphere is balanced by an integral (within the limits of the upper ocean layer) difference between influx of radon due to the decay of ^{226}Ra and loss as a result of the natural decay of ^{222}Rn. The duration of the measurements should have an order of a week or more to satisfy the condition of quasi-stationarity (the period of half-life of radon is equal to 3.85 days), and the effects of horizontal diffusion and advection should be assumed to be small. The dependence of k_g^O on the wind velocity obtained in such a way is shown in Figure 3.4. Estimates from laboratory measurements are also presented here. Despite the large spread in data it can be seen that the gas exchange rate increases with the wind velocity, and, in addition, at $u_a \approx 7$ m/s the regime of gas exchange is changed.

From other indirect methods, a method of determination of k_g^O from measurements of the concentration of the radioactive isotope ^{14}C is worthy of note. As is well known, ^{14}C is formed in the atmosphere due to the action of cosmic rays. Because of this, in the stationary state the flux of ^{14}C through the water–air interface should be balanced by the radioactive decay of ^{14}C in the ocean. This last condition allows us to estimate the global averaged value of k_g^O using an undisturbed (that is, not distorted as a result of nuclear weapon tests in the atmosphere) redistribution of ^{14}C between the ocean and the atmosphere. The estimation of k_g^O obtained by this method is about 6×10^{-4} m/s, though its accuracy is not very high.

The first eddy correlation measurements of the carbon dioxide flux in the surface atmospheric layer were carried out in 1977. The estimates of k_g^O based on these, and on direct recordings of the partial pressure of CO_2 in water and air, led to an unexpected result (see Smith and Jones, 1985): it turned out that the mean value of Q_{CO_2} is close to zero, while the partial pressure of CO_2 in water is markedly higher than in air.

We will return later to an explanation of this fact, but now let us turn to one more important mechanism which operates on wind wave breaking and which, perhaps, is responsible for the enhancement of gas exchange at high wind velocities. By this we mean gas transfer by air bubbles. An analysis of bubble motion in the ocean layer closest to the surface demonstrated (see Memery and Merlivat, 1985) that a flux of gas transported by bubbles is not proportional, generally speaking, to the difference in concentration at the surface c_s^O and in the liquid phase c^O. Nevertheless, for so-called trace gases (small atmospheric admixtures to which CO_2 also belongs) a total flux Q_{CO_2} at the water–air interface can be presented in the form $Q_{CO_2} = (k_g^O + k_b^O)(c_s^O - c_{CO_2}^O)$, where $k_b^O = (c_s^O - c_{CO_2}^O)^{-1} \iint w(\partial\psi/\partial z)Q \, dz \, dr$ is the coefficient of gas transport by bubbles; $\psi = \psi(r, z)$ is the distribution function of bubbles over sizes depending on the depth z; r and $w(r)$ are radius and velocity of a vertical displacement of bubbles; $Q = Q(r, z)$ is the change in the gas content of bubbles during their lifetimes. It is also assumed that bubbles at the depth $z > z'$, where $z' = z'(r)$ is a certain fixed depth, are completely degenerate while at $z \leq z'$ their dimensions remain constant.

We note that at $c_s^O = c_{CO_2}^O$ the gas exchange between bubbles and water will not stop because at concentration $c_{CO_2}^O$, which does not depend on z, the internal gas pressure in bubbles exceeds its partial pressure in water due to surface tension. Formally, this case is met by $k_b^O = \infty$, and the equilibrium state, that is, reduction of Q_{CO_2} to zero, takes place at $k_g^O + k_b^O = 0$ or $k_b^O < 0$. Accordingly, the surface layer has to be oversaturated ($c_{CO_2}^O > c_s^O$), and the gas incoming from bubbles into water has to be balanced by its transfer into the atmosphere. The degree of water oversaturation increases with decreasing gas solubility: for CO_2 it can reach 6.4% (see Memery and

Merlivat, 1985). It is the oversaturation effect of the surface ocean layer in the presence of air bubbles formed in the process of wind wave breaking which explains the above-mentioned result of eddy correlation measurements of CO_2 flux in the surface atmospheric layer.

Gas transport by bubbles determines the difference between exchange coefficients for different gases and, hence, points to the fact of the illegitimacy of using radon as a tracer while examining CO_2 exchange between the ocean and atmosphere. Indeed, on the one hand, radon concentration in the atmosphere is close to zero, so that the presence of bubbles does not affect the strength of radon transfer through the water–air interface; on the other hand, CO_2 concentration in the atmosphere and ocean are comparable and, because of this, gas transport by bubbles has to be important at high wind velocities.

Thus, there is no method among those mentioned above which is free from restrictions preventing, to some extent, the reliable determination of the CO_2 flux at the ocean–atmosphere interface. Data of its measurements under natural conditions are quite limited as well. So it is no wonder that in global carbon cycle models the laboratory estimates of the gas exchange rate are frequently used for determination of the CO_2 flux at the ocean surface.

3.12 Features of small-scale ocean–atmosphere interaction under storm conditions

The following lists the specific features of the small-scale interaction of the ocean and atmosphere under storm conditions: firstly, the formation of an intermediate zone near the water–air interface, with a mixture of finite volumes of water and air; secondly, an abrupt intensification of momentum, heat and humidity exchange as a result of macroscopic transfer. The latter is closely connected with sprays in air and bubbles in water, and they, in turn, with wind-wave breaking.

According to Bortkovskii (1983) there are two types of wind-wave breaking: the *plunging* type, and the *sliding* type. In the first case a wave crest passes over a slope and, being overturned, plunges into the water. In the second case, the breaking mass of water gathers a large number of air bubbles while moving along a wave slope and forms 'white horses' – turbulent flows of water and air mixture on the wave slope. The density of the mixture is less than that of water, and this density difference is sustained by air entrainment, due to which white horses do not degenerate immediately but slide, under gravity, along wave slopes, as over an inclined wall.

The main penetration of air into water occurs due to wave breaking but the contribution of entrainment through the free surface of liquid is also important. As for the formation of sprays this occurs in the following sequence. When the top of an emerging bubble is higher than the free surface, the water flows down from it and the bubble envelope in the vicinity of the top becomes thinner. This thinning continues until the moment when a hole appears. As soon as this occurs the imbalance of the surface tension force causes the hole to expand and shifts the torn envelope off centre. As a result, a ring-shaped rise appears at the hole periphery and, consequently, ring waves arise. Their interference in the centre of the hole is accompanied by the formation of a vertical jet from which one or several drops (sprays) are separated and soar upwards.

This explanation, based on data from filming laboratory experiments is very simplified. Under natural conditions the sprays are also formed by the separation of water particles from wave crests and by the interaction of steep gravitational waves with surface drift current. But available empirical information about the vertical flux and concentration of sprays in the surface atmospheric layer is both meagre and inaccurate. Little wonder that our knowledge of the mechanisms of momentum, heat and humidity transfer under storm conditions still does not go beyond the limits of qualitative presentations, which, briefly, reduce to the following.

In storm winds, the sprays rise from the surface of the water into the air, where they are accelerated by the wind and fall back into the water, thereby imparting the momentum they have gained to the upper water layer. Momentum transfer with air bubbles occurs in much the same manner, but perhaps less efficiently. Also, the breaking waves result in intensive mixing of the upper water layer. These processes together lead to a decrease in the vertical velocity gradients in the near-surface layers of water and air and to an increase in the resistance coefficient of the sea surface. Another factor which increases the resistance coefficient is the appearance of short, steep gravitational waves and foam which increase the effective roughness of the sea surface.

Turning to the description of the mechanisms of heat and humidity exchange between the ocean and the atmosphere under storm conditions we note first that the temperature in the lower part of the surface atmospheric layer can be higher or lower than that of the water surface, and the humidity, as a rule, is less than the saturating humidity. Thus, the temperature of sprays which, at the moment of their separation, is equal to the temperature of the sea surface, and the saturating humidity at the surface of the sprays, will differ from the temperature and humidity of the surface atmospheric layer. These differences determine heat and humidity exchange between sprays and air.

The sublayer within the limits of which the immediate influence of sprays manifests itself is limited by their rise in height (15–20 cm), that is, the thickness of this sublayer is much less than the thickness of the marine surface atmospheric layer. On the other hand, the thickness of the water layer saturated with bubbles under storm conditions is much greater than the thickness of the near-surface water layer in which the eddy fluxes remain approximately constant in the vertical direction, so that the eddy fluxes in the near-surface water layer make up only a part of the full range of fluxes determined, among other things, by the bubble transport.

For reasons that are easy to understand, the effect of bubbles on transfer processes in the near-surface water layer is difficult to estimate, and thus one has to limit oneself to the analysis of the effect of sprays on the structure of the surface atmospheric layer. Such an analysis, within the framework of a thermodynamic model of an isolated drop with subsequent integration over a range of drop sizes was carried out by Bortkovskii (1983). Of course, this approach is justified if the volume concentration of drops does not exceed the critical concentration at which the distance between separate drops is comparable with the thickness of the boundary layer forming on them. Otherwise, the laws of resistance, heat and humidity exchange for separate drops become unusable for their set.

It is well-known that the critical value of volume concentration of spherical particles amounts to ~ 0.02, and the humidity content in the lower part of the surface atmospheric layer at wind velocity 20–25 m/s amounts to 10^{-3}– 10^{-4} g/cm^3. Accordingly, the mean volume concentration of sprays (ratio of humidity content to water density) is one or two orders less than critical. The assumption as to the spherical nature of drops is also justified over the whole range of spray sizes that is typical of the surface atmospheric layer. Hence, the initial prerequisites of the model mentioned above are realistic. This sustains the hope that, after turning to a set of drops characterized by their distribution over size and space, the final aim – the estimation of the integral transfer of momentum, heat and humidity by bubbles – will be attained. We now give the results of model calculations of most interest to us.

According to Bortkovskii (1983) at a drop radius (from 0.003 to 0.005 cm) typical of the surface atmospheric layer the vertical heat and humidity transfer by sprays turns out to be comparable with the eddy transfer. Along with this, a decrease in the radius down to 0.0015 cm results in the fact that humidity transfer by sprays becomes several times larger than the eddy transfer, and the heat transfer becomes negative (air is cooled down by sprays). This is easily explained: small drops quickly reach equilibrium temperature for which the diffusive heat exchange, losses of heat due to

Table 3.2. *Dependence of* C_q^*/C_q *and* C_θ^*/C_θ *on water–air temperature difference and drop radius (according to Bortkovskii, 1983)*

Wind velocity (m/s)	Water–air temperature difference (°C)	Drop radius (cm)			
		0.003		0.005	
		C_q^*/C_q	C_θ^*/C_θ	C_q^*/C_q	C_θ^*/C_θ
20	5	0.97	1.28	0.55	1.29
	1	1.32	0.43	0.76	1.55
	−2	2.01	1.57	0.94	0.40
30	5	1.18	1.08	0.71	1.12
	1	1.85	0.25	0.87	2.23
	−2	2.53	1.80	1.20	0.49

Note: coefficients of heat exchange and evaporation for the environment are assumed to be the same and equal to 1.43×10^{-3} at wind velocity 20 m/s, and 1.58×10^{-3} at wind velocity 30 m/s; the air temperature is referred to the height 10 m.

evaporation and radiative sources and sinks of heat counterbalance each other, and further evaporation of drops occurs due to the influx of heat from the surrounding air. On the other hand, when the radius increases, those drops with an initial temperature higher than the temperature of the environment give up their heat by diffusive heat exchange.

As it turns out, the contribution of sprays to the total momentum flux is relatively small: at a drop radius of 0.003–0.005 cm it amounts to approximately 10% of the total flux. With decreasing drop radius the ratio of momentum transport by sprays to the total transport diminishes; with increasing drop radius it approaches a finite limit.

In order to illustrate the results we compare, following Bortkovskii (1983), the coefficients of heat exchange (C_θ^*) and evaporation (C_q^*) for drops with the usual coefficients of heat exchange(C_θ) and evaporation (C_q). In doing so, we determine coefficients C_θ^*, C_q^* from Equations (3.8.1) but replace the fluxes appearing in (3.8.1) by heat and humidity exchanges due to sprays. The results of the comparison are given in Table 3.2.

As can be seen, an increase in drop radius from 0.003 to 0.005 cm leads to weakening of evaporation that is connected with a decrease in the overall area of the drops' surface, and with intensification of sensible heat exchange representing the difference between the rate of change in drop heat content and losses of heat due to evaporation. The ratio between the coefficients of heat exchange and evaporation changes similarly. Let us pay attention to the sensitivity of C_q^*/C_q and C_θ^*/C_θ to variations in water–air temperature difference and in the humidity difference it controls. Under conditions of thermal inversion when the water–air temperature difference becomes negative, an increase in C_q^*/C_q is explained by a reduction in drop cooling, and by the retention of high values of humidity difference at the drop–environment interface. The opposite situation occurs at unstable stratification: its influence contributes to acceleration of the cooling of drops, to a decrease in humidity difference and, as a result, to a decrease in C_q^*/C_q. But the values of C_q^*/C_q and C_θ^*/C_θ are the most impressive in themselves. They leave no doubt as to the abrupt intensification of heat and moisture exchange under storm conditions.

The contribution of storms to the formation of mean (over a long time period) values of heat exchange and evaporation is determined not only by the intensity of these processes but also by the frequency of storms, by their duration and by temperature and humidity differences at the water–air interface under storm and background (non-storm) conditions. According to data from Ocean Weather Ships (OWS) an increase in heat transfer into the atmosphere manifests itself in autumn, winter and partly in spring. The effect of storms from May to October is not felt in practice. The ratio between total heat fluxes obtained with and without an allowance for storms (this ratio serves as a quantitative measure of storm effects) varies over time and space from 1.38–1.41 in the winter months (OWS A, C, D) to 1.00 in the summer months (OWS D). On average for the year this ratio amounts to 1.32 for OWS A, 1.31 for OWS C, 1.28 for OWS D and 1.15 for OWS E. In other words, in high latitudes (OWS A), and also in the Gulf Stream region (OWS D) and in the North Atlantic Current (OWS C), storm activity determines approximately one-third of the resulting heat and humidity transfer into the atmosphere.

4

Mesoscale ocean–atmosphere interaction

4.1 The planetary boundary layer

The planetary boundary layer (PBL) is the name given to the domain of flow in gaseous and liquid shells of the rotating planet, formed as a result of the simultaneous action of the forces of pressure gradient and turbulent friction and the Coriolis force. We introduce the components of mean velocity u, v and geostrophic wind velocity

$$U_g = -(f\rho)^{-1}\, \partial p/\partial y = G \cos \alpha, \qquad V_g = (f\rho)^{-1}\, \partial p/\partial x = G \sin \alpha,$$

where G is the modulus of the geostrophic wind velocity; α is the angle between an isobar and the axis x; p and ρ are the mean pressure and density; x and y are axes in the Cartesian coordinate system; and we define the thickness h_E of the planetary boundary layer so that the relative departure of mean velocity from geostrophic velocity at level $z = h_E$ does not exceed some preassigned value δ, that is, $G^{-1}[(U_g - u)^2 + (V_g - v)^2]^{1/2}_{z=h_E} \leq \delta$. Then taking into account that the vertical scale of PBL has to depend on the friction velocity u_* and Coriolis parameter f, from dimensional considerations we obtain the Rossby–Montgomery formula

$$h_E = \gamma u_*/|f|, \tag{4.1.1}$$

where γ is a non-dimensional factor, varying according to different authors, in the range 0.1–0.4. At $\gamma = 0.4$ (for the atmosphere this value of γ corresponds to $\delta \approx 20\%$) the PBL thickness amounts to about 1000 m in mid-latitudes, which is an order of magnitude less than the thickness of the troposphere.

Another way of defining h_E was proposed by Charney (1969). He postulated that the PBL remains hydrodynamically stable as long as its effective Reynolds number $Re_E = Gh_E/k_E$, where k_E is a certain effective value of the eddy viscosity coefficient, does not exceed the critical Reynolds number Re_{cr}

for the laminar boundary layer. If by analogy with the laminar boundary layer we assume that $h_E = (2k_E/|f|)^{1/2}$ and solve this equation for k_E, then we will obtain the estimate

$$h_E \geq 2G/f Re_{cr} \qquad (4.1.2)$$

for the PBL thickness.

It is clear that estimates (4.1.1) and (4.1.2) are very rough since they do not take into account the peculiarities of the PBL structure conditioned by the influence of stratification. It is appropriate to note in this connection that there are two types of turbulence: dynamic and convective, generated by Reynolds stresses and buoyancy forces respectively. In the surface atmospheric layer Reynolds stresses dominate at high wind velocities, and buoyancy forces dominate for strong heating of the underlying surface. Dynamic turbulence decreases more rapidly when moving away from the underlying surface than does convective turbulence. As a result, it is convective turbulence which determines the PBL thickness over land. In this case, the upper boundary of the PBL is identified with the lower boundary of high inversion to which the influence of surface heating extends.

At unstable, as well as at stable, stratification, the PBL structure should be determined by the same parameters u_*, $H/\rho c_p$, and β as the structure of the surface atmospheric layer, with only one exception that the Coriolis parameter f is now added. From this parameter set along with the Monin–Obukhov length scale L, one more length scale $\lambda = \kappa u_*/f$ (here the von Karman constant κ is introduced for convenience) and the non-dimensional combination $\mu_0 = \lambda/L = \kappa^2 \beta(-H/\rho c_p)/|f|u_*^2$ (i.e. the non-dimensional stratification parameter) may be compiled. Hence,

$$h_E = \gamma u_* |f|^{-1} \psi(\mu_0), \qquad (4.1.3)$$

where $\psi(\mu_0)$ is a universal function equal to 1 for neutral stratification ($\mu_0 = 0$).

Let us define the function $\psi(\mu_0)$ and the expression for h_E in the regime of strongly unstable stratification. Within the PBL in this case one can distinguish a surface layer, a free convection layer, a mixed layer and an entrainment layer. In the surface layer where dynamic turbulence dominates, the parameters u_*, $H/\rho c_p$, z and β are the determining parameters, and u_* and T_* are characteristic scales of velocity and temperature (see Section 3.8). Accordingly, the non-dimensional vertical gradients of the mean velocity and temperature obtained by normalizing their dimensional expressions to u_* and T_* represent universal functions of the non-dimensional height z/L. From the results of eddy correlation and gradient measurements (see Zilitinkevich,

1970) the surface layer for strongly unstable stratification is limited by height $z \le 0.07|L|$.

The height $z = 0.07|L|$ serves as the lower boundary of the *free convection layer* where dynamic turbulence becomes negligible as compared with convective turbulence. The structure of this layer is determined by three parameters $H/\rho c_p$, z and β, so that the scales of velocity w_* and temperature T_* composed of these parameters take the form $w_* = (z\beta H/\rho c_p)^{1/3}$, $T_* = (H/\rho c_p)/w_*$. According to the Monin–Obukhov similarity theory the non-dimensional profiles of mean velocity and temperature formed by normalizing to w_* and T_* remain constant with height, a condition which is valid up to $z \approx |L|$. The *mixed layer* is situated above this height. Its structure depends neither on u_* nor on z, being determined only by $H/\rho c_p$, β and f. In this case h_E is the characteristic length scale, and $w_* = (h_E\beta H/\rho c_p)^{1/3}$, and $T_* = (H/\rho c_p)/w_*$ are velocity and temperature scales. Accordingly, non-dimensional vertical gradients of mean velocity and temperature represent functions of z/h_E only. The mixed layer extends up to the height $z \approx 0.8h_E$.

In the range of heights from $0.8h_E$ to $1.2h_E$ the *entrainment layer* (an intermediate layer between the PBL and the free atmosphere) is arranged. The structure of this layer is determined by the entrainment rate characterizing the intensity of penetration of turbulence from the free convective layer into the non-turbulent inversion layer, by hydrostatic stability of the latter and by conditions in the free atmosphere.

A rough estimate of the PBL thickness for a strong hydrostatic instability (large negative values of μ_0) can be obtained from dimensional considerations rejecting the entrainment effects and, thereby, assuming that h_E is determined by only three parameters $H/\rho c_p$, β and f. In this case

$$h_E \sim (\beta H/\rho c_p)^{1/2}|f|^{-3/2} \sim \lambda(-\mu_0)^{1/2}. \qquad (4.1.4)$$

It should be emphasized that the multilayer structure is inherent mainly in the convective PBL over land with a typical thickness exceeding the Monin–Obukhov length scale by 50–100 times. The ratio h_E/L is much lower over the ocean (excluding, perhaps, areas of cold deep water formation in the Norwegian and Greenland Seas, and in the Weddell Sea). In such situations, the possibility of ignoring the dynamic turbulence and, therefore, excluding the friction velocity u_* from a number of parameters determining the structure of the free convection layer and mixing layer raises some doubts. These doubts can only be eliminated by turning to a description of the PBL over the ocean within the framework of a more general approach, which includes u_* and h_E in the set of determining parameters.

Under conditions of strongly stable stratification the thickness of the layer containing by turbulence of dynamic origin amounts to about 100 m. An important role in the formation of the structure of the above-mentioned layer is played by the radiating heat influx created by the long-wave emission of the underlying surface and the lowest atmospheric layer. Therefore, when identifying the PBL thickness with the thickness of the surface inversion layer it may be found to be much thicker than the vertical extent of the layer in which the dynamic turbulence predominates. This circumstance should be borne in mind when estimating the expression,

$$h_{\mathrm{E}} \sim u_*^2(-\beta H/\rho c_{\mathrm{p}})^{-1/2}|f|^{-1/2} \sim \lambda\mu_0^{-1/2}, \tag{4.1.5}$$

obtained by Zilitinkevich (1972) from dimensional considerations on the assumption that h_{E} is determined only by $H/\rho c_{\mathrm{p}}$, β and f. It is also necessary to remember that, due to the suppression of turbulence and weakening of mixing, the response of the stable stratified PBL to a change in external parameters becomes slower. As a result, the condition of quasi-stationarity of the PBL implicitly used in the derivation of (4.1.5) may not be fulfilled. There is only one way to eliminate this obstacle – to replace the diagnostic relationship (4.1.5) by the evolution equation for the PBL thickness.

4.2 Problem of closure

We consider the stationary, horizontally homogeneous PBL for parallel, equidistant isobars. We also assume that the horizontal gradients of pressure and density, excluding those cases where the density is involved in a combination with gravity, are equal to their mean values within the PBL. Taking these assumptions into account and neglecting small terms describing the effect of molecular viscosity, the equations for the mean velocity components u, v are written in the form

$$-\frac{1}{\rho}\frac{\partial p}{\partial x} + fv - \frac{\partial}{\partial z}\overline{u'w'} = 0, \tag{4.2.1}$$

$$-\frac{1}{\rho}\frac{\partial p}{\partial y} - fu - \frac{\partial}{\partial z}\overline{v'w'} = 0, \tag{4.2.2}$$

where $-\overline{u'w'}$, $-\overline{v'w'}$ are the components of Reynolds stresses normalized to mean density; other symbols are the same.

We complement (4.2.1) and (4.2.2) with the evolution equation for the mean temperature T

$$\frac{\partial T}{\partial t} = -\frac{\partial}{\partial z}\overline{w'T'}, \tag{4.2.3}$$

and for the mean specific humidity q

$$\frac{\partial q}{\partial t} = -\frac{\partial}{\partial z}\overline{w'q'}, \tag{4.2.4}$$

and also by the continuity equation

$$\frac{\partial u}{\partial x} + \frac{\partial v}{\partial y} + \frac{\partial w}{\partial z} = 0, \tag{4.2.5}$$

by the condition of hydrostatic equilibrium

$$\frac{\partial p}{\partial z} = -g\rho, \tag{4.2.6}$$

and by the equation of state of moist air

$$p = R\rho T(1 + 0.61q). \tag{4.2.7}$$

It should be noted that in (4.2.3) and (4.2.4) we have neglected the sources and sinks of heat and moisture along with molecular diffusion, but have retained, in contrast with (4.2.1) and (4.2.2), terms describing non-stationary effects because otherwise it is impossible to satisfy the condition of vanishing eddy fluxes of heat $\overline{w'T'}$ and moisture $\overline{w'q'}$ vanish at the upper boundary of the PBL.

Let us make one more preliminary remark with regard to the type of averaging used. The most convenient method is averaging over an ensemble (an infinite set of independent realizations) having the following properties: $\overline{\bar{u}_i\bar{u}_j} = \bar{u}_i\bar{u}_j$, $\overline{\bar{u}_iu'_j} = \overline{u'_i\bar{u}_j} = 0$. To realize such averaging in practice is difficult, if at all possible. And one thus has to apply the so-called *ergodic hypothesis*, that is, to assume a flow to be statistically stationary, and the ensemble averaging to be equivalent to averaging over time, because in this case the ensemble-average values do not change with time. Usually, the largest time interval in which the time average approaches its stable stationary value with some tolerated accuracy is chosen as an averaging period. To meet this condition the spectrum of the process to be examined must have a deep minimum separating high-frequency eddy fluctuations from low-frequency oscillations of synoptic origin. Then a selection of any averaging period from the band of the spectral minimum will provide filtration of high-frequency fluctuations and at the same time will not introduce large distortions due to low-frequency oscillations. Such a minimum presents in the spectra of the wind velocity and temperature in the surface atmospheric layers over the horizontally homogeneous underlying surface within the range of periods of the order of an hour; it is absent in the spectra of meteorological elements in the upper part of the PBL and in the surface layer over the horizontally non-homogeneous underlying surface.

Let us turn to the system (4.2.1)–(4.2.7) and note that it contains seven equations and 11 unknowns: u, v, w, p, ρ, T, q, $\overline{u'w'}$, $\overline{v'w'}$, $\overline{w'T'}$ and $\overline{w'q'}$. Thus, the system (4.2.1)–(4.2.7) describing the mean (in the Reynolds' sense) motion in the PBL turns out not to be closed because of the appearance of covariances $\overline{u'w'}$, $\overline{v'w'}$, $\overline{w'T'}$ and $\overline{w'q'}$. When compiling additional equations for these, as was done, for example, in Section 3.10, new unknowns (third moments) appear. The procedure may be continued and at any time new unknowns (the moments of higher orders) will appear. In short, the strict (in the mathematical sense) closure of the equation system for the eddy motion, and, particularly, the motion in the PBL is impossible. This is termed the *problem of closure*. Its solution reduces to the use of some *a priori* assumptions which allow us to express moments of higher order in terms of moments of lower order or by characteristics of the mean flow. Such closures are called closures of the first and second orders respectively.

4.2.1 First-order closure

First-order closure is based on the analogy between the eddy and laminar motions, from which the components $-\overline{u'w'}$ and $-\overline{v'w'}$ of the vertical eddy flux of momentum can be presented as the product of the eddy viscosity coefficient k_M by the vertical gradient of the corresponding component of mean velocity $\partial u/\partial z$, $\partial v/\partial z$, that is, $-\overline{u'w'}_M = k_M \, \partial u/\partial z$, $-\overline{v'w'} = k_M \, \partial v/\partial z$. In exactly the same manner the vertical eddy flux $\overline{w'c'}$ of any other substance with mean concentration c is defined as $-\overline{w'c'} = k_c \, \partial c/\partial z$, where k_c is the eddy diffusion coefficient. Unlike the coefficients of kinematic molecular viscosity ν and diffusion χ, the coefficients k_M and k_c are not proper parameters of fluid and depend directly on the peculiarities of the flow structure. For example, for flows in an open channel the eddy viscosity coefficient changes with depth by the parabolic law, and for a plane jet it changes in proportion to the square root of the distance from the source. Therefore, the above-mentioned expressions for eddy fluxes do not in themselves reflect the peculiarities of the eddy transfer structure and serve only as the embodiment of an idea of the proportionality between the substance transfer in a certain direction and the gradient of the substance transported in the same direction. Thus, the problem should consist of setting a connection between the eddy viscosity and diffusion coefficients and the local characteristics of the mean current.

The first step in this direction was undertaken by Prandtl in 1925. As is well known, the coefficient of the molecular viscosity is proportional to the product of the mean velocity of motion of molecules and the mean length between two subsequent collisions. By analogy with the process of molecular transfer Prandtl postulated that vertical transfer in eddy flow is performed by eddies drifting with velocity w' at distance l. The latter is identified with the mixing length, or, as it is often called, the *turbulence scale* (the mean spatial scale of energy-containing eddies), where the eddy is completely adapted to the conditions of the environment (is mixed with it). Further, considering that the characteristic scale of velocity fluctuations u' and w' are equal to $l\,\partial u/\partial z$, Prandtl obtained $-\overline{u'w'} = l(\partial u^2/\partial z)^2$. From here and from the expression $-\overline{u'w'} = k_M\,\partial u/\partial z$, $k_M = l^2|\partial u/\partial z|$ follows.

To find the new unknown l additional considerations need to be involved. The simplest of these is the proportionality of the turbulence scale in the near-wall region of the flow to the distance z from the wall, that is, $l = \kappa z$, where κ is the non-dimensional constant (the von Karman constant). Such an assumption is not the only possible one. For example, in the region away from a wall where the proportionality between l and z is not valid the von Karman assumption seems to be more justified. In accordance with this the turbulence scale is considered to be a function of the local characteristics of the mean flow, particularly of the first and second derivatives of velocity in the vicinity of the point examined, but not of the velocity itself: the turbulence scale cannot depend on the value of the mean velocity due to the principle of invariance. Then from dimensionality considerations $l = -\kappa \times (\partial u/\partial z)/(\partial^2 u/\partial z^2)$. Generalization of this expression in the case of variable (along height) direction of the mean velocity vector $\mathbf{u}$ was proposed by Rossby (1932). It has the form

$$l = -\kappa \left|\frac{\partial \mathbf{u}}{\partial z}\right| \Big/ \frac{\partial}{\partial z}\left|\frac{\partial \mathbf{u}}{\partial z}\right|.$$

The above-mentioned expressions for the turbulence scale do not take into account the effect of stratification. This restriction can be overcome by including a factor depending on stratification in the right-hand side of the expression. Kazanskii and Monin (1961) used this method having rewritten the Prandtl formula in the form $l = \kappa z(1 - \sigma Rf)^n$, where $Rf = -(\beta H/\rho c_p)/u_*(\partial u/\partial z)$ is the flux Richardson number; $\sigma \approx 12$ and $n \approx 0.35$ are empirical constants. On the other hand, Zilitinkevich and Laikhtman (1965) used the von Karman formula as an initial assumption. They reasoned as follows (see

Zilitinkevich and Laikhtman, 1965, and Zilitinkevich *et al.*, 1967). If the von Karman formula or any equivalent formula $l = -2\kappa(\partial u/\partial z)^2/(\partial/\partial z)(\partial u/\partial z)^2$ is fulfilled for neutral stratification, then for free convection when all the characteristics of turbulence, including the turbulence scale, depend only on the vertical structure of the field of potential temperature θ or buoyancy $\beta\theta$, the expression $l = -2\kappa(\partial/\partial z)(\beta\theta)/(\partial^2/\partial z^2)(\beta\theta)$ pertains. The latter, as with the von Karman formula, is obtained from dimensional considerations assuming that the only determining parameters are the first and second derivatives of the function in question. From $(\partial u/\partial z)^2$ and $\beta\theta$ we compile the combination $\psi^2 = [(\partial u/\partial z)^2 - \alpha_\theta^0\beta\,\partial\theta/\partial z]$ having the necessary asymptotic properties, that is, reducing to $(\partial u/\partial z)^2$ for neutral stratification and to $-\alpha_\theta^0\beta\,\partial\theta/\partial z$ for strongly unstable stratification where α_θ^0 is the reverse of the turbulent Prandtl number (see Section 3.7). Then the expression for the turbulent scale can be presented in the form $l = -2\kappa\psi^2/(\partial\psi^2/\partial z) = -\kappa\psi/(\partial\psi/\partial z)$. Further, when neglecting the effects of non-stationarity, horizontal inhomogeneity and vertical diffusion, then from the budget equation for the kinetic energy of turbulence (see the first relationship in (3.10.15)) it follows that $-\overline{u'w'}(\partial u/\partial z) + \beta\overline{w'\theta'} = \varepsilon$. From here and from definitions $-\overline{u'w'} = k_M\,\partial u/\partial z$, $\overline{w'\theta'} = \alpha_\theta k_M\,\partial\theta/\partial z$, and also from Kolmogorov's relations of approximate similarity $\varepsilon = c_\varepsilon b^{3/2}/l$, $k_M = c_0 b^{1/2}l$, where b is the average kinetic energy of turbulence referred to density, c_ε and c_0 are non-dimensional constants, we obtain $\psi = (c_\varepsilon/c_0)^{1/2}b^{1/2}/l$. Thus, the generalized von Karman formula for the turbulence scale proposed by Zilitinkevich and Laikhtman is finally written in the form

$$l = -\kappa\left(\frac{b^{1/2}}{l}\right)\bigg/\frac{\partial}{\partial z}\left(\frac{b^{1/2}}{l}\right).$$

The concept of mixing length assumes that the eddy viscosity coefficient is equal to zero whenever the gradient of the mean velocity vanishes. In other words, it is suggested that the turbulence is in a state of local equilibrium: generation and dissipation of eddy energy at any point of the flow counterbalance each other, and eddy transfer from neighbouring points and its change over time are absent, or do not play any important role. The fact that this condition is not always valid is obvious. It is enough to recall that in wind tunnels where the turbulence is generated immediately behind a grid and then transported downward along the flow by the mean current according to the concept of mixing length not taking such transport into account, the eddy viscosity coefficient has to take on zero values. As another example: in channels where turbulence is generated mainly in the vicinity of the walls and transported by eddy diffusion to the channel axis, the concept of mixing length neglecting this transport should lead to zero values of the eddy viscosity coefficient at the channel axis. Finally, in the oscillating boundary layer where the intensification of turbulence at the stage of attenuation and the degeneration of turbulence at the stage of amplification of the mean velocity occur, their reproduction within the framework of the concept of mixing length ignoring the effect of non-stationarity on the eddy energy budget is excluded. Thus, the concept of mixing length becomes invalid every time a marked contribution to the eddy energy budget is made by the advective and diffusive transport of eddy energy, and by the pre-history of the process.

The above-mentioned restrictions are a direct consequence of the general hypothesis forming the basis of the *concept of mixing length*, the hypothesis that turbulence is described by the unique length scale and the unique velocity scale. But reality offers us richer possibilities, an example of which is the convective boundary atmospheric layer at the upper boundary of which a negative (opposite to the direction in the surface layer) heat flux is formed as the result of entrainment of warmer air from the inversion layer located above. An identical situation takes place on convective mixing near the base of the upper mixed layer of the ocean where entrainment of colder water from the lower seasonal thermocline is responsible for the appearance of the positive heat flux (that is, again opposite in direction to the surface flux). Such situations are difficult to describe within the framework of the concept of mixing length.

In this respect the integral models or the models of the mixed layer, as they are often called, based on the assumption that all variables within the PBL depth are constant, can be useful. In models of this type the eddy viscosity coefficient is not an unknown quantity to be determined and the closure problem reduces to the parametrization of eddy fluxes at the external boundary of PBL and of its depth.

4.2.2 Second-order closure

The essence of this is that components of Reynolds stresses and eddy fluxes of scalar substances are not approximated by formulae of gradient type which are the basis of the first-order closure (or K-theory as it is also called, in accordance with common designation of eddy exchange coefficients), but they are considered as new unknown quantities derived from equations for the covariations. In Section 3.10 we discussed the method of derivation of equations for the Reynolds stresses. So to avoid repetitions we write Equation (3.10.4), complementing it with terms describing the effect of the Earth's rotation.

For dry air this equation takes the form

$$\frac{\partial}{\partial t}\overline{u_i' u_j'} + \bar{u}_k \frac{\partial}{\partial x_k}\overline{u_i' u_j'} + \left(\overline{u_j' u_k'}\frac{\partial \bar{u}_i}{\partial x_k} + \overline{u_i' u_k'}\frac{\partial \bar{u}_j}{\partial x_k}\right) + \frac{\partial}{\partial x_k}\overline{u_i' u_j' u_k'}$$

$$= -\frac{1}{\rho_0}\left(\frac{\partial}{\partial x_i}\overline{u_j' p'} + \frac{\partial}{\partial x_j}\overline{u_i' p'}\right) + \frac{\overline{p'}}{\rho_0}\left(\frac{\partial u_j'}{\partial x_i} + \frac{\partial u_i'}{\partial x_j}\right) + \beta(\overline{u_i' T'}\delta_{3_i} + \overline{u_j' T'}\delta_{3_j})$$

$$+ v\frac{\partial^2}{\partial x_k^2}\overline{u_i' u_j'} - 2v\overline{\frac{\partial u_i'}{\partial x_k}\frac{\partial u_j'}{\partial x_k}} - f_k(\varepsilon_{jkl}\overline{u_l' u_i'} + \varepsilon_{ikl}\overline{u_l' u_j'}), \quad (4.2.8)$$

where ε_{jkl} is the Levi–Civita symbol defined as $\varepsilon_{123} = \varepsilon_{231} = \varepsilon_{312} = 1$, $\varepsilon_{132} = \varepsilon_{213} = \varepsilon_{321} = -1$ and $\varepsilon_{jkl} = 0$ in all other cases.

Equations for the fluxes of scalar characteristics can be obtained in the same manner as (4.2.8). Namely, we multiply Equation (3.10.3), supplemented with a term describing the effect of the Earth's rotation, by T', and Equation (3.10.11) for dry air, by u_i', then sum and average them. As a result, we obtain the following budget equation for the eddy heat flux:

$$\frac{\partial}{\partial t}\overline{u_i' T'} + \bar{u}_k \frac{\partial}{\partial x_k}\overline{u_i' T'} + \left(\overline{u_i' u_k'}\frac{\partial \bar{T}}{\partial x_k} + \overline{u_k' T'}\frac{\partial \bar{u}_i}{\partial x_k}\right) + \frac{\partial}{\partial x_k}\overline{u_i' u_k' T'}$$

$$= -\frac{1}{\rho_0}\overline{T'\frac{\partial p'}{\partial \bar{x}_i}} + \beta\overline{T'^2}\delta_{3i} + \left(\overline{v T'\frac{\partial^2 u_i'}{\partial x_k^2}} + \chi_T \overline{u_i'\frac{\partial^2 T'}{\partial x_k^2}}\right) - 2\varepsilon_{ijk}\Omega_j\overline{u_k' T'}, \quad (4.2.9)$$

where the first term on the left-hand side describes the change in time, the second term describes the advective transport, the third term describes the interaction of mean and fluctuating motions, the fourth term describes the transfer by velocity fluctuations (the eddy diffusion), and terms on the right-hand side describe, respectively, changes in the eddy heat flux due to correlations of pressure and temperature fluctuations, buoyancy forces, molecular viscosity and diffusion and the Coriolis force.

Equations (4.2.8) and (4.2.9), together with Equations (3.10.2) and (3.10.9) for mean values of the velocity $\bar{u}_i$ and temperature $\bar{T}$ as well as Equation (3.10.3) for temperature dispersion $\overline{T'^2}$ could form a closed system if it were not for terms describing the influence of dissipating processes, correlations of pressure fluctuations, and the eddy transfer. These terms are not directly connected with the variables sought, and should therefore be considered as additional unknowns which have to be presented in terms of the functions sought, or disregarded wherever possible. This, in fact, is the essence of the problem of second-order closure. We consider newly appearing unknown terms in the order in which they are mentioned.

Dissipation terms. These are $v\,\partial^2\overline{u_i'u_j'}/\partial x_k^2$, $2v\overline{(\partial u_i'/\partial x_k)(\partial u_j'/\partial x_k)}$, $v\overline{T'\,\partial^2 u_i'/\partial x_k^2}$ and $\chi_T\overline{u_i'\,\partial^2 T'/\partial x_k^2}$. The first term, as we have already mentioned, describes the influence of molecular diffusion on the evolution of Reynolds stresses; the second term describes the influence of the viscous dissipation. To estimate these terms we introduce characteristic scales of velocity fluctuations $b^{1/2}$, and length l where, as before, b is the kinetic energy of turbulence normalized to the mean density and l is the turbulence scale (the spatial scale of energy-containing eddies). Then $v\,\partial^2\overline{u_i'u_j'}/\partial x_k^2 \sim vb/l^2 \sim (v/b^{1/2}l)(b^{3/2}/l) \sim Re^{-1}(b^{3/2}/l)$, where $Re = b^{1/2}l/v$ is the Reynolds number. From this it follows that at large Reynolds numbers (of order 10^7 in the planetary boundary layer of the atmosphere) the influence of molecular diffusion can be ignored.

Next, for large Reynolds numbers the characteristic spatial scale of energy-containing eddies is much larger than the *Kolmogorov length microscale* $\eta = (v^3/\varepsilon)^{1/4}$ where the dissipation occurs (for an explanation of the sense of ε see below). This means that in the energy spectrum the intervals of energy supply and dissipation are separated from each other by the inertial interval where only cascade energy transfer from large eddies to small ones takes place. A feature of the cascade energy transfer is that any spatial orientation of large-scale eddies is lost in passing to small-scale motions. Hence, small-scale turbulence can be considered as isotropic. In such a case the second of the tensors in which we are interested can be presented in the form $2v\overline{(\partial u_i'/\partial x_k)(\partial u_j'/\partial x_k)} = 2/3\delta_{ij}\varepsilon$, having retained only the diagonal components. Here, $\varepsilon \equiv v\overline{(\partial u_i'/\partial x_k)^2}$ is the average rate of eddy energy dissipation and δ_{ij} is Kronecker's symbol. Vanishing of the non-diagonal components is proved as follows. According to the condition of isotropy all tensors are invariant with respect to any orthogonal transformation (rotation, parallel shift and mirror reflection) of the coordinate system. Therefore, because the tensor $\overline{(\partial u_i'/\partial x_k)(\partial u_j'/\partial x_k)}$ contains an odd number of subscripts, and the mirror reflection is equivalent to a sign change, then owing to the isotropy condition the non-diagonal components should vanish.

We take advantage of the fact that cascade transfer in the inertial interval depends neither on the geometry of the mean flow nor on molecular viscosity, and, inasmuch as it has to be balanced by the sink of the eddy energy in the dissipation interval, then the average dissipation rate ε should not depend on these parameters either. The assumption about the non-dependence of ε on molecular viscosity stated by Kolmogorov (1941) can be interpreted in the following way. Let $\varepsilon = v\overline{(\partial u_i'/\partial x_k)^2}$. Then introducing characteristic scales of velocity fluctuations v_d and length η in the dissipation interval we have $\varepsilon \sim vv_d^2/\eta^2$. Now if v is changed at fixed scales of the velocity fluctuations $b^{1/2}$ and length l in the energy supply interval then v_d and η are adapted in such a way that $(\partial u_i'/\partial x_k)^2 \sim v_d^2/\eta^2$ turns out to be inversely proportional to v, and this means that ε does not depend on v. But, in such a situation, from dimensional considerations $\varepsilon = c_l b^{3/2}/l$, where c_l is a numerical constant.

Finally the third and fourth terms of this group can be rewritten in the form

$$\overline{v T' \frac{\partial^2 u_i'}{\partial x_k^2}} = v \frac{\partial^2}{\partial x_k^2} \overline{u_i' T'} - v \frac{\partial}{\partial x_k}\left(\overline{u_i' \frac{\partial T'}{\partial x_k}}\right) - v \overline{\frac{\partial u_i'}{\partial x_k} \frac{\partial T'}{\partial x_k}},$$

$$\overline{\chi_T u_i' \frac{\partial^2 T'}{\partial x_k^2}} = \chi_T \frac{\partial^2}{\partial x_k^2} \overline{u_i' T'} - \chi_T \frac{\partial}{\partial x_k}\left(\overline{T' \frac{\partial u_i'}{\partial x_k}}\right) - \chi_T \overline{\frac{\partial T'}{\cdot x_k} \frac{\partial u_i}{\partial x_k}},$$

where, as in the preceding case, because of the incommensurability of spatial scales of energy-containing eddies and scales where the molecular viscosity and diffusion become significant, the first two terms on the right-hand sides can be neglected, and the third terms, owing to the condition of isotropy of turbulence, may be assumed to be equal to zero. Thus it turns out that the eddy heat flux is not affected by molecular viscosity and diffusion.

Correlations of pressure fluctuations. In Equation (4.2.8) the term containing correlations of pressure fluctuations consists of two components $(-1/\rho_0)(\partial \overline{u_j' p'}/\partial x_i + \partial \overline{u_i' p'}/\partial x_j)$ and $(\overline{p'/\rho_0})(\partial u_j'/\partial x_i + \partial u_i'/\partial x_j)$. In incompressible fluids $(\partial u_i'/\partial x_i = 0)$ the second component does not have an effect on the kinetic energy of turbulence and only redistributes it between different components of the Reynolds stress tensor and, thereby, promotes their isotropization. As for the first component, the volume integral over any finite domain of a turbulent flow within the interior of the laminar flow is equal to zero. Hence the first component describes the redistribution of Reynolds stresses in space, that is, it can be parametrized by a gradient-type formula.

Taking into account both these circumstances, most authors following Rotta (1951) combine the first component with the term describing eddy diffusion (Reynolds stresses transfer by velocity fluctuations), and approximate the second one in the form

$$(\overline{p'/\rho_0})(\partial u_j'/\partial x_i + \partial u_i'/\partial x_j) = -c_2 b^{1/2} l^{-1}(\overline{u_i' u_j'} - \tfrac{2}{3}\delta_{ij} b),$$

assuming it to be proportional to the degree of anisotropy (difference of the terms in parentheses on the right). This expression, as can easily be seen, ensures attenuation of all components of the Reynolds stress tensor and possesses all the necessary properties of tensor symmetry.

Eddy diffusion. Eddy diffusion includes two terms describing, respectively, the diffusion transfer $(-1/\rho_0)(\partial \overline{u_j' p'}/\partial x_i + \partial \overline{u_i' p'}/\partial x_j)$ created by pressure fluctuations, and the divergence $\partial(\overline{u_i'u_j'u_k'})/\partial x_k$ of Reynolds stress transfer. The second term, while integrating over volume restricted either by a laminar flow or by a homogeneous isotropic flow, vanishes and, hence, much like the first term, describes the redistribution of Reynolds stresses inside the turbulent flow domain. Therefore, for an approximate combination of these terms the following expression of gradient type is usually used:

$$\frac{\partial}{\partial x_k}\overline{u_i'u_j'u_k'} - \frac{1}{\rho_0}\left(\frac{\partial}{\partial x_i}\overline{u_j'p'} + \frac{\partial}{\partial x_j}\overline{u_i'p'}\right) = c_3\frac{\partial}{\partial x_k}\left(b^{1/2}l\frac{\partial}{\partial x_k}\overline{u_i'u_j'}\right).$$

The latter has the disadvantage that it does not satisfy the condition of tensor symmetry for $\overline{u_i', u_j', u_k'}$. This disadvantage is absent in the expression

$$\frac{\partial}{\partial x_k}\overline{u_i'u_j'u_k'} - \frac{1}{\rho_0}\left(\frac{\partial}{\partial x_i}\overline{u_j'p'} + \frac{\partial}{\partial x_j}\overline{u_i'p'}\right)$$

$$= c_4\frac{\partial}{\partial x_k}\left[b^{1/2}l\left(\frac{\partial}{\partial x_k}\overline{u_i'u_j'} + \frac{\partial}{\partial x_j}\overline{u_i'u_k'} + \frac{\partial}{\partial x_i}\overline{u_j'u_k'}\right)\right].$$

Here and above c_3 and c_4 are numerical constants.

But even this simple expression is much more complicated than the first one and, more important, its realization does not lead to better results.

Terms in the budget equations for the eddy heat flux, temperature dispersion, and turbulent energy containing correlations of pressure fluctuations and the transfer by velocity fluctuations, are approximated similarly.

Determination of the turbulence scale. To complete the equation system it is necessary to determine the turbulence scale. For this purpose, one uses either hypothetical relations of the type mentioned in the first-order closures, or evolution equations for the turbulence scale or any combination $b^m l^n$ with various values of powers m and n. It will be recalled that the budget equation (3.10.6) for turbulent energy determines the kinetic energy of turbulence, and so its inclusion in the combination $b^m l^n$ does not increase the number of unknowns. The last-named version of closure is preferable because it eliminates arbitrariness when choosing an expression for the turbulence scale in the first case, and eliminates difficulties when specifying the boundary condition for the turbulence scale at the upper boundary of the PBL in the second case. We examine it, assuming the turbulent energy dissipation rate to be a dependent variable.

At large Reynolds numbers when the turbulence can be considered to be locally isotropic the average rate ε of turbulent energy dissipation is equal to the product of the kinematic molecular viscosity coefficient v and the dispersion of fluctuations of the relative vorticity $(\partial u_i'/\partial x_k)^2$ (see (3.10.6)). We take advantage of this circumstance when deriving the equation for $\varepsilon = v(\partial u_i'/\partial x_k)^2$. Applying the operator $\partial/\partial x_k$ to (3.10.3) and then multiplying it by $v(\partial u_i'/\partial x_k)$ and averaging we obtain the following equation:

$$\frac{\partial \varepsilon}{\partial t} + \bar{u}_k \frac{\partial \varepsilon}{\partial x_k} + \frac{\partial}{\partial x_k}\overline{u_k'\varepsilon'} + 2v\overline{u_k' \frac{\partial u_i'}{\partial x_k} \frac{\partial}{\partial x_k}\left(\frac{\partial \bar{u}_i}{\partial x_k}\right)} = v\frac{\partial^2 \varepsilon}{\partial x_k^2} - 2\overline{\left(v\frac{\partial^2 u_i'}{\partial x_k^2}\right)^2}. \quad (4.2.10)$$

Here we have assumed that $\partial \bar{u}_k/\partial x_k = 0$, $\partial u_k'/\partial x_k = 0$ (fluid is incompressible), $\partial(\partial p'/\partial x_i)/\partial x_k = 0$ (curl of gradient is equal to zero), $(\partial u_i'/\partial x_k)(\partial T'/\partial x_k) = 0$ (the turbulence is locally isotropic), and takes into account the equality

$$\overline{\frac{\partial u_i'}{\partial x_k}\frac{\partial^3 u_i}{\partial x_k^3}} = \frac{1}{2}\left[\frac{\partial^2}{\partial x_k^2}\overline{\left(\frac{\partial u_i'}{\partial x_k}\right)^2} - 2\overline{\left(\frac{\partial^2 u_i'}{\partial x_k^2}\right)^2}\right].$$

In (4.2.10) the first term on the left-hand side describes the change in time of the average rate of turbulent energy dissipation; the second – an advective transport; the third – the divergence of the transfer of turbulent energy dissipation fluctuations $\varepsilon' = v(\partial u'/\partial x_k)^2$ by velocity fluctuations (the eddy diffusion); the fourth – the production at the expense of interaction of mean and fluctuating motions, and terms on the right-hand side describe, respectively, the molecular diffusion and destruction due to molecular viscosity. The last five of the above-mentioned terms are of order $b^2 l^{-2}$, and the average rate of the turbulent energy dissipation is of order $b^{3/2}l^{-1}$ so that a change of ε in time occurs with a time scale $b^{-1/2}l$ which is equal to a characteristic time scale of energy containing eddies.

Equation (4.2.10) contains three new unknowns: the eddy diffusion, as well as the source (production) and the sink (destruction) of ε, which should be presented in terms of the variables sought. There are no general recommendations as there are no obvious physical grounds for a universal dependence between them. Because of this we have to compromise again and supplement Equation (4.2.10) with hypothetical relations selected, to some extent, from considerations of simplicity and convenience of realization. In particular, it can be suggested that the source and sink of ε cannot be considered separately because at large Reynolds numbers they both approach zero. The most general representation of the sum of these terms and the eddy flux $-\overline{u_k'\varepsilon'}$ takes the form

$$-2v\overline{u_k'\frac{\partial u_i'}{\partial x_k}\frac{\partial}{\partial x_k}\left(\frac{\partial \bar{u}_i}{\partial x_k}\right)} - 2\overline{\left(v\frac{\partial^2 u_i'}{\partial x_k^2}\right)^2} = \left(c_5\frac{G}{\varepsilon} - c_6\right)\frac{\varepsilon^2}{b}; \qquad -\overline{u_k'\varepsilon'} = \alpha_\varepsilon k_M \frac{\partial \varepsilon}{\partial x_k},$$

where G is the turbulent energy generation defined by Equation (3.10.6); α_ε is the ratio between the coefficients of the eddy diffusion for ε and eddy viscosity; c_5 and c_6 are empirical constants.

The disadvantages of this method of closure are the controversy surrounding the assumed expressions for the production and destruction of ε and the uncertainty of the numerical constants. These circumstances, and also the fact that the solution of the problem of second-order closure gives the feeling of an art confused with scientific principles cause some uneasiness. The question inevitably comes to mind as to whether there is a criterion of correctness of the parametrization of correlations containing pressure fluctuations, and of other unknowns in the equations for Reynolds stresses and eddy fluxes. Such a criterion, known under the name of *a condition of realizability* (possibility from the physical viewpoint), was formulated by Schumann in 1977. It requires that all the characteristics which have to be non-negative (say, dispersion) to always be the same, and the correlation coefficients in the modulus should not exceed 1. As was shown by Lumley (1980) an attempt to satisfy the above requirements results in extremely complicated schemes of parametrization which are hardly useful in actual calculations. And because of this, one is usually of the opinion that at the present stage of research development it is more expedient to use simpler parametrizations, for which the condition of realizability can be upset but unsatisfactory results will be obtained only in extreme cases.

4.3 Laws of resistance and heat and humidity exchange

We consider the stationary, horizontally homogeneous, stratified PBL. Under stationary conditions the conclusion of constancy with height of the sensible and latent heat eddy fluxes follows from Equations (4.2.3) and (4.2.4). Using this fact and the *principle of self-similarity*, that is, independence of the developed turbulence regime on molecular constants and characteristics of the underlying surface, Kazansky and Monin (1961) came to the conclusion that the PBL structure should be determined only by u_*, $H/\rho c_p$, E/ρ, β and f. From this parameter set, two independent combinations with the dimension of length are formulated: the modified Monin–Obukhov length scale $L = u_*^2/[\kappa^2(\beta T_* + 0.61gq_*)]$, where $T_* = (-H/\rho c_p)/\kappa u_*$, $q_* = (-E/\rho)/\kappa u_*$ are temperature and specific humidity scales (see Section 3.9), and the thickness scale $\lambda = \kappa u_*/|f|$ of the neutrally stratified PBL (see Section 4.1) and, respectively, the non-dimensional stratification parameter $\mu_0 = \lambda/L$. Thus, if we choose u_*/κ, T_*, q_* and λ as characteristic scales of the mean velocity, temperature, specific humidity and length, respectively, then according to the principle of self-similarity all the non-dimensional statistical characteristics of the developed turbulence regime of the PBL have to be universal functions of two arguments: the non-dimensional height $\xi = z/L$ and the stratification parameter μ_0. In particular, general expressions for *defects in velocity, temperature and specific humidity* (differences in their values at an arbitrary level z and at the upper boundary h_E of the PBL) will take the form

$$
\left.
\begin{aligned}
u(z) - G \cos \alpha &= (u_*/\kappa)\psi_u(\xi, \mu_0), \\
v(z) - G \sin \alpha &= (u_*/\kappa)\psi_v(\xi, \mu_0) \text{ sign } f, \\
\theta(z) - \theta_h &= T_*\psi_\theta(\xi, \mu_0), \\
q(z) - q_h &= q_*\psi_q(\xi, \mu_0),
\end{aligned}
\right\} \tag{4.3.1}
$$

where ψ_u, ψ_v, ψ_θ and ψ_q are universal functions vanishing at $z = h_E$; θ_h and q_h are the potential temperature and specific humidity at this level; G and α are moduli of the geostrophic wind velocity and the angle between an isobar and the x axis (see Section 4.1).

Let us turn now to a derivation of the laws of resistance and heat and humidity exchange connecting characteristics (u_*, α, T_*, q_*) with the external parameters of the PBL. For this purpose we take advantage of the procedure of an asymptotic matching which allows for the existence of the so-called overlapping region where Equations (4.3.1) for the outer part of the PBL at small ξ coincide with the expressions

$$
\left.
\begin{aligned}
u(z) &= (u_*/\kappa) \ln z/z_0, \quad v(z) = 0, \quad \theta(z) - \theta_0 = (T_*/\alpha_\theta^0) \ln z/z_0, \\
q(z) - q_0 &= (q_*/\alpha_q^0) \ln z/z_0,
\end{aligned}
\right\} \tag{4.3.2}
$$

for the surface atmospheric layer at large z/z_0. Carrying out the given requirement (combining Equations (4.3.1) and (4.3.2)) we obtain

$$
\left.
\begin{aligned}
\ln \frac{\lambda}{z} - \frac{\kappa G \cos \alpha}{u_*} &= \psi_u(\xi, \mu_0) - \ln \xi, \\
-\frac{\kappa G \sin \alpha}{u_*} &= \psi_v(\xi, \mu_0) \text{ sign } f, \\
\ln \frac{\lambda}{z_0} - \frac{\alpha_\theta^0(\theta_h - \theta_0)}{T_*} &= \alpha_\theta^0\psi_\theta(\xi, \mu_0) - \ln \xi, \\
\ln \frac{\lambda}{z_0} - \frac{\alpha_q^0(q_h - q_0)}{q_*} &= \alpha_q^0\psi_q(\xi, \mu_0) - \ln \xi,
\end{aligned}
\right\} \tag{4.3.3}
$$

where the factor *sign f* in the second formula, as well as in Equation (4.3.1), takes into account the opposite directions of rotation of the velocity vector in the Northern and Southern Hemispheres.

Further, because the right-hand sides of Equations (4.3.3) are functions of ξ, but the left-hand sides are not, then to satisfy these equalities it is necessary for their right-hand sides to approach non-zero finite limits depending only on μ_0 at small values of the argument ξ. Therefore, we introduce the following definitions:

$$\lim_{\xi \to 0} [\psi_u(\xi, \mu_0) - \ln \xi] \equiv B(\mu_0) + \ln \kappa,$$

$$\lim_{\xi \to 0} \psi_v(\xi, \mu_0) \equiv A(\mu_0),$$

$$\lim_{\xi \to 0} [\alpha_\theta^0 \psi_\theta(\xi, \mu_0) - \ln \xi] \equiv C(\mu_0) + \ln \kappa,$$

$$\lim_{\xi \to 0} [\alpha_q^0 \psi_q(\xi, \mu_0) - \ln \xi] \equiv D(\mu_0) + \ln \kappa,$$

where A, B, C and D are the non-dimensional universal functions of the argument μ_0. Then taking into account the facts that $\ln \lambda/z_0 = \ln(\kappa u_*/G)(G/|f|z_0) = \ln (\kappa u_*/G) + \ln Ro$, where $Ro = G/|f|z_0$ is the Rossby number, and that on the basis of the second equality in (4.3.3)

$$\cos \alpha = (1 - \sin^2 \alpha)^{1/2} = (u_*/\kappa G)[\kappa^2(u_*/G)^{-2} - A^2(\mu_0)]^{1/2}$$

we find

$$\left. \begin{aligned} \ln Ro &= B(\mu_0) - \ln \frac{u_*}{G} + \sqrt{\left[\frac{\kappa^2}{(u_*/G)^2} - A^2(\mu_0) \right]}, \\ \sin \alpha &= -\frac{A(\mu_0)}{\kappa} \frac{u_*}{G} \operatorname{sign} f, \\ \alpha_q^0 (\theta_h - \theta_0)/T_* &= [\ln(Ro \, u_*/G) - C(\mu_0)], \\ \alpha_\theta^0 (q_h - q_0)/q_* &= [\ln(Ro \, u_*/G) - D(\mu_0)]. \end{aligned} \right\} \quad (4.3.4)$$

The above relationships are in fact the laws of resistance and heat and humidity exchange for the PBL. With their help and knowing the form of the functions A, B, C and D, one can set a connection between the geostrophical friction coefficient u_*/G, the angle α, the coefficients of heat exchange $T_*/(\theta_h - \theta_0)$ and evaporation $q_*/(q_h - q_0)$, as well as the stratification parameter $\mu_0 = \kappa^3(u_*/G)^{-1}(\beta T_* + 0.61 g q_*)/|f|G$ on the one hand, and external parameters G, $(\theta_h - \theta_0)$, $(q_h - q_0)$, f and z_0 of the planetary boundary layer on the other.

The law of resistance for the neutrally stratified PBL was first obtained by Kazanskii and Monin (1961). The relations (4.3.4) for the stratified PBL were recommended by Zilitinkevich (1970). Later they were repeatedly verified. As a result it was shown that experimental data generally confirm the universal character of dependencies $A(\mu_0)$, $B(\mu_0)$, $C(\mu_0)$ and $D(\mu_0)$ despite the inevitable spread.

Following Zilitinkevich (1970) we consider the asymptotic properties of the functions A, B, C and D for stable ($\mu_0 < 0$) and unstable ($\mu_0 > 0$) stratification. According to (4.2.1), (4.2.2) and considerations presented in Section 4.1 the divergence of the vertical eddy momentum flux in the outer part of the PBL is of the order of $u_*^2/|f|h_E$, where h_E is the PBL thickness defined depending on conditions of hydrostatic stability by Equation (4.1.4) or Equation (4.1.5). Taking this circumstance into account as well as the fact that $h_E \sim (\beta H/\rho c_p)^{1/2}|f|^{-3/2}$ for the strongly unstable stratification and that $\beta H/\rho c_p$ and μ_0 are connected to each other by the relationship $\beta H/\rho c_p = (|f|u_*^2/\kappa^2)(-\mu_0)$, we have the following expressions for defects in the velocity components in the outer part of the PBL:

$$\left.\begin{aligned}
u(z) - G\cos\alpha &= (u_*/\kappa)(-\mu_0)^{-1/2}\psi_u(z/h_E), \\
v(z) - G\sin\alpha &= (u_*/\kappa)(-\mu_0)^{-1/2}\psi_v(z/h_E),
\end{aligned}\right\} \tag{4.3.5}$$

where ψ_u, ψ_v are universal functions which differ from those in (4.3.1).

We match (4.3.5) with respective asymptotic formulae

$$\left.\begin{aligned}
u(z) &= (u_*/\kappa)[a_u + c_u(z/-L)^{-1/3} - \ln(z_0/-L)], \\
v(z) &= 0,
\end{aligned}\right\} \tag{4.3.6}$$

complying with the surface atmospheric layer for the free convection regime (see (3.8.7)), where a_u and c_u are non-dimensional constants. As a result we arrive at the same law of resistance as in (4.3.4) but now

$$A(\mu_0) = (-\mu_0)^{-1/2}M_1, \qquad B(\mu_0) = (-\mu_0)^{-1/2}M_2 + \ln(-\mu_0)/\kappa - a_u.$$

For strongly stable stratification when $h_E \sim u_*^2(-\beta H/\rho c_p)^{-1/2}|f|^{-1/2}$ the expressions for velocity defects in the outer part of the PBL take the form

$$\left.\begin{aligned}
u(z) - G\cos\alpha &= (u_*/\kappa)\mu_0^{1/2}\psi_u(z/h_E), \\
v(z) - G\sin\alpha &= (u_*/\kappa)\mu_0^{1/2}\psi_v(z/h_E).
\end{aligned}\right\} \tag{4.3.7}$$

Combining them with the asymptotic formulae

$$u(z) = (u_*/\kappa)[b_u(z/L) - \ln(z_0/L)], \qquad v(z) = 0, \tag{4.3.8}$$

complying with the surface atmospheric layer for the strong stability regime (see (3.8.8)), we arrive again at the first two relations in (4.3.4) but now

$$A(\mu_0) = \mu_0^{1/2}N_1, \qquad B(\mu_0) = \mu_0^{1/2}N_2 + \ln(\mu_0/\kappa).$$

Similarly, the expressions for defects of temperature and specific humidity in the outer part of the PBL in the regimes of strong instability and strong stability are written in the form

$$\left.\begin{array}{l} \theta(z) - \theta_h = T_*\psi_\theta(z/h_\mathrm{E}), \\ q(z) - q_h = q_*\psi_q(z/h_\mathrm{E}), \end{array}\right\} \tag{4.3.9}$$

where, depending on conditions of stratification, h_E is given either in the form of (4.1.4) or in the form of (4.1.5). Matching these expressions with their respective expressions

$$\left.\begin{array}{l} \theta(z) - \theta_0 = T_*[a_\theta + c_\theta(z/-L)^{-1/3} - (1/\alpha_\theta^0)\ln(z_0/-L)], \\ q(z) - q_0 = q_*[a_q + c_q(z/-L)^{-1/3} - (1/\alpha_q^0)\ln(z_0/-L)], \end{array}\right\} \tag{4.3.10}$$

for the surface atmospheric layer in the regime of free convection and

$$\left.\begin{array}{l} \theta(z) - \theta_0 = T_*[b_\theta(z/L) - (1/\alpha_\theta^0)\ln(z_0/L)], \\ q(z) - q_0 = q_*[b_q(z/L) - (1/\alpha_q^0)\ln(z_0/L)], \end{array}\right\} \tag{4.3.11}$$

in the regime of strong stability we obtain the last two relations in (4.3.4). As this takes place it turns out that $C(\mu_0) = \ln(-\mu_0/\kappa) + \alpha_\theta^0 M_3 - \alpha_\theta^0 a_\theta$, $D(\mu_0) = \ln(-\mu_0/\kappa) + \alpha_q^0 M_4 - \alpha_q^0 a_q$ in the first case and $C(\mu_0) = \ln(\mu_0/\kappa) + \alpha_\theta^0 N_3$, $D(\mu_0) = \ln(\mu_0/\kappa) + \alpha_q^0 N_4$ in the second case. Here M_3, M_4, N_3, N_4, as well as M_1, M_2, N_1, N_2, are the numerical constants derived from experimental data. We note that experimental estimates of the universal functions A, B, C and D are subjected to large scattering (see Figure 4.1), caused, firstly, by the approximate character of the theory ignoring the influence of non-stationarity and horizontal inhomogeneity of real fields of wind velocity, air temperature and humidity, and, secondly, by errors of handling of experimental data arising from indeterminacy in selection of the PBL thickness. We pay attention also to the strong sensitivity of the universal functions A, B, C and D to variations in the parameter μ_0, which dictates the necessity of considering the effects of stratification when calculating the characteristics of the interaction between the PBL and the underlying surface.

4.4 System of planetary boundary layers of the ocean and atmosphere

Meteorological characteristics (the geostrophic wind velocity, temperature, specific humidity and net radiation flux) at the upper boundary of the atmospheric PBL, and hydrological characteristics (the geostrophic current velocity, temperature and salinity) at the lower boundary of the ocean PBL) serve as external parameters for a system of two PBLs. It is these parameters which determine one or other structures of the PBL system, conditions at the interface and internal connections between the separate elements.

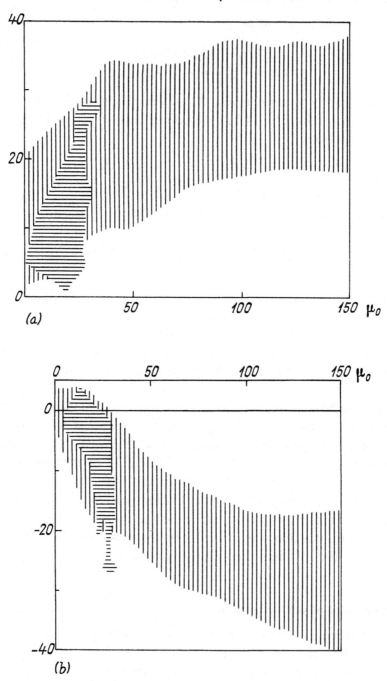

Figure 4.1 Universal functions *A*, *B* and *C* for the stable stratified planetary boundary layer. Vertically shaded regions are experimental estimates by Yamada (1976); horizontally shaded regions are estimates by Nieuwstadt (1981).

(continued)

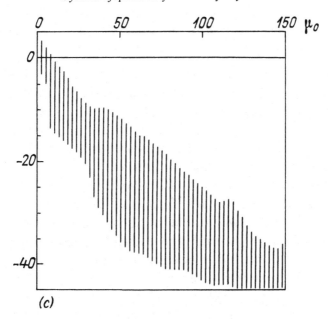

(c)

Figure 4.1 (*continued*).

Before starting the discussion of theoretical models of the PBL system we note that the external parameters listed above are determined, in turn, by the processes of interaction between the atmospheric PBL and free atmosphere, and between the ocean PBL and deep ocean. But because the eddy fluxes of various substances decrease as they move away from the interface of two environments and vanish at the outer boundaries of the PBLs, then interaction between the boundary layers, the free atmosphere and the deep ocean is realized not by the eddy transfer but, rather, by processes with different origins. The ordered vertical motions and the penetration of turbulence into a neighbouring non-turbulent domain (entrainment) serve as connecting links in the system of the atmospheric PBL–free atmosphere, and the ocean PBL–deep ocean. It should also be borne in mind that the relaxation time for velocity fields in the PBL system is much less than in the free atmosphere and especially in the deep ocean. Therefore, when examining the dynamics of two PBLs their isolation from the ocean–atmosphere system and subsequent analysis without consideration of the mechanisms of feedbacks with the free atmosphere and with the deep ocean turns out to be justified in a number of cases. One discusses the known autonomy of the system of the atmospheric and ocean planetary boundary layers in these terms.

4.4.1 *Theoretical models using* a priori *information on the magnitude and profile of the eddy viscosity coefficient*

When analysing observational data on the wind and ice drift in the period of the 'Fram' drift, Nansen noticed that drifting ice in the Central Arctic deviated to the right of the wind direction by 30–40°. He explained this fact by the Earth's rotation and indicated that because the upper ocean layer is driven to motion by ice the currents induced in the ocean should deviate from the wind direction even more than 30–40°. Theoretical research on this problem was initiated by Ekman (1902), who obtained results conforming to Nansen's qualitative conclusions. Ekman (1905) presented a theoretical confirmation of the clockwise (in the Northern Hemisphere) rotation of the drift current velocity with depth and adduced some arguments in favour of using the results obtained to describe wind velocity changes with height over land. This idea was later realized by Akerblom (1908) and independently by Exner (1912). It is advisable to dwell at length on Exner's work because he was the first to combine the planetary boundary layers of the ocean and atmosphere in a single dynamic system.

Under stationary conditions and parallel and equidistant isobars the motion in the PBL is described by the following equations:

$$\frac{d}{dz_i} k_{Mi} \frac{du_i}{dz_i} + f(v_i - V_{gi}) = 0; \qquad \frac{d}{dz_i} k_{Mi} \frac{du_i}{dz_i} - f(u_i - U_{gi}) = 0, \quad (4.4.1)$$

where u_i, v_i are components of the wind velocity or drift current; z_i is the vertical coordinate oriented vertically upward in the atmosphere and downward in the ocean (the origin is located at the free ocean surface); the subscript i takes on values 1, 2 while $i = 1$ complies with the atmospheric PBL, $i = 2$ complies with the oceanic PBL; all other designations are the same.

Equations (4.1.1) are solved on the assumption that velocity components V_{g2}, U_{g2} of the geostrophic current are equal to zero, coefficients k_{Mi} of eddy viscosity in each PBL remain constant with height and fixed, and the following conditions are fulfilled at the ocean–atmosphere interface and at considerable distance from it:

$$\left. \begin{array}{c} u_i = u_2, \qquad v_1 = v_2, \qquad k_{M1}\rho_1 \dfrac{du_1}{dz_1} = -k_{M2}\rho_2 \dfrac{du_2}{dz_2}, \\[3mm] k_{M1}\rho_1 \dfrac{dv_1}{dz_1} = -k_{M2}\rho_2 \dfrac{dv_2}{dz_2} \quad \text{at } z_{1,2} = 0, \end{array} \right\} \quad (4.4.2)$$

$$u_i \Rightarrow U_{gi}, \qquad v_i \Rightarrow V_{gi} \text{ at } z_i \Rightarrow \infty. \quad (4.4.3)$$

Boundary conditions (4.4.2), as known, mean continuity of the velocity and momentum flux at the free ocean surface; conditions (4.4.3) describe the boundedness of the velocity at a sufficiently large height within the limits of the respective PBL.

The final expressions for the vertical profile of the wind and current velocity take the form

$$u_i = U_{gi} + e^{-a_i z_i}[(u_0 - U_{gi})\cos a_i z_i + (v_0 - V_{gi})\sin a_i z_i], \atop v_i = V_{gi} + e^{-a_i z_i}[(v_0 - V_{gi})\cos a_i z_i - (u_0 - U_{gi})\sin a_i z_i], } \quad (4.4.4)$$

where u_0, v_0 are the current velocity components at the ocean surface defined by the equalities

$$u_0 = U_{gi}\left(1 + \frac{\rho_2}{\rho_1}\sqrt{\left(\frac{a_1}{a_2}\right)}\right)^{-1}, \quad v_0 = V_{gi}\left(1 + \frac{\rho_2}{\rho_1}\sqrt{\left(\frac{a_1}{a_2}\right)}\right)^{-1}, \atop a_i = (2k_{Mi})^{-1/2}|f|^{1/2}. } \quad (4.4.5)$$

Taking advantage of estimates of the eddy viscosity coefficients existing at that time Koschmider (1938) calculated the change of wind velocity in the atmosphere and of drift current velocity in the ocean. It turned out that the velocity distribution in the system of two PBLs was described by the double Ekman spiral, and the current velocity at the water–air interface was not zero and coincided in direction with the geostrophic wind (see Figure 4.2). From this, in particular, follows the conclusion about the anticlockwise rotation of the wind velocity vector with height in the next-to-surface atmospheric layer.

Further development of the theory was achieved by Shvets (1939) and Hesselberg (1954). To obtain better agreement between calculated results and experimental data they took into account the change in the atmospheric pressure gradient with height. In doing so, Hesselberg, unlike Shvets, assigned the slip condition as the boundary condition at the water–air interface. Both authors considered the eddy viscosity coefficient to be independent of height, and preassigned. Calculated changes of wind velocity with height were found to be qualitatively close to observational changes: the rapid increase of wind velocity was marked within the first several hundred metres, then between 500 and 1000 metres it fell or even terminated altogether, and at larger heights the velocity increase commenced again.

The generalization of the Exner model in the case where the ocean surface is covered with freely drifting ice was presented by Laikhtman (1958). In this case the continuity condition for the momentum flux at the water–air interface was replaced by the budget equation for forces acting upon a unit

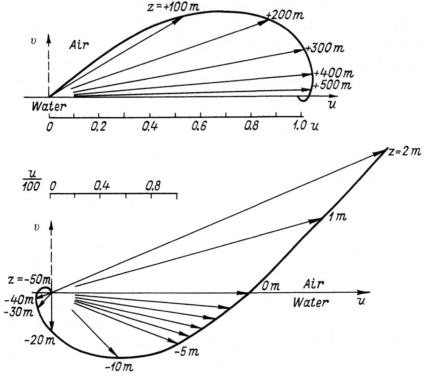

Figure 4.2 Double Ekman spiral. The lower spiral shows a 50-fold gain.

(that is, referred to a unit area) ice mass, i.e.

$$
\left.
\begin{aligned}
fm_1 v_0 + \rho_1 k_{M1} \frac{du_1}{dz_1}\bigg|_{z_1=0} + \rho_2 k_{M2} \frac{du_2}{dz_2}\bigg|_{z_2=0} &= 0, \\
-fm_1 u_0 + \rho_1 k_{M1} \frac{dv_1}{dz_1}\bigg|_{z_1=0} + \rho_2 k_{M2} \frac{dv_2}{dz_2}\bigg|_{z_2=0} &= 0.
\end{aligned}
\right\}
\tag{4.4.6}
$$

It is suggested here that the ocean surface is horizontal and the eddy viscosity coefficients k_{M1}, k_{M2} do not vanish at the air–ice and ice–water interfaces.

When guiding the x axis along the direction of the geostrophic wind ($V_{g1} = 0$) and assuming that u_0, $v_0 \ll U_{g1}$, then the expressions for the components of the ice drift velocity will take the form

$$
\left.
\begin{aligned}
u_0 &= 2U_{g1}(\rho_1/\rho_2) \frac{(a_2/a_1)^{1/2}(1 + m_1 a_2/\rho_2)}{1 + (1 + 2m_1 a_2/\rho_2)^2}, \\
v_0 &= -2U_{g1}(\rho_1/\rho_2) \frac{(a_2/a_1)^{1/2} m_1 a_2/\rho_2}{1 + (1 + 2m_1 a_2/\rho_2)^2}.
\end{aligned}
\right\}
\tag{4.4.7}
$$

As may be seen, $v_0/u_0 = -(m_1 a_2/\rho_2)/(1 + m_1 a_2/\rho_2) < 0$. This means that the ice drift in a deep sea is always directed to the side of high pressure, and hence, in the presence of ice cover, the left turn of wind in the surface atmospheric layer will be greater than that predicted by Exner's model.

4.4.2 Simplest closed models

Equations (4.4.1) are not closed. Besides the components of the wind and drift current velocities they contain two more unknowns: eddy viscosity coefficients in the atmospheric and ocean planetary boundary layers. To close the system it is necessary to supplement the equations of motion with relations connecting eddy viscosity coefficients with the velocity distribution. For this purpose Laikhtman (1958) proposed to use the integral turbulent energy budget equation

$$\int_0^{h_i} k_{\mathrm{M}i}\left[\left(\frac{\mathrm{d}u_i}{\mathrm{d}z_i}\right)^2 + \left(\frac{\mathrm{d}v_i}{\mathrm{d}z_i}\right)^2 - g\frac{\mathrm{d}\ln S_i}{\mathrm{d}z_i}\right]\mathrm{d}z_i - \hat{\varepsilon}_i h_i = 0, \qquad (4.4.8)$$

where S_i is the potential air temperature θ, or water density ρ_2; h_i is the thickness of the ith PBL; $\hat{\varepsilon}$ is the average (within the PBL) turbulent energy dissipation rate.

We note that in (4.4.8) the diffusive turbulent energy fluxes at the outer boundaries of the PBLs and at the water–air interface are assumed to be equal to zero. Besides the obvious advantages of using this equation to determine the eddy viscosity coefficient, it has its own disadvantages: the appearance of two new unknown parameters ($\hat{\varepsilon}_i, h_i$) and the necessity of approximating the vertical distribution of the eddy viscosity coefficient by some function of height.

In the simplest closed models two methods of determining $\hat{\varepsilon}$ are the most commonly used. In the first, average dissipation is assumed to be in proportion to the integral (over the PBL) turbulent energy production of shear and convective origin; the second reduces to the assumption that $\hat{\varepsilon}_i$ is a function of the eddy viscosity coefficient, averaged within the PBL, and of the PBL thickness. With this assumption we have, on the basis of dimensionality considerations,

$$\hat{\varepsilon}_i = c_\varepsilon k_{\mathrm{M}i}^3/h_i^4, \qquad (4.4.9)$$

where c_ε is a numerical constant estimated from experimental data.

As for h_i, it can be identified by the height at which the derivative of the mean velocity modulus becomes zero initially:

$$\frac{d}{\partial z_i}(u_i^2 + v_i^2)^{1/2}|_{z=h_i} = 0, \tag{4.4.10a}$$

or by the thickness of the friction layer at the upper boundary of which the Ekman condition is fulfilled:

$$[(u_i - U_{gi})^2 + (v_i - V_{gi})^2]_{z=h}^{1/2} = e^{-\pi}(U_{gi}^2 + V_{gi}^2)^{1/2}, \tag{4.4.10b}$$

In the absence of the geostrophic current the last condition reduces to the form

$$(u_2^2 + v_2^2)^{1/2}|_{z=h_2} = e^{-\pi}(u_0^2 + v_0^2)^{1/2}. \tag{4.4.10c}$$

The system of Equations (4.4.1)–(4.4.3) and (4.4.8)–(4.4.10) given above is closed. Its solution at $k_{Mi} = const$ was obtained by Kagan (1971). The formulae for the eddy viscosity coefficients in the neutrally stratified PBLs of the atmosphere and the ocean have the form

$$k_{M1} \sim G_1^2/|f|; \qquad k_{M2} \sim (\rho_1/\rho_2)G_1^2/|f|. \tag{4.4.11}$$

It follows from here and (4.4.5) that the wind coefficient (ratio between the modulus c_0 of the surface drift velocity and the modulus G_1 of the geostrophic wind velocity) is independent of wind velocity and approximately equals $(\rho_1/\rho_2)^{1/2}$.

The model examined correctly reproduces the relationships between different characteristics of the PBL system and external parameters. But to obtain, with its help, an acceptable agreement between calculated and observed values of characteristics sought is not possible due to the use of eddy viscosity coefficients which are assumed to be constant with height in the atmosphere and with depth in the ocean. Because of this, subsequent improvement of the model has followed the path of replacing the average eddy viscosity coefficients by some function of the vertical coordinate. The sensitivity of the solution to the choice of different changes in eddy viscosity coefficients along the vertical was examined by Kagan (1971). In this work it was shown that replacement of an eddy viscosity coefficient variable over depth in the ocean by its average value does not have any effect on the wind field structure, and only changes the current field structure. This change manifests itself in a small increase in the current velocity in the upper, and a decrease in the lower, part of the friction layer. At the same time use of the average eddy viscosity coefficient in the atmosphere results in an increase in the drift current velocity by 50–60%.

The angle between the direction of the surface drift current and the surface isobar turns out to be quite sensitive to the approximation of the eddy viscosity coefficient in the atmospheric PBL. If for a constant (over height) eddy viscosity coefficient the surface drift current is strictly directed along the isobar (coincidence with the isobar takes place only at constant eddy viscosity coefficients in the atmosphere and the ocean), then for a variable eddy viscosity coefficient in the atmosphere the surface drift current is directed to the side of high pressure, making an angle of 30° to the isobar.

Kagan (1971) has attempted to find an optimal variant of approximation of the vertical distribution of the eddy viscosity coefficient in the atmospheric and ocean PBLs. This requires observed data obtained under strictly controlled conditions, and in view of their almost complete absence he has restricted himself to comparison of calculated and observed values of the wind coefficient. Unfortunately, the observational data for the wind coefficient vary over a rather wide range – from 0.0092 according to Hela, up to 0.0328 according to Mohn – that is explained by errors in determination of the surface current velocity identified from data on the drift of various objects. In other words, available estimates, as a matter of fact, describe not the surface current but, rather, a current in a layer of finite (and different in every separate case) thickness. The accuracy of wind velocity measurements used on determination of the wind coefficient also leaves much to be desired (see Section 1.4). Therefore, from this, quite rich, set of estimates of the wind coefficient, only two estimates, with most statistics and free from the above-mentioned disadvantages, were chosen. These estimates, from Hughes (1956) and Tomczak (1963), were found from measured current velocity values obtained with the help of drifting postcards several mm thick and geostrophic wind velocity values calculated from data on atmospheric surface pressure. Eventually, it was ascertained that approximation of the vertical profile of the eddy viscosity coefficient by a scheme 'with a knee', that is, its approximation by a linear function of height in the surface layer and constant with height in the rest of the atmospheric PBL, and presetting the eddy viscosity coefficient to be constant with depth in the ocean PBL, guarantees quite acceptable accuracy in the calculation of the wind coefficient: according to observed data it changes from 0.0192 through 0.0226; the calculated results yield 0.025.

An ideologically close model of the atmospheric and oceanic PBL system was developed by Yoshihara (1968). Here (as well as in all the other models listed) the atmospheric and ocean PBLs were considered to be horizontally homogeneous. Meanwhile, there are many proofs that in zones of atmospheric and hydrological fronts the horizontal gradients of characteristics can differ markedly from zero. For example, in the zone of the Sargasso Sea hydrological front the temperature difference amounts to more than 1 °C per 10 km. Some parts of the front are known where the temperature difference may be even as much as 3 °C per 100 m. The first model of the atmospheric and oceanic PBL system, taking into account the horizontal inhomogeneity of fields of meteorological and hydrological characteristics, was proposed by Kagan (1971). In this work a stationary solution to the equations of motion, state, heat transport and turbulent energy budget in both environments, and also the budget equations for water vapour in the atmosphere, and for salinity in the ocean was obtained. In doing so the horizontal temperature and salinity gradients were considered to be constant over the vertical, and fixed. The continuity conditions for the velocity, temperature and momentum were accepted at the ocean surface. The surface temperature was determined from the heat balance condition. It was also considered that at the upper boundary of the atmospheric PBL, and at the lower boundary of the ocean PBL the wind and current become geostrophic, and the air temperature and humidity, as well as the water temperature and salinity, take on fixed values. The problem has been solved for constant eddy viscosity coefficients in both media.

The model described was then generalized for the case when the ocean surface is covered by freely drifting ice (see Kagan, 1971). The solution obtained allowed one to explain a number of peculiarities of the thermal and dynamic ocean–atmosphere interactions in the presence of ice cover, and, particularly, intensification of the inversion in the atmosphere in winter, which, as it turns out, is conditioned by the shielding influence of ice contributing to a decrease in the heat influx to the upper ice surface from below, and by a drop in its temperature. However, due to the approximate description of real processes and the assumption of constant eddy viscosity coefficients this explanation has a rather qualitative character: in winter this assumption leads to an overestimation of the temperature at the upper ice surface, and in summer, when the upper ice surface temperature becomes equal to the water-freezing temperature, to considerable (almost seven times) underestimation of heat expenses for ice melting. Because of this, in an improved version of the model, the vertical distribution of the eddy viscosity coefficient in the atmospheric PBL was approximated by the scheme 'with a knee' (see Kagan, 1971). Test results demonstrated that the model reproduces seasonal changes in the ice surface temperature, drift velocity and intensity of ice growth and melting to good precision. The annual mean values of the ice growth or melting rate, and the ice drift velocity, turned out to be equal to 57 cm/year and 10 cm/s; observed data yield 30–50 cm/year and 8 cm/s respectively.

We also mention the non-stationary model of the atmospheric ocean PBL system developed by Pandolfo (1969). Here, as in Kagan (1971), the horizontal gradients of temperature and humidity in the atmosphere and the horizontal gradients of temperature and salinity in the ocean were considered to be prescribed, though arbitrary, functions of the vertical coordinate, and instead of the turbulent energy budget equation the Prandtl formula was applied. The modification reduced to accounting for the effects of stratification and wind waves. A detailed discussion of this model can be found in Kagan (1971).

4.4.3 Semiempirical models not using a priori information on the magnitude and profile of the eddy viscosity coefficient

A general feature of all the models discussed is the fact that one or another approximation of vertical profiles of the eddy viscosity coefficient was used in their construction. It was shown in a review article by Zilitinkevich *et al.* (1967) that, due to the strong variability of the eddy viscosity coefficient in the atmosphere, even a good approximation does not provide a reliable representation of the vertical structure of the atmospheric PBL under all possible conditions. The same can be said with respect to the ocean PBL, the only difference being that information about the eddy viscosity coefficient in the ocean is more limited. A few attempts at estimating the eddy viscosity coefficient from data of direct determinations of Reynolds fluxes in the ocean are detailed in the literature. Panteleyev (1960) was the first to undertake such an attempt. He demonstrated that the eddy viscosity coefficient reaches a maximum at a depth of 10–15 m. Gongwer and Finkle (1960) showed that it decreases below this depth, and in the range of depths from 100–150 m to 400 m it remains practically constant with depth. A report with published data on the magnitude of the eddy viscosity coefficient is contained in Neumann and Pirson (1966), where it is stated that the eddy viscosity coefficient can change, depending on conditions, by two orders of magnitude – from 10 to 10^3 cm^2/s. Thus, data available at present allow determination of only the order and rough form of its profile, and it is natural that, even in spite of the simplicity and clearness of the models where the eddy viscosity coefficient is considered to be a prescribed function of the vertical coordinate, these models are unsatisfactory.

A theoretical model of the system of the atmospheric and ocean PBLs which does not use *a priori* assumptions of the magnitude and profile of the eddy viscosity coefficient was developed by Laikhtman (1966). We dwell at length on this not only because it is the most efficient of all the models mentioned above but also to show a very elegant method of reducing a two-layer problem to a single-layer one.

When closing the equations of motion by the budget equation for the turbulent energy and by relations of approximate similarity which are common in semiempirical turbulence theory, the equation set for the atmospheric and ocean PBLs takes the form

$$
\left.
\begin{aligned}
&\frac{d}{dz_i} k_{Mi} \frac{du_i}{dz_i} + f(v_i - V_{gi}) = 0, \qquad \frac{d}{dz_i} k_{Mi} \frac{dv_i}{dz_i} - f(u_i - U_{gi}) = 0, \\
&k_{Mi} \left[\left(\frac{du_i}{dz_i}\right)^2 + \left(\frac{dv_i}{dz_i}\right)^2 - g\frac{d \ln S_i}{dz_i} \right] - c\frac{b_i^2}{k_{Mi}} + \alpha_b \frac{d}{dz_i} k_{Mi} \frac{db_i}{dz_i} = 0, \\
&k_{Mi} = b_i \left(\frac{\kappa c^{1/2}}{u_{*i}} z_{0i} + \kappa c^{1/4} \int_{z_{0i}}^{z_i} \frac{dz_i}{b_i^{1/2}} \right).
\end{aligned}
\right\} \quad (4.4.12)
$$

Here the last expression is obtained by transformation of the approximate similarity relationship for the eddy viscosity coefficient and the generalized von Karman formula for the turbulence scale (see Section 4.2.1); α_b and c are universal constants with numerical values equal to 0.73 and 0.046 respectively; all another symbols are the same.

The turbulent energy budget equation (the third equation in (4.4.12)) includes the function S_i, which, as has been already noted, represents either the potential air temperature θ (at $i = 1$), or the sea water density ρ_2 (at $i = 2$). In order not to complicate the problem by the determination of a new unknown function it can be assumed that $k_{Mi} dS_i/dz_i = const$ or advantage can be taken of any interpolation formula for dS_i/dz_i. In both cases the condition $dS_i/dz_i \neq 0$ means that the vertical structure of the atmospheric and ocean PBLs described by Equation (4.4.12) depends on stratification in both media.

Equations (4.4.12) are supplemented by the following boundary conditions:

$$
u_i \Rightarrow U_{gi}, \qquad v_i \Rightarrow V_{gi}, \qquad b_i \Rightarrow 0 \text{ at } z_i \Rightarrow \infty, \qquad (4.4.13)
$$

$$
\left.
\begin{aligned}
&k_{M1}\rho_1\, du_1/dz_1 = -k_{M2}\rho_2\, du_2/dz_2, \qquad k_{M1}\rho_1\, dv_1/dz_1 = -k_{M2}\rho_2\, dv_2/dz_2, \\
&u_1 = u_2, \qquad v_1 = v_2, \qquad b_i = c^{-1/2}u_{*i}^2 \quad (i = 1, 2) \text{ at } z_i = z_{0i},
\end{aligned}
\right\}
$$

$$(4.4.14)$$

where the last condition follows from the turbulent energy budget equation on the assumption that in the vicinity of the water–air interface the production of turbulent energy is balanced by dissipation.

If we now introduce the non-dimensional variables with subscript n,

$$
\left.
\begin{aligned}
&(u_i, v_i) = (-1)^{i+1}(u_{*i}/\kappa)(u_{ni}, v_{ni}), \qquad z_i = (\kappa u_{*i}/|f|)z_{ni}, \\
&k_{Mi} = (\kappa^2 u_{*i}^2/|f|)k_{Mni}, \qquad b_i = c^{-1/2}u_{*i}^2 b_{ni},
\end{aligned}
\right\} \quad (4.4.15)
$$

and then change from velocities to stresses,[1] system (4.4.12)–(4.4.14) will take the form

$$
\left.
\begin{array}{l}
\dfrac{\mathrm{d}^2 \eta_{ni}}{\mathrm{d}z_{ni}^2} + \dfrac{\sigma_{ni}}{k_{Mni}} = 0, \qquad \dfrac{\mathrm{d}^2 \sigma_{ni}}{\mathrm{d}z_{ni}^2} - \dfrac{\eta_{ni}}{k_{Mni}} = 0, \\[2mm]
\dfrac{\eta_{ni}^2 + \sigma_{ni}^2}{k_{Mni}} - \mu_{0i} k_{Mni} \dfrac{\mathrm{d}S_{ni}}{\mathrm{d}z_{ni}} - \dfrac{b_{ni}^2}{k_{Mni}} - \beta \dfrac{\mathrm{d}}{\mathrm{d}z_{ni}} k_{Mni} \dfrac{\mathrm{d}b_{ni}}{\mathrm{d}z_{ni}} = 0,
\end{array}
\right\} \tag{4.4.16}
$$

$$
\left.
\begin{array}{l}
k_{Mni} = b_{ni}\left(z_{0ni} + \displaystyle\int_{z_{0ni}}^{z_{ni}} b_{ni}^{-1/2}\, \mathrm{d}z_{ni} \right), \\[3mm]
\eta_{ni}, \sigma_{ni}, b_{ni} \Rightarrow 0 \ \text{at}\ z_{ni} \Rightarrow \infty,
\end{array}
\right\} \tag{4.4.17}
$$

$$
\left.
\begin{array}{l}
\eta_{n1} = (\rho_2 u_{*2}^2 / \rho_1 u_{*1}^2)\eta_{n2}, \qquad \sigma_{n1} = (\rho_2 u_{*2}^2 / \rho_1 u_{*1}^2)\sigma_{n2}, \\[2mm]
-\dfrac{\mathrm{d}\eta_{n1}}{\mathrm{d}z_{ni}} + \dfrac{\kappa V_{g1}}{u_{*1}} = \left(\dfrac{\rho_1}{\rho_2}\right)^{1/2}\left(\dfrac{\mathrm{d}\eta_{n2}}{\mathrm{d}z_{n2}} + \dfrac{\kappa V_{g2}}{u_{*2}}\right), \\[3mm]
\dfrac{\mathrm{d}\sigma_{n1}}{\mathrm{d}z_{ni}} + \dfrac{\kappa U_{g1}}{u_{*1}} = -\left(\dfrac{\rho_1}{\rho_2}\right)^{1/2}\left(\dfrac{\mathrm{d}\sigma_{n2}}{\mathrm{d}z_{n2}} - \dfrac{\kappa U_{g2}}{u_{*2}}\right), \\[3mm]
b_{ni} \Rightarrow 1 \ \text{at}\ z_{ni} \Rightarrow 0,
\end{array}
\right\} \tag{4.4.18}
$$

where $\eta_{ni} = k_{Mni}\, \mathrm{d}u_{ni}/\mathrm{d}z_{ni}$, $\sigma_{ni} = k_{Mni}\, \mathrm{d}v_{ni}/\mathrm{d}z_{ni}$ are non-dimensional components of the eddy friction stress; $\mu_{0i} = -\kappa^2(g/S_i)(\pi_{0i}/\rho_i c_i)/|f|u_{*i}^2$; π_{0i} is the eddy flux of sensible heat (at $i = 1$), or of mass (at $i = 2$) at the ocean surface; $\rho_i c_i$ is the air or water volume heat capacity; β is a universal constant composed of α_b and c and equal to 0.54.

We note that the set (4.4.16)–(4.4.18) looks very much like the analogous system for the atmospheric PBL for which the numerical solution is given in Laikhtman (1970). The only distinction between the problem here and the one in Laikhtman (1970) is in the formulation of boundary conditions at $z_{ni} \Rightarrow 0$ (in Laikhtman, 1970, it is assumed that $\eta_n = 1$, $\sigma_n = 0$, $b_n = 1$ instead of (4.4.18)). To determine the universal functions of the two-layer problem (4.4.16)–(4.4.18), an ancillary coordinate system (x', y', z_{ni}), is introduced where the direction of axis x' and scales u_{*i} are selected to satisfy conditions $\eta_{n1} = \eta_{n2} = 1$, $\sigma_{n1} = \sigma_{n2} = 0$. In accordance with the first two equalities in (4.4.18) this means that $(\rho_2 u_{*2}^2 / \rho_1 u_{*1}^2) = 1$ and that the x' axis is directed along the surface stress and makes a certain angle α with the

---

[1] The use of this technique, proposed by Monin (1950), has the advantage that it allows, firstly, a reduction by two ($\kappa V_{gi}/u_{*i}$, $\kappa U_{gi}/u_{*i}$) in the amount of non-dimensional variables and, secondly, to carry over the boundary conditions (4.4.14) from the roughness level z_{0i} to the level $z_i = 0$. Such a replacement of the lower limit of integration is justified because for small z_i the friction stress and the turbulent energy do not in practice change with height.

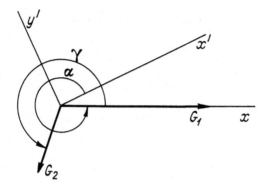

Figure 4.3 Frame of reference (for angles) and wind and current velocity vectors in the planetary boundary layers of the atmosphere and ocean.

x axis of the standard coordinate system (see Figure 4.3). The friction velocity u_{*i} and the angle α between the direction of the geostrophic wind and surface stress is determined with the help of the, still unused, condition of velocity continuity at the water–air interface (third and fourth relations in (4.4.18)). As a result we obtain the following expressions for the geostrophic friction coefficient $\chi = u_{*1}/\kappa G_1$ and the angle α:

$$
\left.
\begin{aligned}
\chi &= (1 - 2n \cos \gamma + n^2)^{1/2} \left[\left(\frac{d\eta_{n1}}{dz_{n1}} + \sqrt{\left(\frac{\rho_1}{\rho_2} \right)} \frac{d\eta_{n2}}{dz_{n2}} \right)^2 \right. \\
&\quad \left. + \left(\frac{d\sigma_{n1}}{dz_{n1}} + \sqrt{\left(\frac{\rho_1}{\rho_2} \right)} \frac{d\sigma_{n2}}{dz_{n2}} \right)^2 \right]^{-1/2} \quad \text{at } z_{n1} = z_{0n1}, z_{n2} = z_{0n2}, \\
\tan \alpha &= -\frac{1 - n \ (N \sin \gamma + \cos \gamma)}{N - n \ (N \cos \gamma - \sin \gamma)},
\end{aligned}
\right\} \quad (4.4.19)
$$

where γ is the angle between the geostrophic wind (isobar) and geostrophic current (see Figure 4.3);

$$
n = G_2/G_1; \qquad N = \left(\frac{d\sigma_{n1}}{dz_{n1}} + \sqrt{\left(\frac{\rho_1}{\rho_2} \right)} \frac{d\sigma_{n2}}{dz_{n2}} \right) \Bigg/ \left(\frac{d\eta_{n1}}{dz_{n1}} + \sqrt{\left(\frac{\rho_1}{\rho_2} \right)} \frac{d\eta_{n2}}{dz_{n2}} \right)
$$

$$
\text{at } z_{n1} = z_{0n1}, \qquad z_{n2} = z_{0n2}.
$$

We note also that if the roughness parameter z_{01} of the ocean surface is fixed then $z_{0n1} = (\kappa^2 \chi Ro)^{-1}$; if Charnock's formula, $z_{01} = mu_{*1}^2/g$, is assigned then $z_{0n1} = m(G_1|f|/g)\chi$. Here $Ro = G_1/|f|z_{01}$ is the Rossby number; m is the numerical factor. Accordingly, for z_{0n2} we have for both cases $z_{0n2} = (\sqrt{(\rho_1/\rho_2)}\kappa^2 \chi Ro)^{-1}$ and $z_{0n2} = m\sqrt{(\rho_1/\rho_2)}(G_1|f|/g)\chi$.

The dependence of χ and α on the non-dimensional parameter $mG_1|f|/g$

and on the stratification parameter μ_0 is presented by Laikhtman (1970) in tabular form so it is easy to find u_{*1} and then, with the help of the relation $(\rho_2 u_{*2}^2/\rho_1 u_{*1}^2) = 1$, to calculate u_{*2} and the characteristic scales of all variables sought and defined by Equations (4.4.15). After that the vertical distributions of the wind velocity, current velocity and the eddy viscosity coefficients are restored by using the expressions

$$\left.\begin{aligned}
u_1 &= G_1 \cos\alpha + \chi G_1 \, d\sigma_{n1}/dz_{n1}, \\
v_1 &= G_1 \sin\alpha - \chi G_1 \, d\eta_{n1}/dz_{n1}, \\
u_2 &= G_2 \cos(\alpha + \gamma) - (\rho_1/\rho_2)^{1/2}\chi G_1 \, d\sigma_{n2}/dz_{n2}, \\
v_2 &= G_2 \sin(\alpha + \gamma) + (\rho_1/\rho_2)^{1/2}\chi G_1 \, d\eta_{n2}/dz_{n2}, \\
k_{M1} &= \kappa^4(\chi^2 G_1^2/|f|)k_{Mn1}, \\
k_{M2} &= (\rho_1/\rho_2)\kappa^4(\chi^2 G_1^2/|f|)k_{Mn2}.
\end{aligned}\right\} \qquad (4.4.20)$$

Tests showed quite satisfactory coincidence of the calculated and observed dependencies of the resistance coefficient $C_u = (u_{*1}/u_{10})^2$ on wind velocity (here u_{10} is the wind velocity at a height of 10 m). The solution sensitivity to the choice of numerical constants was examined in detail by Radikevich (1968). He also calculated, on the basis of the model in question, seasonal mean fields of wind velocity, drift current, eddy viscosity coefficients and momentum, heat and water vapour fluxes in the North Atlantic.

The model of the ice drift proposed by Laikhtman (1986) was the natural generalization of the model discussed. Its essence is the following. The system of equations for the atmospheric and ocean PBLs, by means of proper rotation of the coordinate axes, reduces to the system of equations for the atmospheric PBL for which the solution is known. The matching of the equations is performed by the equations of motion for ice and the continuity condition for velocity at the water–ice and ice–air interfaces. As a result the ice drift characteristics are determined by the solution of a system of four transcendental equations. The analysis of this system of equations showed that for neutral stratification in both media the ice drift velocity and direction are functions of three non-dimensional parameters: $m_1|f|/\kappa^2\rho_1 G_1$, $G_1/|f|z_{01}$ and z_{01}/z_{02}. Control calculations performed for fixed values of listed parameters demonstrate that the ice drift coefficient c_0/G_1 and the angle between the ice drift direction and isobar are insensitive in practice to variations of z_{01}/z_{02}. From the two remaining parameters the first one $(m_1|f|/\kappa^2\rho_1 G_1)$ determines the velocity and has almost no effect on the direction of ice drift; the second one $(G_1/|f|z_{01})$ does the opposite.

The description of the models of the atmospheric and ocean PBL system would be incomplete if we did not note the following two circumstances. As is known, the water–air interface oscillates near its mean position in such a way that the different points of the horizontal plane in the vicinity of the mean sea level will be in air and in water simultaneously. The further the considered plane is located from the free surface the larger the changes in the ratio between areas of water and air. If we now recollect the difference between statistical regimes of turbulence in both media, then it becomes absolutely unclear what meaning we should put on the operation of averaging in the zone of water and air contact. It is clear only that the regularities of the processes in this zone cannot be described in the same way as for the atmospheric and ocean PBLs.

In this connection, in a number of works (see, for example, Laikhtman *et al.*, 1968) the concept of the *transitional layer* has been introduced, within the limits of which the interface oscillates. Since within this layer all physical characteristics are subjected to considerable changes it cannot be 'contracted to plane' and we cannot expect the continuity conditions to be met. Obviously, this obstacle can be bypassed by using the moveable coordinates consistent with the water–air interface configuration (this is justified only in the case where wind waves do not break), or by identification of the transitional layer with the two-phase liquid layer in the regime of the developed turbulence. Only the future will show whether these, or some other more radical methods, will be successful or not.

Next, in the models examined of the atmospheric and ocean PBL system it was suggested that the turbulent energy in the upper ocean layer is generated only due to the interaction of Reynolds stresses with shears of mean velocity. Because of this the condition of vanishing of the diffusive turbulent energy flux was taken into account as a boundary condition at the ocean surface. There is another viewpoint, which was presented for the first time by Kitaigorodskii (1970), according to which the main mechanism for the production of turbulent energy in the upper ocean layer is diffusion of turbulent energy from the transitional layer where the energy of wind waves transforms into turbulent energy as a result of their breaking. Thus, there are two alternative ideas about the mechanism of turbulent energy generation in the upper ocean layer. Is it crucial to elucidate where the truth lies? Certainly, it is most convenient to take an intermediate position stipulating immediately that both the above-mentioned ideas are equivalent. But let us try to understand which arguments lead to a revision of the traditional presentations. There are two such arguments: (1) either the energy E_w transmitted from wind to waves is equal to the energy E_d used to supply drift currents, or at the developed waves it is much larger than the latter; (2) within the framework of Ekman's model of the PBL (shear flow) it is impossible to obtain an estimate of the turbulent energy dissipation typical of the upper layer, and therefore, if all energy transmitted from wind to waves in the regime of developed waves dissipates, then only turbulence of wave origin is able to provide a proper level of dissipation. We have doubts about the second argument, though the first one also needs verification.

So, we will consider Ekman's model of the PBL. Within the framework of this model drift currents are described by the following equations:

$$
\left.
\begin{aligned}
&\frac{d}{dz_2} k_{M2} \frac{du_2}{dz_2} + fv_2 = 0, \qquad \frac{d}{dz_2} k_{M2} \frac{dv_2}{dz_2} - fu_2 = 0, \\
&k_{M2}\, du_2/dz_2 = -(\rho_1/\rho_2)u_{*1}, \qquad k_{M2}\, dv_2/dz_2 = 0 \text{ at } z_2 \Rightarrow 0, \\
&u_2, v_2 \Rightarrow 0 \text{ at } z_2 \Rightarrow \infty,
\end{aligned}
\right\} \qquad (4.4.21)
$$

where it is assumed that the x axis is directed along the tangential wind stress.

We derive an energy equation and then integrate it over z_2 from 0 to h_2. As a result we obtain

$$
\frac{\rho_1}{\rho_2} u_0 u_{*1}^2 - \int_0^{h_2} k_{M2}\left[\left(\frac{du_2}{dz_2}\right)^2 + \left(\frac{dv_2}{dz_2}\right)^2\right] dz_2 = 0.
$$

Here the first term on the left describes the energy supply from wind to current; the second term describes turbulent energy production in the ocean PBL.

We will assume that the ocean is stratified neutrally and that the diffusive turbulent energy flux at the ocean surface is equal to zero. Then from the turbulent energy budget equation (4.4.8) it follows that

$$
\hat{\varepsilon}_2 = (\rho_1/\rho_2)u_0 u_{*1}^2/h_2. \qquad (4.4.22)
$$

Substitution of the expression $h_2 = 0.1(\rho_1/\rho_2)^{1/2}C_u u_a/|f|$, as well as the relation $u_{*1}^2 = C_u u_a^2$ and values of $\rho_1/\rho_2 = 1.3 \times 10^{-3}$, $C_u = 1.2 \times 10^{-3}$, $u_a = 10\text{--}15$ m/s, $|f| = 10^{-4}$ s^{-1}, into (4.4.22) yields the estimate $\hat{\varepsilon}_2 \approx (2.5\text{--}5.5) \times 10^{-2}$ cm^2/s^3. The typical value of $\hat{\varepsilon}_2$ for the upper ocean layer (see Phillips, 1977) has the same order of magnitude. The order of $\hat{\varepsilon}_2$ cannot be changed, even considering the fact that only part (40–90%) of the total momentum flux τ is consumed to supply the drift current (see Kitaigorodskii, 1970). Not only the magnitude but also the vertical distribution of the turbulent energy dissipation obtained within the framework of Ekman's model of the PBL is in good agreement with experimental data. All this points to the fact that the large production and strong dissipation of the turbulent energy observed in the upper ocean layer are not necessarily connected with the existence of the diffusive turbulent energy flux from the transitional layer. Moreover, this flux can be considered as equal to zero if one assumes, following Stewart and Grant (1962), that all small-scale turbulence appearing in the transitional layer due to wind-wave breaking completely dissipates here.

However, the momentum transfer from wind to waves, and from them (due to breaking) to the drift currents does not always occur locally: such a transfer takes place only in short waves, while long waves can carry their momentum a long distance before dissipating. Because of this the curious situation where two diametrically opposed viewpoints lead to similar results in estimation of turbulent energy dissipation arises.

5

Large-scale ocean–atmosphere interaction

5.1 Classification of climatic system models

No one doubts that the study of any physical phenomenon should be based on observational data where possible. But let us imagine that there are no such data, or not enough of them. What to do then? Of course, one can always put off the solution of the problem in hand until better times when the volume of empirical information will be sufficient and conclusions following its analysis will obtain credit. But what if we cannot afford the luxury of carrying out costly field measurements, or the time scale of phenomena to be examined is great (say, comparable with the duration of one generation or even the whole of humanity)? Finally, what should we do if the results of the phenomenon that we need to know are not permissible in principle (for example, in the case where there is a problem of clarifying the consequences of a nuclear conflict)? In such situations there is only one answer – to use mathematical models of the phenomena to be investigated.

Mathematical models of the climatic system represent the complex of hydrothermodynamic equations describing the state of separate subsystems complemented by proper initial and boundary conditions and parametrizations for those physical processes which, whether owing to imperfections in our knowledge, or owing to the limitations of available technical and economical means, or both of these, cannot be solved explicitly. All the current models of the climatic system are divided into analytical and numerical, into deterministic and stochastic, into hydrothermodynamic and thermodynamic (energy balance), into zero-dimensional, 0.5-dimensional (box), one-dimensional, two-dimensional (including zonal) and three-dimensional. This, as any other, classification is conventional to a certain extent. Its purpose is only to classify models by method of solution (analytical and numerical), by whether an ensemble of climatic system states are taken into account or not (stochastic and deterministic), by the way of reproducing the ordered large-scale circulation (immediately, in hydrothermodynamical models, and in parametrized form, in thermodynamical ones), and, finally, by the number of independent variables – spatial coordinates (zero-dimensional, 0.5-dimensional, etc., models).

We explain in more detail the principle of model division into deterministic and stochastic. As has already been mentioned (see Section 1.2), the climatic system has an extremely wide spectrum of time scales from a fraction of a second, for turbulence, to hundreds of millions of years for the cyclic reorganization of convective motions in the Earth's mantle and for continental drift. Now there is no one acceptable model that can simulate this diversity of temporal variability of the climatic system and take into account the real geography of continents and oceans. Apparently, there will be no such model in the near future. We should take this into consideration and compromise by reducing the range of time scales studied or replacing a direct description of processes which cannot be resolved by a parametric one. This is precisely what happens in short-term processes. As for long-term processes, their determining parameters are usually fixed. This last circumstance can lead to distortion in the response of slow inertial links of the climatic system to external forcing.

But when rejecting the fixing of parameters of inertial links of the climatic system it not only creates the problem of closure, similarly to that in turbulence theory, but also leads to the appearance of additional terms in model equations that describe random disturbances. Let $\mathbf{x} = (x_i)$ be a vector of a fast component state, and $\mathbf{y} = (y_i)$ be a vector of a slow component state of the climatic system (say, the atmosphere and the ocean), and the first one of them is specified by the typical time scale t_{0x}, the second by t_{0y} connected with t_{0x} by the inequality $t_{0x} \ll t_{0y}$. The latter means that in the spectrum of climatic variability the fast and slow components are separated relative to each other. In this case the evolution of the climatic set can be described by the following set of equations:

$$\frac{\mathrm{d}x_i}{\mathrm{d}t} = f_i(\mathbf{x}, \mathbf{y}), \qquad\qquad (5.1.1)$$

$$\frac{\mathrm{d}y_i}{\mathrm{d}t} = g_i(\mathbf{x}, t), \qquad\qquad (5.1.2)$$

where f_i and g_i are known non-linear functions of $\mathbf{x}$ and $\mathbf{y}$.

We consider the evolution of the climatic system on time scales $t \gg t_{0x}$. It is natural in this case to attempt to reduce the set (5.1.1) and (5.1.2) to one equation for the slow variable $\mathbf{y}$, averaging Equation (5.1.2) in time with a period much larger than t_{0x} but much smaller than t_{0y}. In doing so we obtain

$$\frac{\overline{dy_i}}{dt} = \overline{g_i(\mathbf{y}, \mathbf{x})}, \tag{5.1.3}$$

where the overbar symbolizes time averaging.

We replace averaging over time by averaging over an ensemble of all possible realizations of the fast variable $\mathbf{x}$ at a fixed value of the slow variable $\mathbf{y}$ and parametrize the variable $\mathbf{x}$ appearing in $\langle g_i(\mathbf{y}, \mathbf{x}) \rangle$ (here angle brackets mean probability averaging) in terms of $\mathbf{y}$. First we recollect that $\overline{g_i(\mathbf{y}, \mathbf{x})}$ differs from $\langle g_i(\mathbf{y}, \mathbf{x}) \rangle$ by the fluctuation term g_i' describing the contribution to the dispersion of the variable $\mathbf{y}$ from all high-frequency disturbances of the variable $\mathbf{x}$ and disturbances from the range $t_{0x} \ll t \ll t_{0y}$ of the spectral minimum, that is, disturbances which are determined by the interaction of the fast and slow components and, together with high-frequency distur- bances, form white noise. Thus, parametrizing $\langle g_i(\mathbf{y}, \mathbf{x}) \rangle$ in the form $\langle g_i(\mathbf{y}, \mathbf{x}) \rangle = \tilde{g}_i(\mathbf{y}) + g_i'$, we arrive at the equation

$$dy_i/dt = \tilde{g}_i(\mathbf{y}) + g_i'. \tag{5.1.4}$$

From (5.1.4) it follows that the evolution of the slow component of the climatic system from any initial state will be described not by one curve in a phase space of parameters y_i, but by a family of curves complying with all possible random disturbances g_i'. It is precisely this fact which is the main difference between deterministic and stochastic models. The former does not take into account the existence of random disturbances at all, the latter takes it into account in one form or another.

Let us discuss the basic ideas and principles of construction of deterministic models which have played an extremely important role in the climate theory and have still not exhausted their usefulness. One of the restrictions of these models is the difficulty of interpreting results obtained in the absence of numerical experiments carried out for various combinations of internal and external model parameters. The similarity theory of the circulation of planetary atmospheres developed by Golitsyn (1973) and then modified by Zilitinkevich and Monin (1977) as applied to the case of the global ocean– atmosphere interaction may be quite useful in this respect.

5.2 Similarity theory for global ocean–atmosphere interaction

As has been mentioned, the primary cause of atmospheric motions is the flux $q = (1 - \alpha_p)(\pi a^2 / 4\pi a^2) S_0$ of the absorbed short-wave solar radiation (here S_0 is the solar constant, α_p is the planetary albedo, see below). Besides q, a set of the external parameters determining the atmospheric circulation includes the radius a of the planet, the angular velocity Ω of its rotation, the mass $m_A = p_s/g$ of the atmospheric column of a unit cross-section, the gravity g related to m_A and pressure p_s at the low boundary of the atmosphere by the relationship $g = p_s/m_A$, the specific heat of air at constant pressure c_p, the specific heat of air at constant volume c_v defined by the difference $(c_p - c_v) = R/\mu$ (where R is the universal gas constant), the relative air molecular mass μ, and, finally, the Stefan–Boltzmann constant σ or its reduced value $f\sigma$ (here f is the emittance) determining the outgoing flux of long-wave radiation for a prescribed temperature.

Golitsyn suggested that these eight parameters, which all, with the exception of μ, are dimensional, completely determine the global properties of the atmosphere. But because the dimensional parameters contain four independent dimensions (length, time, mass and temperature), then, on the basis of the π-theorem of the dimensional analysis, it is possible to form only the following three dimensionless combinations:

$$\Pi_\Omega = \frac{a}{2L} = \frac{a\Omega}{c_r}, \qquad \Pi_g = \frac{H}{a} = \frac{c_r^2}{\kappa g a},$$

$$\Pi_m = \frac{\tau_p}{\tau_r} = (f\sigma)^{3/8} c_p^{-3/2} q^{5/8} \frac{a}{m_A},$$

where $L = c_r/2\Omega$ and $H = c_r^2/\kappa g$ are the horizontal scale of synoptic disturbances and the height of the homogeneous atmosphere; $c_r = (\kappa R T_r/\mu)^{1/2}$ is the sound velocity; $\kappa = c_p/c_v$ is the ratio of heat capacities; $T_r = (q/f\sigma)^{1/4}$ is the average temperature of the outgoing emission; $\tau_p = a/c_r$ is the relaxation time of pressure disturbances; $\tau_r = c_p m_A T_r/q$ is the relaxation time of the local radiative equilibrium.

We agree that the angular velocity Ω of planetary rotation is only included in all formulae via Π_Ω; the gravity g – only via Π_g; and mass m_A of the atmospheric column of the unit cross-section – only via Π_m. We also exclude from the examination the dimensionless parameter κ, assuming it to be equal to a numerical constant. Then any characteristic F of the atmospheric circulation has to have the form

$$F = F_0(f\sigma, c_p, q, a)\psi_F(\Pi_\Omega, \Pi_g, \Pi_m),$$

where F_0 is the combination with dimension F from the parameters $f\sigma$, c_p, q and a; ψ_F is the dimensionless function of non-dimensional combinations Π_Ω, Π_g and Π_m.

Let $\Pi_\Omega \ll 1$, $\Pi_g \ll 1$ and $\Pi_m \ll 1$. In this case, according to Golitsyn, the function ψ_F can be replaced by its limiting value (equal to unity) corresponding to zero values of Π_Ω, Π_g and Π_m. The assumption about the existence of the finite limit of the function ψ_F (in other words, *complete self-similarity* by parameters Π_Ω, Π_g and Π_m) represents one more hypothesis forming the basis of the theory. An application of this hypothesis allows us to obtain a simple formula for the total kinetic energy E of the atmospheric circulation. Indeed, for the given assumptions,

$$E = 2\pi(f\sigma)^{1/8}c_p^{-1/2}q^{7/8}a^3, \tag{5.2.1}$$

where the factor 2π is introduced for the sake of convenience.

On the other hand, $E = \frac{1}{2}(4\pi a^2 m_A)V^2$, where V is the root-mean-square velocity of atmospheric motion. Hence, in accordance with (5.2.1)

$$V = (2\pi)^{1/2}m_A^{-1/2}(f\sigma)^{1/8}c_p^{-1/2}q^{7/8}a^2. \tag{5.2.2}$$

Next, if the advective heat transport in the atmosphere is balanced by the long-wave radiation emitted into space, that is,

$$m_A c_p V \frac{\delta T}{\pi a/2} \approx (f\sigma)T_r^4, \tag{5.2.3}$$

where δT is the characteristic equator-to-pole temperature difference, then after substitution of the expressions for V and T_r into (5.2.3) we have

$$\delta T = \frac{1}{4}(2\pi)^{1/2}m_A^{-1/2}(f\sigma)^{-1/8}c_p^{-1/2}q^{1/8}a^{-1}. \tag{5.2.4}$$

We note now that for the Earth's conditions the inequality $\Pi_\Omega \ll 1$ is not valid (for the Earth, $\Pi_\Omega = 1.43$). Nevertheless, an agreement between estimates obtained with the help of (5.2.1), (5.2.2) and (5.2.4) and observed data turned out to be satisfactory. This demonstrates that the similarity theory may be used in this case, too.

In order to take into account the existence of two media – the atmosphere and ocean – and the processes of interaction between them Zilitinkevich and Monin proposed modification of the similarity theory as follows. Firstly, they replaced Equation (5.2.3) by the heat budget equation for the ocean–atmosphere system averaged over time throughout the annual cycle and over an area of the sphere surface between the equator and some fixed latitude (say, $\varphi = \pi/4$). This equation, on the assumption that there is no heat transport through the equator, is written in the form

$$(MHT_A + MHT_O) = R_A, \tag{5.2.5}$$

where

$$MHT_A = c_p m_A \frac{\overline{vT}}{a}$$

is the meridional heat transport in the atmosphere at latitude $\varphi = \pi/4$, referred to a unit area of the sphere belt $0 < \varphi < \pi/4$; MHT_O is the meridional heat transport in the ocean normalized in the same way; R_A is the average (in the examined belt) net radiation flux at the upper boundary of the atmosphere; v is the meridional component of wind velocity; an overbar means averaging over the circle of latitude $\varphi = \pi/4$.

Second, the net radiation flux $R_A(\varphi) = F^\downarrow(\varphi) - F^\uparrow(\varphi)$ at the upper boundary of the atmosphere is set proportional to the difference in the temperature T_s at the underlying surface for $\varphi = \pi/4$ and its global average value T_0. Substitution of the expressions for the absorbed short-wave solar radiation flux $F^\downarrow(\varphi) = (4/\pi)q\cos\varphi$ and for the outgoing long-wave radiation flux $F^\uparrow(\varphi) = q(1 + b_0(T_s - T_0)/T_0)$ into the relationship $R_A(\varphi) \sim (T_s - T_0)$ yields

$$T_s - T_0 = \delta T_0(\cos\varphi - \pi/4), \tag{5.2.6}$$

$$R_A(\varphi) = q\left(1 - \frac{\pi b_0}{4}\frac{\delta T_0}{T_0}\right)\left(\frac{4}{\pi}\cos\varphi - 1\right). \tag{5.2.7}$$

Here $\delta T_0 = (T_E - T_P)$ is the annual mean difference of the underlying surface temperatures at the equator (T_E) and the pole (T_P); $b_0 = 2.6$ is the numerical constant.

At $\varphi = 0$ and $\varphi = \pi/2$ from (5.2.6) it follows that

$$T_E - T_0 = \delta T_0(1 - \pi/4); \qquad T_P - T_0 = -\frac{\pi}{4}\delta T_0, \tag{5.2.8}$$

and from (5.2.7), after averaging over the domain $0 < \varphi < \pi/4$ and using the ratio

$$T_r/T_0 = 0.879, \tag{5.2.9}$$

connecting the average temperature T_r of the outgoing emission with the global average temperature T_0 of the underlying surface , it appears that

$$R_A = b_1\left(1 - b_2\frac{\delta T_0}{T_r}\right), \tag{5.2.10}$$

where $b_1 = 0.16$ and $b_2 = 1.8$ are the new numerical constants; values of these and all other constants mentioned below are borrowed from Zilitinkevich and Monin (1977).

Thirdly, when describing the meridional heat transport MHT_A in the atmosphere Zilitinkevich and Monin suggested that the main contribution to the average product $\overline{vT}$ is made by the correlation of synoptic velocity and temperature disturbances, and that the basic mechanism of these disturbances in the stratified atmosphere is the baroclinic instability of the zonal flow. In this case $\overline{vT} \sim \sigma_T \sigma_v$, where $\sigma_v \sim L\hat{u}/a$, $\sigma_T \sim L\delta T_0/a$ are the root mean-square disturbances of the velocity and temperature; $L \sim \Pi_T^{1/2} c_r/\Omega$ is the horizontal scale of baroclinic disturbances proportional to Rossby's deformation radius; $\Pi_T = ((\kappa - 1)/\kappa)(1 - \gamma/\gamma_a) \approx 0.1$ is the non-dimensional combination describing the vertical atmospheric stratification; γ_a and γ are the adiabatic and actual vertical temperature gradients in the atmosphere.

Substituting $\overline{vT}$ into MHT_A and using the definitions of Π_m and Π_Ω we obtain

$$MHT_A = \frac{b_3 \Pi_T}{\Pi_m \Pi_\Omega^2} \frac{\hat{u}\delta T_0}{c_r T_r} q, \tag{5.2.11}$$

where $b_3 \approx 12$ is one more numerical constant.

The mass-weighted vertical average zonal wind velocity $\hat{u}$ appears in (5.2.11). To find it, the difference analogue of the thermal wind formula

$$\hat{u} = \frac{b_4 c_r}{\Pi_\Omega} \frac{\delta T_0}{T_r} \tag{5.2.12}$$

is used. It follows from the geostrophic relation with an allowance for the hydrostatic equation, the Clapeyron equation and the approximate equality $\delta p/p \approx \delta T/T$; here p is the atmospheric pressure, $b_4 = 0.35$ is the numerical constant.

Finally, in estimating the meridional heat transport MHT_O in the ocean it is assumed that on average throughout the year the total amount of heat entering the ocean area being examined has to be balanced by the meridional transport, that is, $(2/3)Q_0 = MHT_O$ where Q_0 is the resulting heat flux at the ocean–atmosphere interface referred to the unit area $2\pi a^2 \sin \pi/4$ of the zonal belt of $0 < \varphi < \pi/4$; the factor $2/3$ is the fraction of the ocean in the zonal belt (the relative ocean area in the domain being examined).

It is also anticipated that the seasonal and latitude variations of water temperature encompass not the entire ocean thickness but only the upper active layer, and that the vertical difference in temperatures at the surface and at the low boundary of the ocean active layer in the zonal belt $0 < \varphi < \pi/4$ is proportional to the horizontal temperature difference $\delta T_w = (T_E - T_I)$, where T_I is the sea water freezing temperature. Considering these assumptions the vertical heat flux is defined as $Q_0/\rho_0 c_0 = k_H \delta T_w/h$, where h is the thickness of the active layer; k_H is the effective coefficient of the vertical heat diffusion including the proportionality factor between the vertical and horizontal temperature differences in the ocean; $\rho_0 c_0$ is the volume heat capacity of sea water.

The expression for $Q_0/\rho_0 c_0$ written above contains two new unknowns k_H and h. The first one is found from the condition of the existence of strongly stable stratification within the active layer. In this case the flux Richardson's number

$$Rf = \frac{\beta Q_0}{\rho_0 c_0}\left(\frac{u_{*w}^4}{k_M}\right)^{-1}$$

and the ratio $\alpha_H = k_H/k_M$ approach their limiting values, and the combination $\alpha_H Rf$ – to the constant Rf_0, which is approximately equal to 5×10^{-4}. Here u_{*w} is the friction velocity in the next-to-surface ocean layer, k_M is the coefficient of vertical viscosity, $\beta = g\alpha_T$ is the buoyancy parameter, α_T is the coefficient of sea water thermal expansion.

The equality $\alpha_H Rf = Rf_0$, together with the equation for Q_0 yields $k_H = (Rf_0 h/\beta\delta T_w)^{1/2}u_{*w}^2$. To determine the second unknown the equation $\partial h/\partial t = b_5 k_H/h$ (here $b_5 = 17$ is the numerical constant) is used. This equation describes the growth of the active layer thickness due to the entrainment of the water from the deep layer during the cold six-month season.

In accordance with this equation the characteristic value of h is

$$h = (b_5 k_H t_0)^{1/2}, \tag{5.2.13}$$

where t_0 is the annual period.

Now when h is known the equations for k_H, $Q_0/\rho_0 c_0$ and MHT_0 can be rewritten in the form

$$\left.\begin{aligned}
k_H &= \left(\frac{Rf_0 u_{*w}^4}{\beta\delta T_w}\right)^{2/3}(b_5 t_0)^{1/3}, \\[2mm]
Q_0/\rho_0 c_0 &= \left(\frac{Rf_0 u_{*w}^4}{\beta b_5 t_0}\right)^{1/3}(\delta T_w)^{2/3}, \\[2mm]
MHT_0 &= \tfrac{2}{3}\rho_0 c_0\left(\frac{Rf_0 u_{*w}^4}{\beta b_5 t_0}\right)^{1/3}(\delta T_w)^{2/3}.
\end{aligned}\right\} \tag{5.2.14}$$

It only remains to find an expression for u_{*w} and a relation between δT_w and δT_0. The following equality is used to estimate u_{*w}:

$$u_{*w} = \left(C_u \frac{\rho}{\rho_0}\hat{u}^2\right)^{1/2}, \tag{5.2.15}$$

arising from the continuity condition for momentum fluxes at the ocean–atmosphere interface. Here C_u is the resistance coefficient of the sea surface; ρ is the air density.

As for the relationship between δT_w and δT_0, the condition $\delta T_w = \delta T_0$ is apparently valid in the absence, and $\delta T_w = (T_E - T_I)$ in the presence, of ice in the polar latitudes. This condition, together with (5.2.5) and (5.2.7)–(5.2.15), forms a closed set to determine the climatic characteristics of the atmosphere $(T_0, \delta T_0, T_E, T_P, \hat{u}, MHT_A)$ and the ocean $(\delta T_w, u_{*w}, k_H, MHT_O, Q_0, h)$. For example, for the present-day period, when the Antarctic is covered with ice and the Arctic is almost isolated from the World Ocean, the theory yields the following estimates: $T_0 = 17\,°C$, $\delta T_0 = 36\,°C$, $T_E = 27\,°C$, $T_P = -9\,°C$, $\hat{u} = 10$ m/s, $MHT_A = 244$ W/m^2, $\delta T_w = 27\,°C$, $u_{*w} = 1.4$ cm/s, $k_H = 1.7$ cm^2/s, $MHT_O = 84$ W^2/m, $Q_0 = 126$ W/m^2, $h = 300$ m. In general, these are in good agreement with the empirical data.

We turn now to a discussion of general ideas on the construction of deterministic models. We start with the simplest ones in the hierarchy of the model position assigned for the *low-parametric models*, that is, the models describing the climatic system by means of a small number of parameters.

5.3 Zero-dimensional models

The simplest model is a model with one parameter – the temperature T_r of the outgoing terrestrial emission. It is defined by the condition of radiative equilibrium of the planet characterizing a balance between the absorbed short-wave solar radiation flux $\pi a^2 S_0 (1 - \alpha_p)$ and the outgoing long-wave emission flux $4\pi a^2 f\sigma T_r^4$, that is,

$$\tfrac{1}{4} S_0 (1 - \alpha_p) = f\sigma T_r^4, \tag{5.3.1}$$

where, along with the known symbols, α_p is the mean weighted planetary albedo defined by the expression $\alpha_p = \tfrac{1}{2} \int_{-1}^{1} S(x)\alpha(x)\,dx$; $x = \sin\varphi$; $\alpha(x)$ is the local albedo at latitude φ (or at fixed x); $S(x)$ is the annual mean distribution of solar radiation at the upper atmospheric boundary normalized so that the integral of $S(x)$ from 0 to 1 is equal to 1. The following formula serves as a good approximation for $S(x)$ (see North *et al.*, 1981)

$$S(x) \approx 1 + S_2 P_2(x), \tag{5.3.2}$$

where $S_2 = -0.477$ is a numerical constant; $P_2(x) = (3x^2 - 1)/2$ is the second Legendre polynomial. In accordance with (5.3.2) the annual mean distribution of solar radiation at the upper atmospheric boundary is a parabolic function of x, having zero derivative at the equator ($x = 0$) and taken to be 0.523 at the pole ($x = 1$).

From (5.3.1) it follows for $\alpha_p = 0.30$ that $T_r = 254.6$ K, that is, the temperature of the outgoing terrestrial emission turns out to be much less than the observed global average temperature of the surface atmospheric layer, amounting to for 287.4 K. The cause of this discrepancy is clear: it is the greenhouse effect of the atmosphere which is not taken into account, and which is formed by absorption of the long-wave emission of the Earth's surface and of the adjoining air layer primarily by water vapour and carbon dioxide. To describe the feedback between the flux I of the outgoing long-wave emission and the average global air surface temperature T_0 we take advantage of the approximation $I = A + BT_0$ (see Budyko, 1969) where $A = 203.3$ W/m^2, $B = 2.09$ W/m^2 °C are empirical constants which implicitly take into account the effects of cloudiness and active admixtures in the atmosphere of the Northern Hemisphere. Equating the flux I to the flux of the absorbed solar radiation we have

$$\tfrac{1}{4}S_0(1 - \alpha_p) = A + BT_0, \tag{5.3.3}$$

from which with $\alpha_p = 0.30$ and the above-mentioned values of the numerical constants A and B we obtain the estimate $T_0 = 14.97$ °C which in practice coincides with the observed average (for the Northern Hemisphere) surface air temperature $T_0 = 14.9$ °C.

We now consider the influence of the feedback between the planetary albedo and surface air temperature. In doing so we assume that the planetary albedo is equal to 0.30 if the planet has no snow and ice, and 0.62 if it is covered with snow and ice. After substitution of these values into (5.3.3) we find that $T_0 = 15$ °C in the first, and $T_0 = -36.4$ °C in the second case. The Earth occupies an intermediate position between these two extreme cases. We assume that the southern boundary of the ice–snow cover is consistent with $x = x_s$, and that $x_s = 1$ at $T_0 > 15$ °C, $x_s = 0$ at $T_0 < -15$ °C, and $x_s = 1 + (T_0 - 15)/30$ at -15 °C $\le T_0 \le 15$ °C. Then designating the local albedo as α_j for $0 < x < x_s$ we obtain from the definition of α_p and (5.3.3)

$$H_0(x_s(T_0)) \equiv 1 - \alpha_p = 1 - \int_0^1 \alpha(x)S(x)\,dx$$

$$= 1 - \alpha_i + (\alpha_i - \alpha_j)[x_s - \tfrac{1}{2}S_2(x_s - x_s^3)], \tag{5.3.4}$$

where $H_0(x_s(T_0))$ is the global coefficient of the short-wave solar radiation absorption (the planetary co-albedo).

For the present-day value of the solar constant S_0 Equation (5.3.3) has three solutions. The first one complies with the present climate, the second one – with the climate when about 30% of the planet's area is covered by ice, the third one – with the climate when the planet is completely covered by ice (the case of the *white Earth*).

We examine the linear stability of all the above-mentioned solutions of Equation (5.3.3). First we write down a proper non-stationary heat budget

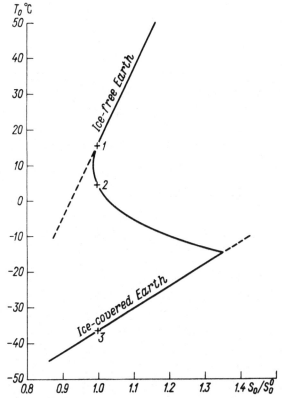

Figure 5.1 Dependence of the global average surface air temperature on S_0/S_0^0 for the zero-dimensional model, according to North *et al.* (1981).

equation. Considering (5.3.4), this takes the form

$$c\frac{\mathrm{d}T_0}{\mathrm{d}t} + I(T_0) = \tfrac{1}{4}S_0H_0(x_s(T_0))$$
(5.3.5)

and in the limit (as $t \Rightarrow \infty$) reduces to an algebraic (with respect to T_0) equation which for the above-mentioned approximations of I and $H_0(x_s(T_0))$ describes the dependence of T_0 on S_0/S_0^0 presented in Figure 5.1. Here S_0^0 is the present-day value of the solar constant, c is the heat capacity of the climatic system in question.

We define T_0 as $T_0^0 + \delta T_0$, where T_0^0 is the solution of the stationary equation; δT_0 is a small departure from it. Then we express $I(T_0)$ and $H(x_s(T_0))$ as a Taylor series expansion keeping only the first terms containing δT_0. Then after a substitution $T_0 = T_0^0 + \delta T_0$; $I(T_0) = I(T_0^0) + (\partial I/\partial T_0)_{T_0^0}\,\delta T_0$; $H(x_s(T_0)) = H_0(x_s(T_0^0)) + (\partial H_0/\partial T_0)_{T_0^0}\,\delta T_0$ into (5.3.5) and subtraction of the stationary equation

$$I(T_0^0) = \tfrac{1}{4}S_0H_0(x_s(T_0^0)),$$
(5.3.6)

we have

$$c \frac{\mathrm{d}}{\mathrm{d}t} \delta T_0 + \left[\left(\frac{\partial I}{\partial T} \right)_{T_0^0} - \tfrac{1}{4} S_0 \left(\frac{\partial H_0}{\partial T_0} \right)_{T_0^0} \right] \delta T_0 = 0. \tag{5.3.7}$$

We transform the expression in square brackets and to do this we differentiate (5.3.6) with respect to T_0. As a result we have

$$(\partial I/\partial T_0) = (H_0/4)(\partial S_0/\partial T_0) + (S_0/4)(\partial H_0/\partial T_0)$$

whence $[(\partial I/\partial T_0) - (S_0/4)(\partial H_0/\partial T_0)]_{T_0^0} = (H_0/4)(\partial S_0/\partial T_0)_{T_0^0}$. Substitution of this relationship into (5.3.7) yields

$$c \frac{\mathrm{d}}{\mathrm{d}t} \delta T_0 + \frac{1}{4} \left(\frac{\partial S_0}{\partial T_0} \right)_{T_0^0} H_0 \delta T_0 = 0. \tag{5.3.8}$$

This ordinary first-order differential equation has an exponential solution with an index $(H_0/4c)(\partial S_0/\partial T_0)_{T_0^0}$. Hence, the tendency for δT_0 towards zero (stability), or to recede from it (instability), should be determined exclusively by the sign of the combination $(H_0/4c)(\partial S_0/\partial T_0)_{T_0^0}$. But because c and H_0 are always positive then the stability or instability of the stationary solution has to depend on the sign of the derivative $(\partial S_0/\partial T_0)_{T_0^0}$. Thus the following theorem applies (see Cahalan and North, 1979): *the stationary state of the climate system described by Equation (5.3.3) will be stable when* $(\partial S_0/\partial T_0)_{T_0^0} > 0$, *and unstable when* $(\partial S_0/\partial T_0)_{T_0^0} < 0$. The last inequality is valid if the negative feedback between the outgoing long-wave emission and surface air temperature is weaker than the positive feedback between the planetary albedo and surface air temperature, that is, if $(\partial I/\partial T_0)_{T_0^0} < \tfrac{1}{4} S_0(\partial H_0/\partial T_0)_{T_0^0}$ (see Equation (5.3.7)).

So we arrive at the following two conclusions on the basis of the considerations mentioned above: (1) the climatic regime complying with point 2 in Figure 5.1 is unstable in the sense that for small changes in the solar constant it converts into the regime of the ice-free Earth or into the regime of the ice-covered (white) Earth; (2) the climatic regimes which have global average surface air temperature above 10 °C or below −15 °C are stable with respect to small variations in the solar constant. All this is applied to both the present-day climate which is characterized by partial glaciation of the Earth and climates of the ice-covered and ice-free Earth.

One-dimensional models of a latitudinal structure where only one dependent variable – surface air temperature – is considered to be a function of latitude are close, in the ideological sense, to the zero-dimensional models. Therefore, to cover the succession of ideas we now discuss models, and only after this do we consider the richer, from the viewpoint of their physical content, 0.5-dimensional (box) models.

5.4 One-dimensional models

We distinguish the zonal belt of the unit meridional extension in the climatic system. The respective heat budget equation for annual mean conditions has the form

$$I(x) + \nabla \cdot MHT(x) = \tfrac{1}{4}S_0 S(x) a(x, x_s), \qquad (5.4.1)$$

where the first term on the left-hand side describes, as before, the outgoing long-wave emission, the second term describes the divergence of the meridional heat transport integrated over the thickness of the atmosphere and the ocean; the term on the right-hand side describes the short-wave solar radiation absorbed by the climatic system; $a(x) = [1 - \alpha(x)]$ is the local absorption coefficient; other symbols are the same.

We supplement Equation (5.4.1) with representation of each function in terms of the annual mean zonally averaged surface air temperature $T(x)$ and appropriate boundary conditions at the poles. We assume that the meridional heat transport is zero at the equator and the poles. To determine $I(x)$ and $a(x)$ we use the approximations proposed by Budyko (1969), that is, we assume that $I(x) = A + BT(x)$, $a(x, x_s) = a_i$ at $T(x) < T_s$ and $a(x, x_s) = a_j$ at $T(x) > T_s$, where, as in the preceding section, A and B are numerical constants, a_i and a_j are absorption coefficients for the short-wave solar radiation in ice-covered and ice-free latitude zones; $T_s = -10\,°C$ is the annual mean surface air temperature at the boundary ($x = x_s$) between these latitude zones.

Before dealing with parametrization of the meridional heat transport divergence, we consider, according to North *et al.* (1981), two extreme cases where $MHT(x)$ is equal to infinity and zero. In the first case the surface air temperature at all latitudes must remain identical, that is, the one-dimensional model will reproduce a situation similar to the one described by the zero-dimensional model. The respective solution complying with the one-dimensional model has to coincide with that shown in Figure 5.1, the only difference being that in the range of temperature changes from $-15\,°C$ to $+15\,°C$ the solid curve in Figure 5.1 is now replaced by the dashed lines which are an extension of the solid lines representing solutions in the regime of the ice-free (dark) and ice-covered (white) Earth. The physical interpretation of the broken curve at the phase plane $(T, S_0/S_0^0)$ is apparent: for a decrease in the solar constant there is a reduction of the surface air temperature of the isothermal planet. The latter remains dark up to the point where the surface air temperature reaches $-10\,°C$, following which the planet suddenly becomes white. The new (appropriate for the regime of the white planet) surface air temperature is $-44\,°C$.

In the second limiting case (zero meridional heat transport) Equation (5.4.1) reduces to the form

$$A + BT(x) = \tfrac{1}{4}S_0 S(x) a(x, x_s). \tag{5.4.2}$$

If $a(x, x_s)$ at $x = x_s$ takes an average value between a_i and a_j, then, applying Equation (5.4.2) to the boundary between ice-covered and ice-free latitude zones, we obtain $S_0(x_s) = 4(A + BT_s)/S(x_s)\bar{a}$, where $\bar{a} = (a_i + a_j)/2$.

This relation defines the value of the solar constant S_0 providing the setting of the southern ice boundary at one fixed latitude or another. Calculated results presented in Figure 5.2 demonstrate quite a curious situation: if we agree with current theories on the Sun's evolution and, in accordance with these, assume that the solar constant at the early stages of the Earth's geological history was less by 20–40% than its present-day value, then the question arises as to why the whole Earth was not covered by ice as predicted, at present, by all the available one-dimensional climatic models? Several explanations can be proposed for this paradox, among them water shortage for creation of the ice cover, and the existence of unconsidered negative feedbacks determined, for example, by a change in cloudiness and gaseous composition of the atmosphere, etc.

We note in this connection that the results of the calculation were obtained

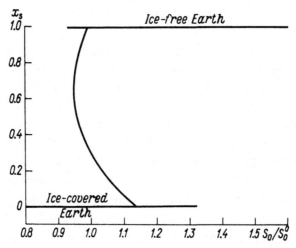

Figure 5.2 Dependence of the southern ice edge location on S_0/S_0^0 for the one-dimensional model (transport approximation), according to North *et al.* (1981).

with an implicit assumption of the universality of the numerical constants A and B which, as a matter of fact, have to be changed on the geological time scale at least, due to changes in the content of absorbing substances in the atmosphere. It is also necessary to take into account the fact that one-dimensional models with zero meridional heat transport are not able to reproduce the real climatic system. Indeed, it may be seen from Figure 5.2 that the present position of the southern boundary of the polar glaciation ($x_s = 0.95$) is met by the solar constant which exceeds its contemporary value by 70%. Moreover, for the given values S_0/S_0^0 and x_s the global average surface air temperature turns out to be equal to 93 °C, and at $x = x_s$ a temperature discontinuity of the order of 50 °C occurs (see (5.4.2)). It is clear that without taking the meridional heat transport into account it is impossible to eliminate discrepancies in the observed data.

Two methods for parametrization of the meridional heat transport divergence within the framework of one-dimensional models are known. The first one (the so-called *transport approximation*) was proposed by Budyko (1969) (see also Budyko, 1980), the second one (the *diffusive approximation*) was proposed by Faegre (1972) and North (1975). We discuss these parametrizations at length in order to establish their peculiarities and general features.

Transport approximation. For annual mean conditions the meridional heat transport in the climatic system is determined only by the net radiation flux at the upper atmospheric boundary. On the other hand, it has to be correlated with the average (over the atmospheric thickness) air temperature. But the vertical variability of the average air temperature is small compared to the horizontal variability, so that the vertical average temperature has to be correlated with the surface air temperature. In other words, there should be a relationship between the meridional heat transport and the surface air temperature distribution. These considerations serve as the basis for approximation of the meridional heat transport divergence in the form

$$\nabla \cdot MHT(x) = \beta[T(x) - T_0], \tag{5.4.3}$$

where, as before, T_0 is the global average surface air temperature; $\beta = 3.75 \times 10^4$ W/m² °C is an empirical constant.

Substitution of (5.4.3), and the definition of $I(x)$, into (5.4.1) yields

$$A + BT(x) + \beta[T(x) - T_0] = \tfrac{1}{4}S_0 S(x) a(x, x_s). \tag{5.4.4}$$

From here, using the expression

$$A + BT_0 = \tfrac{1}{4}S_0 H_0(x_s), \tag{5.4.5}$$

found by integration of (5.4.4) over x from 0 to 1, we obtain the relationship for the determination of the annual mean zonally averaged temperature in the surface atmospheric layer. This has the form

$$T(x) = -AB^{-1} + \tfrac{1}{4}S_0[S(x)a(x, x_s) + \beta B^{-1}H_0(x_s)]B^{-1}(1 + \beta B^{-1})^{-1}. \quad (5.4.6)$$

It remains to find the location x_s of the southern boundary of the polar glaciation. Using the condition $T(x) = T_s$ at $x = x_s$ we obtain

$$T_s(x) = -AB^{-1} + \tfrac{1}{4}S_0[S(x_s)a(x_s, x_s) + \beta B^{-1}H_0(x_s)]B^{-1}(1 + \beta B^{-1})^{-1}.$$
$$(5.4.7)$$

where $a(x_s, x_s) = (a_i + a_j)/2$ is the average absorption coefficient.

The dependence of x_s on S_0/S_0^0 derived from (5.4.7) is shown in Figure 5.2. We pay attention to the jump-like change of the southern boundary of the polar glaciation after it reaches latitude 50°. The following explains this fact: at the boundary of the polar glaciation ($x = x_s$) the flux of the outgoing long-wave radiation remains constant for fixed T_s. Because of this, when S_0/S_0^0 decreases, the southern boundary moves to the equator (x_s decreases) and, therefore, $S(x)$ increases. Simultaneously, a decrease in the global average surface air temperature occurs and, following this an increase occurs in the meridional heat transport divergence. But as soon as S_0/S_0^0 becomes sufficiently small the meridional heat transport divergence increases more than the flux of the absorbed short-wave solar radiation and the polar glaciation propagates rapidly to the equator.

The reverse sequence of events takes place with an increase in S_0/S_0^0. Nevertheless, the rapid transition from the regime of partial glaciation to the limiting regime (ice-free Earth) is preserved in this case as well.

Diffusive approximation. When identifying the heat content with a passive scalar characteristic the annual mean meridional heat transport will be proportional to the temperature gradient, or, in terms of $T(x)$, to $-(1 - x^2)^{1/2}\,dT(x)/dx$ with a numerical factor called the phenomenological coefficient of heat diffusion. Accordingly, the meridional heat transport divergence will be equal to

$$\nabla \cdot MHT(x) = -\frac{d}{dx}D(1 - x^2)\frac{dT}{dx},$$

where the factor D represents the product of the heat diffusion coefficient by $1/a^2$.

We take advantage of this approximation and rewrite Equation (5.4.1) in the form

$$-\frac{d}{dx} D(1 - x^2) \frac{dT}{dx} + A + BT = \tfrac{1}{4} S_0 S(x) a(x, x_s). \tag{5.4.8}$$

Let the conditions of vanishing of the meridional heat transport be boundary conditions at the equator and the pole, that is,

$$-D(1 - x^2)^{1/2} \frac{dT}{dx}\bigg|_{x = 0,1} = 0. \tag{5.4.9}$$

The first of these conditions is equivalent to an assumption of the symmetry of the surface air temperature field in both hemispheres.

We will find a solution of Equation (5.4.8) with $D = const$ in the form of a series in Legendre polynomials $P_n(x)$. First we recollect that the Legendre polynomials are eigenfunctions of the diffusive operator

$$-\frac{d}{dx} (1 - x^2) \frac{dP_n(x)}{dx} = n(n + 1)P_n(x)$$

and that only those that have even index n satisfy the boundary conditions (5.4.9). Hence, the solution of (5.4.8), (5.4.9) can be presented in the form $T(x) = \sum_n T_n P_n(x)$, $n = 0, 2, \ldots$. We substitute this expansion into (5.4.8), then multiply it by $P_n(x)$ and integrate over x from 0 to 1. Then, using the orthogonality condition $\int_0^1 P_n(x)P_m(x)\, dx = (2n + 1)^{-1}\delta_{nm}$, where δ_{nm} is the Kronecker symbol, we obtain the following relationship for the expansion coefficient T_n:

$$[n(n + 1)D + B]T_n + \delta_{0n} A = \tfrac{1}{4} S_0 H_n, \tag{5.4.10}$$

where $H_n \equiv H_n(x_s) = (2n + 1) \int_0^1 P_n(x)S(x)a(x, x_s)\, dx$ is the polynomial function of the argument x_s tabulated for $a(x, x_s) = a_0 + a_2 P_2(x)$ by North (1975).

Specifically, when $n = 0$ the relationship (5.4.10) reduces to (5.4.5), from where, for the above-mentioned values of the numerical constants A and B and function $H_0(s)$, $T_0 = 14.97\,°C$ follows. For $n = 2$ the value of the function $H_n(x_s)$ is equal to $H_2 \approx -0.5$. Assuming, according to North *et al.* (1981), that $DB^{-1} = 0.310$ we have $T_2 = -28\,°C$ from (5.4.10). If we restrict ourselves to the first two terms of the expansion $T \approx T_0 + T_2 P_2(x)$ – the *two-mode approximation* – and substitute the values of T_0 and T_2, then the agreement between calculated and observed meridional distributions of the annual mean surface air temperature will be good. Of course, the fact that the diffusive parametrization is in close agreement with observed data is impressive. But it should be remembered that this is the result of a lucky choice of parameters of the two-mode approximation rather than of the advantages of the parametrization adopted. By the way, the diffusive parametrization of the meridional heat transport within the limits of the two-mode approximation coincides with that proposed by Budyko (1969). Indeed, substituting the definition of T into the expression for the diffusive operator we obtain with $D = const$ and $n = 2$

$$-\frac{d}{dx}\left[D(1 - x^2)\frac{dT}{dx}\right] = 6DT_2P_2(x) = 6D[T(x) - T_0],$$

from here and from the definition of T it is found that $\beta = 6D$.

To determine x_s we follow North (1975). Instead of $T(x)$ we introduce a new dependent variable – the outgoing long-wave emission flux $I(x)$. Then Equation (5.4.8) is rewritten in the form

$$-\frac{d}{dx}D'(1 - x^2)\frac{dI}{dx} + I = \tfrac{1}{4}S_0S(x)a(x, x_s), \tag{5.4.11}$$

where $D' = D/B$.

Next, we present I in the form of an even Legendre polynomial series and make the same transformations as in the derivation of the expression for T, that is, we substitute the expansion into (5.4.11), and then multiply the equation obtained by $P_n(x)$ and integrate over x from 0 to 1. As a result we will have $[n(n + 1)D' + 1]I_n = (S_0/4)H_n(x_s)$. Substitution of this relationship into $I = \sum_n I_nP_n(x)$, $n = 0, 2, \ldots,$ yields, at $x = x_s$,

$$I(x_s) = \tfrac{1}{4}S_0 \sum_n \frac{H_n(x_s)P_n(x_s)}{n(n + 1)D' + 1}, \quad n = 0, 2, \ldots, \tag{5.4.12}$$

from which the position of the southern boundary of the polar glaciation is found.

The dependence of x_s on S_0^0 is shown in Figure 5.3. Under its construction in expansion (5.4.12) the terms with subscript n from 0 to 6 inclusive have been retained. It can be seen that in the vicinity of $S_0/S_0^0 = 1$ the function x_s has several stationary solutions. In particular, there are five solutions for $S_0/S_0^0 \approx 0.98$, which agree, respectively, with the ice-free Earth ($x_s = 1$), small ($x_s \approx 0.99$), intermediate ($x_s \approx 0.88$), and large ($x_s \approx 0.30$) area of the polar glaciation, and with the ice-covered Earth ($x_s \approx 0$). The stability of these solutions can be verified similarly as for the zero-dimensional model, that is, by means of the addition of the term $c\, \partial I(x, t)/\partial t$ with the heat capacity coefficient c to the left-hand side of Equation (5.4.11) and linearization of the non-stationary equation with respect to small disturbances of the stationary solution. Having performed such transformations North *et al.* (1981) showed that the climatic regime corresponding to the intermediate area of the polar glaciation, as well

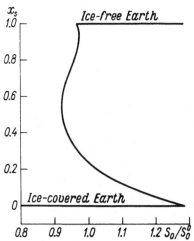

Figure 5.3 Dependence of the southern ice edge location on S_0/S_0^0 for the one-dimensional model (diffusion approximation), according to North *et al.* (1981).

as the regimes of ice-free and ice-covered Earth, are stable, and that the remaining regimes belonging to segments of the curve in Figure 5.3 with negative values of the derivatives $\partial(S_0/S_0^0)/\partial x_s$ are unstable. This suggests that the last two regimes cannot be realized in the process of modern climate evolution.

The most striking difference between Figure 5.2 and Figure 5.3 takes place in the regime of total glaciation when $S_0/S_0^0 > 1.1$ and in the regime of ice-free Earth when $S_0/S_0^0 < 1.1$: the diffusive approximation predicts the branching of solutions, the transport approximation predicts their merging. Also, on the diffusive approximation the derivative $\partial(S_0/S_0^0)/\partial x_s$ for $x_s \approx 0.95$ changes sign, so a gradual increase in S_0/S_0^0 from $S_0/S_0^0 = 1$ is accompanied at first by smooth and then, after the derivative $\partial(S_0/S_0^0)/\partial x$ passes the point of changing sign, by an abrupt increase in x_s (decrease in the polar glaciation area) reminiscent of a jump-like transition to the regime of ice-free Earth. According to North *et al.* (1981), the marked peculiarity of the solution disappears when taking into account the dependence of the diffusion coefficient on the meridional temperature gradient and when retaining higher modes in the expansion (5.4.12), and therefore it should be qualified only as artefact, stipulated by the limitation of assumptions forming the basis of the model.

5.5 0.5-dimensional (box) models

In Section 1.1 we showed that the evolution of the surface air temperature is determined not only by one (inherent to the atmosphere) time scale but depends crucially on the time scales of the upper mixed layer (UML) and deep ocean layer (DOL) interacting with each other and with the atmosphere. This consideration is the basis of the box thermodynamic model suggested by Kagan *et al.* (1990) and is intended to simulate the seasonal variability of the climatic system of the Northern Hemisphere.

The seasonal time scale of variability predetermines the necessity of singling out the UML as an independent subsystem with all consequences arising with respect to the description of the processes at the UML–DOL interface.

We approximate the seasonal thermocline separating the UML and the DOL by a temperature jump, and single out two periods within the annual cycle – periods of falling and rising of the lower boundary of the UML. We will also suppose that an increase in UML thickness occurs exclusively owing to the entrainment of colder water from the DOL. The last condition is equivalent to the assumption that the eddy heat flux at the lower boundary of the UML differs from zero (it is used for heating water entrained from below) when UML thickness increases, and otherwise is equal to zero when entrainment is absent. Vanishing of this heat flux means that in a period of rising of the lower boundary of the UML any heat exchange between the UML and DOL stops and within the UML a new (shallower) thermocline is formed preventing heat penetration into the lower part of the UML. But isolation of part of the UML and its joining with the DOL are tantamount to the assignment of an equivalent heat flux at the upper boundary of the DOL defined by the condition of heat conservation in the UML–DOL system. In other words, within the framework of the adopted parametrization for the heat exchange at the UML–DOL interface, the heat transfer from the UML into the DOL occurs only in the period of rising, and is absent in the period of falling, of the lower boundary of the UML.

Another factor controlling the thermal regime of the DOL when considering this layer as a whole is an ordered vertical transfer of cold deep water created by upwelling. According to Munk (1966) the balance of these two factors – the heat transfer from above and the cold transfer from below – provides the existence of the main thermocline in the ocean. But owing to the law of mass conservation the vertical velocity integrated over the ocean area must vanish. Hence it is necessary to suggest the existence in the ocean of a downwelling area with a degenerate main thermocline. From general considerations confirmed by the results of laboratory experiments using revolving tanks, it is clear that the most intense downwelling has to be timed to an area of cold deep water formation and that this area and the area of upwelling must be connected to each other by the cold water transport from the first area into the second one in the deep layer, and by an opposite transport of warm water in the surface layer. These are the major prerequisites used in the design of the model in question.

The conditional concepts as to the character of the meridional circulation in the ocean form the basis of the model. That is, it is suggested that the area of upwelling is located in temperate and low latitudes and that in this area the upwelling is regulated by the inflow of cold deep waters from high latitudes and by the sinking of warm surface waters. Calculated results of the meridional heat transport in the ocean (see Section 2.3) and observed spatial distributions of the potential temperature, salinity and phosphate concentration obtained within the framework of GEOSECS and TTO programmes testify to such a meridional circulation cell.

Three areas can be singled out in the ocean – an area of temperate and low latitudes (hereinafter the upwelling area), the area of cold deep water formation in high latitudes and the polar ocean area covered with ice. It is assumed that in the upwelling area, where ocean warming occurs, the UML and DOL are separated by the clearly marked seasonal thermocline preventing heat exchange between them (in the polar ocean this function is performed by the halocline), and in the area of cold deep water formation both layers unite as a result of strong cooling accompanied by the development of intensive convection. Similarly, two zonal areas are singled out in the atmosphere: the area of temperate and low latitudes stretching over the upwelling area and part of the land contiguous with it, and the area of high latitudes located over the remaining part of the land, over the area of cold deep water formation and the polar ocean. As a result the atmosphere and the ocean are represented in the form of a system of seven boxes (Figure 5.4): two atmospheric boxes (northern and southern) and five ocean boxes (the UML and DOL in the upwelling area and in the polar ocean, and the area of cold deep water formation).

The evolutionary equations describing the heat budget in each separate box have the form

$$\frac{d\theta_1}{dt} = F_{T1}^A + \frac{2^{\kappa_1}}{c_p m_A} \sum_{i=1}^{4} (f_1 Q_{T1i}^{LA} + (1 - f_1) Q_{T1i}^{OA}), \qquad (5.5.1)$$

$$\frac{d\theta_2}{dt} = F_{T2}^A + \frac{2^{\kappa_1}}{c_p m_A} \sum_{i=1}^{4} (f_2 Q_{T2i}^{LA} + f_1 Q_{T2i}^{IA} + (1 - f_2 - f_1) Q_{T2i}^{OA}), \quad (5.5.2)$$

$$\frac{d}{dt} T_1 h_1 - T_1 w_e = -(F_{T1}^O + w T_1) - Q_{h_1 - 0}^T + \sum_{i=1}^{4} Q_{T1i}^{OS}/\rho_0 c_0 + J, \quad (5.5.3)$$

$$\frac{d}{dt} T_2(H - h_1) + T_2 w_e = -(F_{T2}^O - w T_0) + Q_{h_1 + 0}^T - J, \qquad (5.5.4)$$

$$\frac{d}{dt} T_0 H = (s_1^O/s_0^O)[(F_{T1}^O + F_{T2}^O) + w(T_2 - T_1)] + \sum_{i=1}^{4} Q_{T0i}^{OS}/\rho_0 c_0$$

$$- \beta_1 L_1 (M_1 + M_s)/\rho_0 c_0 s_0^O + D_0^T/s_0^O, \tag{5.5.5}$$

$$\frac{d}{dt} T_{-1} h_{-1} = Q_{-1}^{OS}/\rho_0 c_0 - Q_{h-1}^T + D_{-1}^T/s_{-1}^O, \tag{5.5.6}$$

$$\frac{d}{dt} T_{-2}(H - h_{-1}) = Q_{h-1}^T + D_{-2}^T/s_{-1}^O, \tag{5.5.7}$$

where the first two equations refer to the southern and northern atmospheric boxes, respectively, the third and fourth refer to the UML and DOL in the upwelling area, the fifth refers to the area of cold deep water formation, and the two last refer to the UML and DOL in the polar ocean.

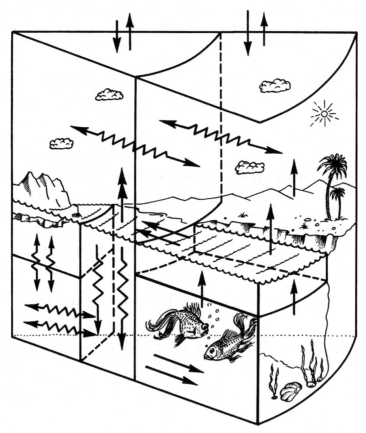

Figure 5.4 Schematic representation of the climatic system in the 0.5-dimensional box model.

We now explain the separate terms in these equations and their meanings. Terms on the left-hand sides of the Equations (5.5.1) and (5.5.2) describe changes in time of the mass-weighted, vertically averaged air potential temperature θ referred to the mean level ($p_s/2$) in the atmosphere; the first terms on the right-hand sides – the meridional transport of sensible heat, the second terms on the right-hand sides – heat sources and sinks in the atmosphere; the first terms on the left-hand sides of Equations (5.5.3)– (5.5.7) – heat content changes in the ocean area examined (here T_1 and T_2 are the UML and DOL temperature in the upwelling area, T_0 is the temperature in the cold deep water formation area, T_{-1} and T_{-2} are the temperatures of the UML and DOL in the polar ocean, h_1 and h_{-1} are UML thickness in the upwelling area and in the polar ocean); the second terms on the left-hand sides describe the effect of the entrainment at the UML–DOL interface; the first terms on the right-hand sides of Equations (5.5.3)–(5.5.5) describe the diffusive and advective heat exchange between the upwelling area and the area of cold deep water formation; the second terms in (5.5.3) and (5.5.4) describe the eddy heat flux at the upper boundary of the UML and the equivalent heat flux at the upper boundary of the DOL in the upwelling area; the third term in (5.5.3) and the second term in (5.5.5) describe the heat exchange at the ocean–atmosphere interface; the penultimate term on the right-hand side of (5.5.5) describes the change of heat content in the area of cold deep water formation due to melting ice and snow exported from the polar ocean; the first two terms on the right-hand side of (5.5.6) describe the heat flux at the water–ice interface and the heat exchange between the UML and DOL in the polar ocean; the last terms in (5.5.5)–(5.5.7) describe the change in the heat content created by trapping of water from a neighbouring ocean area as the result of displacement of the southern boundary of sea ice. Here, in addition to the symbols already specified, H is the ocean depth; M_I and M_s are ice and snow masses in the polar ocean; $f_1 = s_1^L/s_1$ and $f_2 = s_2^L/s_2$ is the ratio between the area s_1^L, s_2^L of land in the southern and northern boxes and the area s_1 and s_2 of these boxes (fractions of continents),[1] $f_i = s_{-1}^O/s_2$ and s_{-1}^O are the fraction and area of sea ice in the northern box, s_0^O is the area of the cold deep water formation domain connected to area s_1^O of the upwelling domain and to the common ocean area s^O in the Northern Hemisphere by the relation $s_0^O = s^O - s_1^O - s_{-1}^O$; $m_A = p_s/g$ is the mass of the unit atmospheric column; p_s is the surface atmospheric pressure; g is gravity; β_I is the dimensional parameter assumed as proportional to M_I with

---

[1] The land in each box is represented in the form of a 'segment' of constant extension in the zonal direction with the northern boundary of the land 'segment' in the northern box coinciding with the latitude circle at 71.6 °N.

proportionality factor β_1'; $\kappa_1 = R/c_p$; R and c_p are the gas constant and the heat capacity of air; ρ_0 and c_0 are density and heat capacity of sea water; subscripts A and O indicate belonging to the atmosphere and the ocean; double subscripts LA, IA, OA, LS and OS indicate belonging to the atmosphere over land, to the atmosphere over sea ice and over the ocean, to the land and ocean surfaces respectively; the meaning of the remaining symbols will be further clarified.

To find M_I and M_s we picture the ice cover in the polar ocean area as a film of finite thickness whose surface can be covered by snow or melted water, and single out three periods within the year cycle. *Winter*: there is snow on the ice surface; the temperature T_s of the active surface (in this case the snow surface) is less than the water freezing temperature T_{SO}; the growth and melting of ice occurs at the ice–water interface. *Spring*: the ice surface is covered by a mixture of snow and melted water; $T_s = T_{SO}$; melting or growth of ice occurs at its low surface. *Summer* and *autumn*. The ice surface is covered by a layer of melted water; the temperature of this layer remains equal to T_{SO}; melting and growth of ice occurs on both surfaces. The beginning and the end of this period are identified, respectively, with periods when the thickness of the snow and melted water layers are zero.

In these situations the evolution equations for the ice mass $M_I = \rho_I h_I s_{-1}^O$, snow $M_s = \rho_s h_s s_{-1}^O$, and melted water $M_w = \rho_w h_w s_{-1}^O$ are written in the form

$$
\frac{dM_I}{dt} = \begin{cases}
-\beta M_I - \dfrac{s_{-1}^O}{L_I}(Q_I - Q_{-1}^{OS}) & \text{for } T_s < T_{SO};\ M_s > 0;\ M_w = 0, \\[2ex]
-\beta M_I - \dfrac{s_{-1}^O}{L_I}(Q_I - Q_{-1}^{OS}) & \text{for } T_s = T_{SO};\ M_s > 0;\ M_w > 0, \quad (5.5.8) \\[2ex]
-\beta M_I - \dfrac{s_{-1}^O}{L_I}(Q_I - Q_{-1}^{OS}) + \dfrac{dM_I'}{dt} & \text{for } T_s = T_{SO};\ M_s = 0;\ M_w > 0.
\end{cases}
$$

$$
\frac{dM_s}{dt} = \begin{cases}
-\beta_1 M_s + s_{-1}^O(P_2 + Q_{-14}^{IS}/L_*) & \text{for } T_s < T_{SO};\ M_s > 0;\ M_w = 0, \\[1ex]
-\beta_1 M_s + dM_s'/dt & \text{for } T_s = T_{SO};\ M_s > 0;\ M_w > 0, \quad (5.5.9) \\[1ex]
0 & \text{for } T_s = T_{SO};\ M_s = 0;\ M_w > 0.
\end{cases}
$$

$$
\frac{dM_w}{dt} = \begin{cases}
0 & \text{for } T_s < T_{SO};\ M_s > 0;\ M_w = 0, \\[1ex]
-\beta_1 M_w + s_{-1}^O(P_2 + Q_{-14}^{IS}/L_*) - dM_s'/dt & \text{for } T_s = T_{SO};\ M_s > 0;\ M_w > 0, \\[1ex]
-\beta_1 M_w + s_{-1}^O(P_2 + Q_{-14}^{IS}/L_*) - dM_I'/dt & \text{for } T_s = T_{SO};\ M_s = 0;\ M_w > 0.
\end{cases}
$$

$$(5.5.10)$$

where the first terms on the right-hand sides describe the ice, snow, and melted water export from the polar basin; the second term in (5.5.8) describes the growth or melting of ice at its low surface; the second terms in the first equation of (5.5.9) and in the two last equations of (5.5.10) describe the increase or decrease in snow and melted water masses due to precipitation and evaporation; the remaining terms describe the change in ice, snow and melted water masses owing to water phase transitions at the upper surface of the ice–snow cover. Here h_I, h_s, and h_w are the ice, snow and melted water layer thicknesses; ρ_I, ρ_s, and ρ_w are their densities, P_2 and Q^{IS}_{-14} are the precipitation and latent heat flux; L_* is the heat of sublimation equal to $L + L_I$ when $T_s < T_{SO}$, and to L when $T_s \geq T_{SO}$; L and L_I are latent heat of condensation and melting; the double subscript IS indicates belonging to the upper surface of the ice–snow cover.

The required values of T_s, dM'_s/dt and dM'_I/dt are obtained from the heat budget equation for the upper surface:

$$\sum_{i=1}^{4} Q^{IS}_{-1} - Q_I = \begin{cases} 0 & \text{for } T_s < T_{SO};\ M_s > 0;\ M_w = 0, \\[2mm] -\dfrac{L_I}{s^{O}_{-1}}\dfrac{dM'_s}{dt} & \text{for } T_s = T_{SO};\ M_s > 0;\ M_w > 0, \\[2mm] -\dfrac{L_I}{s^{O}_{-1}}\dfrac{dM'_I}{dt} & \text{for } T_s = T_{SO};\ M_s = 0;\ M_w > 0, \end{cases} \tag{5.5.11}$$

where, as in (5.5.3) and (5.5.5), the term containing the symbol $\sum$ represents the sum of fluxes of the absorbed short-wave ($i = 1$) and long-wave ($i = 2$) radiation, as well as of sensible ($i = 3$) and latent ($i = 4$) heat; the second term on the left-hand side of (5.5.11) describes the vertical heat flux in the ice–snow cover determined by the condition of the quasi-stationarity of the temperature in the form

$$Q_I = \begin{cases} (T_s - T_{-1})(h_I/\lambda_I + h_s/\lambda_s)^{-1} & \text{for } T_s < T_{SO};\ M_s > 0;\ M_w = 0, \\[2mm] (T_s - T_{-1})(h_I/\lambda_I + h_s/\lambda_s + h_w/\lambda_w)^{-1} & \text{for } T_s = T_{SO};\ M_s > 0;\ M_w > 0, \\[2mm] (T_s - T_{-1})(h_I/\lambda_I)^{-1} & \text{for } T_s = T_{SO};\ M_s = 0;\ M_w > 0. \end{cases} \tag{5.5.12}$$

Here λ_I, λ_s and λ_w are ice, snow and liquid water thermal conductivity.

We determine the vertical heat flux Q^{OS}_{-1} at the low ice surface. For this purpose we assume that the UML thickness in the polar ocean remains constant in time, and temperature T_{-1} is equal to the sea water freezing temperature. Then on the basis of (5.5.6) we have

$$Q^{OS}_{-1}/\rho_0 c_0 = Q^{T}_{h-1} - D^{T}_{-1}/s^{O}_{-1}. \tag{5.5.13}$$

This equation is supplemented by the expressions

$$Q^{T}_{h-1} = \kappa_0(T_{-1} - T_{-2}), \tag{5.5.14}$$

$$D^T_{-1} = \begin{cases} \gamma_0 h_{-1}(T_0 - T_{-1})\, ds^o_{-1}/dt & \text{for } ds^o_{-1}/dt > 0, \\ 0 & \text{for } ds^o_{-1}/dt \leq 0, \end{cases}$$

$$D^I_{-2} = \begin{cases} (H - h_{-1})(T_0 - T_{-2})\, ds^o_{-1}/dt & \text{for } ds^o_{-1}/dt > 0, \\ 0 & \text{for } ds^o_{-1}/dt \leq 0, \end{cases} \qquad (5.5.15)$$

$$D^T_0 = \begin{cases} 0 & \text{for } ds^o_{-1}/dt > 0, \\ -[(H - h_{-1})(T_0 - T_{-2}) + (1 + \gamma_0)h_{-1}(T_0 - T_{-1})]|ds^o_{-1}/dt| \\ \hspace{5cm} \text{at } ds^o_{-1}/dt \leq 0, \end{cases}$$

for the heat flux Q^T_{h-1} at the UML–DOL interface and the rate of heat content change in the polar ocean (D^T_{-1}, D^T_{-2}) and in the area of cold deep water formation (D^T_0) due to trapping of water from the neighbouring areas, and also by the relationship

$$h_I/h_{Im} = 1 - \exp[-(s^o_{-1}/s^o_{-1m})^{1/2}], \qquad (5.5.16)$$

connecting the sea ice thickness h_I and area s^o_{-1} and complying with the conditions $h_I \Rightarrow 0$ as $s^o_{-1} \Rightarrow 0$ and $h_I \Rightarrow h_{Im}$ as $s^o_{-1} \Rightarrow \infty$, where h_{Im} and s^o_{-1m} are limiting values of the sea ice thickness and area (the latter in this case is equal to the ocean area s^o in the Northern Hemisphere); κ_0 is a dimensional factor; γ_0 is the non-dimensional constant characterizing the ratio between the amount of heat trapped by the polar ocean from the area of cold deep water formation and the amount of heat consumed for sea ice melting (it is assumed that the remaining trapped heat remains in the area of cold deep water formation and is expended immediately in an increase in the heat exchange between the ocean and the atmosphere in the northern box as this takes place in reality when leads are available.

We direct our attention to the determination of the thickness h_1 of the UML, as well as of the eddy heat flux $Q^T_{h_1-0}$ at the low boundary of the UML, and the equivalent heat flux $Q^T_{h_1+0}$ at the upper boundary of the DOL in the upwelling area. When it is assumed, as was mentioned above, that the flux $Q^T_{h_1-0}$ determined by entrainment of water from the DOL into the UML differs from zero only in periods of UML deepening, and that cutting of part of the UML accompanying the formation of a new thermocline and its integration with the DOL are tantamount to an assignment of equivalent flux $Q^T_{h_1+0}$ at the upper boundary of the DOL defined by the condition of heat conservation in the UML–DOL system, then

$$Q^T_{h_1-0} = \begin{cases} (T_1 - T_2)w_e & \text{for } w_e > 0, \\ 0 & \text{for } w_e \leq 0, \end{cases} \qquad (5.5.17)$$

$$Q_{h_1+0}^{\mathrm{T}} = \begin{cases} 0 & \text{for } w_e > 0, \\ -(T_1 - T_2)w_e & \text{for } w_e \leq 0, \end{cases} \tag{5.5.18}$$

where $w_e = (dh_1/dt + w)$ is the entrainment velocity, w is the velocity of vertical motions in the upwelling area.

To estimate h_1 we use the turbulent energy budget equation integrated within UML limits. At the same time we suppose that the integral dissipation and generation of turbulent energy of mechanical and convective origin are proportional to each other, and that the turbulent energy of mechanical origin does not propagate beyond the limits of the Ekman boundary layer. Then

$$\left.\begin{aligned} \frac{dh_1}{dt} + w &= -\frac{C_1}{T_1 - T_2} - \left[1 - \frac{C_2}{C_1}\mathrm{Rf}^{-1}\delta\left(\frac{h_1}{h_E}\right)\right]\sum_{i=1}^{4} Q_{T1i}^{\mathrm{OS}}/\rho_0 c_0 \\ &\quad\text{for } \mathrm{Rf} \leq \frac{C_2}{C_1}\delta\left(\frac{h_1}{h_E}\right), \\ \frac{h_1}{h_E} &= 1 - \frac{C_1}{C_2}\mathrm{Rf} \quad \text{for } \mathrm{Rf} \geq 0, \; w_e < 0, \end{aligned}\right\} \tag{5.5.19}$$

where $\mathrm{Rf} = -ga_T(h_1/u_*^3)\sum_{i=1}^{4} Q_{T1i}^{\mathrm{OS}}/\rho_0 c_0$ is the flux Richardson number; $h_E = u_{*1}/C_3|f|$ is the thickness of the Ekman boundary layer in the ocean; $\delta(h_1/h_E)$ is the function equal to 1 as $h_1 < h_E$, and otherwise to 0; u_{*1} is the wind friction velocity; f is the Coriolis parameter, α_T is the coefficient of sea-water thermal expansion; C_1, C_2 and C_3 are non-dimensional numerical constants.

The velocity w of vertical motions in the upwelling area appearing here and above can be considered as a model parameter. But it is possible to express it differently in terms of the meridional temperature difference in the system of ocean boxes (say, by means of $(T_2 - T_1)$), that is, to assume, following Stommel (1960), that not only the processes of cooling in the area of cold deep water formation participate in the formation of the deep circulation but also the processes which control the dynamics of currents in temperate and low latitudes. In this case

$$w = C_4(T_2 - T_0)/a\delta\varphi^0, \tag{5.5.20}$$

where a is the Earth's radius; $\delta\varphi^0$ is a difference in latitudes of the ocean boxes; C_4 is a dimensional factor.

As is known, in box ocean models the spatial variability of the UML is usually neglected. Meanwhile, the comparison of mean (over the North Atlantic area) values of temperature in the UML calculated from local and space-averaged data show that this assumption leads to systematic understating of the UML temperature. To eliminate this disadvantage we parametrize the spatial correlation $J = h_1\langle(\sum_{i=1}^{4} Q_{T1i}^{\mathrm{OS}}/\rho_0 c_0 - Q_{h_1-0}^{\mathrm{T}})\rangle$ (here angle brackets signify spatial averaging) in the form

$$J = C_5 h_1(T_1 - T_2), \tag{5.5.21}$$

where C_5 is one more dimensional constant. We also add the same term (but with inverse sign) in the equation for the DOL in order to ensure the condition of heat conservation in the system of ocean boxes.

To parametrize the diffusive heat transport F_T^O in the ocean determined by the correlation between synoptic disturbances of the meridional components of velocity and temperature, we assume that the meridional and zonal components of the current velocity are proportional to each other (this assumption is justified for blocking of zonal flows), and that the meridional mass transport from the UML into the area of cold deep water formation, and from this area to the DOL, should be associated with w by the condition of mass conservation. Then

$$\begin{pmatrix} F^O \\ F_{T2}^O \end{pmatrix} = \frac{w\Lambda}{a\delta\varphi^O} \begin{pmatrix} C_6(T_1 - T_0) \\ C_6'(T_2 - T_0) \end{pmatrix}, \tag{5.5.22}$$

where Λ is the baroclinic radius of deformation, C_6 and C_6' are non-dimensional numerical constants.

We parametrize the meridional heat transport F_T^A in the atmosphere in accordance with Stone (1972). Normalizing this transport to the atmospheric box area we have

$$\begin{pmatrix} F_{T1}^A \\ F_{T2}^A \end{pmatrix} = \begin{pmatrix} -C_7 \dfrac{\cos\varphi_\Gamma}{\sin\varphi_\Gamma} \\ \\ C_7 \dfrac{\cos\varphi_\Gamma}{1 - \sin\varphi_\Gamma} \end{pmatrix} |\theta_1 - \theta_2| \frac{\theta_1 - \theta_2}{a\delta\varphi^A}, \tag{5.5.23}$$

where $\delta\varphi^A$ is the difference in latitude of the atmospheric boxes; φ_Γ is the latitude of the boundary between them; C_7 is a dimensional constant.

Heat sources and sinks $\sum_{i=1}^{4} Q_i^A$ (the first superscripts are omitted here) are presented in the form

$$\left. \begin{array}{ll} Q_1^A = \chi(1 - \alpha_A)Q, & Q_2^A = \sigma[\nu T_s^4 - (\nu^\downarrow + \nu^\uparrow)T_A^4], \\ Q_3^A = -Q_3^S, & Q_4^A = L_* P, \end{array} \right\} \tag{5.5.24}$$

where $Q = S_0/4$ is the flux of the short-wave solar radiation at the upper atmospheric boundary; S_0 is the solar constant; T_s and T_A are values of the absolute temperature at the underlying surface (in the case of the ocean surface T_s is equal to T_1 or T_0) and at the mean level in the atmosphere; χ is the short-wave radiation absorption coefficient; $v^\uparrow$ and $v^\downarrow$ are upward and downward long-wave radiation emissivity; v is the coefficient of long-wave radiation absorption; α_A is the atmospheric albedo; σ is the Stefan–Boltzmann constant; as before, P is the amount of precipitation falling per unit area of the underlying surface per unit of time.

The unknown function P is obtained from the condition of instantaneous precipitation of the moisture excess determined by the difference between evaporation and the meridional water vapour transport by means of synoptic disturbances. This condition is written in the form

$$\begin{pmatrix} P_2 \\ P_2 \end{pmatrix} = m_A \begin{pmatrix} -C_9' \dfrac{\cos \varphi_r}{\sin \varphi_r} \\[2mm] C_9 \dfrac{\cos \varphi_r}{1 - \sin \varphi_r} \end{pmatrix} |\theta_1 - \theta_2| \dfrac{q_1 - q_2}{a \delta \varphi^A}$$
$$- \begin{pmatrix} L^{-1}(f_1 Q_{14}^{LS} + (1 - f_1)Q_{14}^{OS}) \\ L_*^{-1}(f_2 Q_{24}^{LS} + f_1 Q_{-14}^{IS}) + L^{-1}(1 - f_2 - f_1)Q_{04}^{OS}) \end{pmatrix}, \quad (5.5.25)$$

where C_9 is a dimensional constant; the composition of precipitation is identified by the temperature of the underlying surface: if it is less than the water freezing temperature T_{SO} then it is snow that falls, otherwise it is rain.

To find q we use the standard power dependence of specific humidity on pressure p. As a result, we arrive at the relationship

$$\begin{pmatrix} q_1 \\ q_2 \end{pmatrix} = \begin{pmatrix} q_{a1}(1 + \kappa_{21})^{-1} \\ q_{a2}(1 + \kappa_{22})^{-1} \end{pmatrix}, \quad (5.5.26)$$

where $q_{a1} = f_1 q_{a1}^L + (1 - f_1)q_{a1}^O$, $q_{a2} = f_2 q_{a2}^L + f_1 q_{a2}^I + (1 - f_2 - f_1)q_{a2}^O$ are the average (within the limits of southern and northern boxes) values of specific humidity at some fixed level (say, at $p/p_s = 0.9985$) in the surface atmospheric layer; superscripts L, I and O correspond, respectively, to the air surface layer over land, sea ice and ocean; κ_{21} and κ_{22} are numerical constants. Over any type of underlying surface q_a can be found using the expression $q_a = r_a q_m(T_a)$, where $q_m(T_a)$ is the maximum specific humidity defined by the Clausius–Clapeyron equation; T_a and r_a are the temperature and specific humidity at the level $p/p_s = 0.9985$.

The relationship (5.5.26) includes one more unknown, T_a. We will assume that the potential temperature is a linear function of pressure. Then the absolute temperature T_a at the level $p/p_s = 0.9985$ is defined by the expression

$$T_a = (p/p_s)^{\kappa_1}[2^{\kappa_1 + 1}(1 - p/p_s)T_A + (1 - 2p/p_s)T_s],$$

representing the local dependence between the surface air temperature and the underlying surface temperature. This dependence is valid over land, ice and the upwelling area but is violated over the relatively small area of cold deep water formation, an effect associated with cold air transport from a neighbouring continent and with deep ocean mixing. This peculiarity can be considered in the framework of a box model if the surface air temperature over the area of cold deep water formation is assumed to be described by the interpolation formula:

$$T_a = \beta_a[f_2 T_a^L + f_1 T_a^I + (1 - f_2 - f_1)T_a^O] + (1 - \beta_a)T_a^O, \qquad (5.5.27)$$

where β_a is the numerical factor equal to zero as $T_s^L > 273$ K, and 1 as $T_s^L \leq 273$ K.

We now consider the heat fluxes Q_i^S appearing in (5.5.3), (5.5.5), (5.5.9), (5.5.10), (5.5.19) and (5.5.24). We discuss the flux Q_1^S of the short-wave solar radiation absorbed by the underlying layer, as well as the net flux Q_2^S of the long-wave radiation at the underlying surface, and eddy fluxes of sensible Q_3^S, and latent Q_4^S heat. Let us represent them in the form

$$\left. \begin{aligned} &Q_1^S = (1 - \chi)(1 - \alpha_s)(1 - \alpha_A)Q, \; Q_2^S = \sigma(v^\downarrow T_A^4 - T_s^4), \\ &Q_3^S = \rho_a C_u^{1/2} u_*(T_a - T_s), \\ &Q_4^S = \rho_a L_* C_u^{1/2} u_* r_s(c_1(T_a - T_s) - c_2), \end{aligned} \right\} \qquad (5.5.28)$$

where $c_1 = \mathrm{d}q_m/\mathrm{d}|t|_{T=T_a}$, $c_2 = (1 - r_a/r_s)q_m(T_a)$; α_s is the albedo of the underlying surface; C_u is the resistance coefficient; ρ_a is the air density.

We define the relative humidity (r_s) at the underlying surface and the relative humidity in the surface atmospheric layer as $r_s = W/W_c$ and $r_a = r_* W/W_c$, where W is the moisture content and W_c is its critical value for which evaporation is equal to potential evapotranspiration (evaporability); r_* is a numerical constant, and then we recollect that the climatic system contains four types of underlying surface: the sea surface, the sea ice surface, the snow-covered ice and the snow-free land surfaces. We assume that $W/W_c = 1$ for the first three types of underlying surface and that the critical moisture content W_c and soil moisture capacity W_0 are connected with each other by the relation $W_c = 0.75W_0$ (see Manabe and Bryan, 1969). Next, using data presented by Korzun (1974) we assume that in the case where the temperature T_s of the land surface is higher than the water freezing temperature T_{SO} and the land moisture content W is less than the moisture capacity W_0, a certain part of precipitation (say, $\gamma_R(W/W_0)(P/\rho_0)$, here P/ρ_0 is the thickness of the precipitation layer in water equivalent; ρ_0 is the fresh water density, γ_R is a numerical factor) is consumed for run-off formation. Meanwhile the remaining part is used for land moisture content change. But if $W \geq W_0$, then the whole moisture excess resulting from the difference between precipitation and evaporation is transformed into run-off. Hence, for the southern box, where the temperature of the land surface is always higher than T_{SO}, we have

$$\frac{dW_1}{dt} = \begin{cases} \left(1 - \gamma_{\text{R1}} \dfrac{W_1}{W_{01}}\right) \dfrac{P_1}{\rho_0} + \dfrac{Q_{14}^{\text{LS}}}{\rho_0 L} & \text{for } W_1/W_{01} < 1, \\ 0 & \text{for } W_1/W_{01} \geq 1, \end{cases}$$

$$R_1 = \begin{cases} \gamma_{\text{R1}} \dfrac{W_1}{W_{01}} \dfrac{P_1}{\rho_0} & \text{for } W_1/W_{01} < 1, \\ \dfrac{P_1}{\rho_0} + \dfrac{Q_{14}^{\text{LS}}}{\rho_0 L} & \text{for } W_1/W_{01} \geq 1, \end{cases} \tag{5.5.29}$$

where $Q_{14}^{\text{LS}} = (W_1/W_{c1})LE_1^{\text{LS}}$ as $W_1 < W_{c1}$, or $Q_{14}^{\text{LS}} = LE_1^{\text{LS}}$ as $W_1 \geq W_{c1}$ is the latent heat flux at the land surface in the southern box; E_1^{LS} is the potential evapotranspiration; R_1 is the run-off referred to a unit area of the land surface.

We now take into account the fact that the land surface in the southern box belongs to different watersheds. Because of this R_1 can be presented as the sum of the run-off R_{12} over the watershed s_{12}^{L} of the Arctic Ocean and the run-off R_{11} over the watershed s_{11}^{L} of all other oceans. Meanwhile,

$$R_{11} = R_1 \frac{s_{11}^{\text{L}}}{s_1^{\text{L}}}, \qquad R_{12} = R_1 \frac{s_{12}^{\text{L}}}{s_1^{\text{L}}}, \tag{5.5.30}$$

where $s_1^{\text{L}} = (s_{11}^{\text{L}} + s_{12}^{\text{L}})$ is the land area in the southern box.

To estimate s_{11}^{L} and s_{12}^{L} we assume that the land area s_2^{L} in the northern box falls to the Arctic Ocean watershed only. Since the total area s_{A} of this basin consisting of s_{12}^{L} and s_2^{L} is known, then $s_{11}^{\text{L}} = s^{\text{L}} - s_{\text{A}}$, $s_{12}^{\text{L}} = s_{\text{A}} - s_2^{\text{L}}$, where it is assumed that $s_2^{\text{L}} < s_{\text{A}}$. Substitution of these relations into (5.5.30) yields the final expressions for the determination of R_{11} and R_{12} by the value of R_1 obtained.

The land surface in the northern box may be covered with snow or it may be free from snow, and the temperature of the land surface may be higher or lower than the water freezing temperature T_{SO}. Below we describe three periods in the annual run-off cycle:

Winter. There is snow on the land; the temperature of the underlying surface (the snow cover in this case) is lower than T_{SO}; there is no water filtration into the soil; the soil moisture content does not change; the run-off is equal to the water inflow from the southern box. In this case

$$\frac{dW_2}{dt} = 0; \qquad R_2 = R_1 s_{12}^L / s_2^L; \qquad \frac{dM_s^L}{dt} = s_2^L P_2 + s_2^L Q_{24}^{LS}/L_*$$

$$\text{for } T_s < T_{SO}, M_s^L > 0, \quad (5.5.31)$$

where $M_s^L = \rho_s h_s s_2^L$ is the mass of snow at the land surface in the northern box.

Spring high water. The temperature T_s of the snow cover is equal to T_{SO}; the soil moisture content does not change as before; the run-off is formed due to snow melting, precipitation, evaporation and water inflow from the southern box, that is,

$$\left.\begin{aligned} &\frac{dW_2}{dt} = 0; \qquad R_2 = R_1 \frac{s_{12}^L}{s_2^L} - \frac{1}{\rho_0 s_2^L}\frac{dM_s^L}{dt} + \frac{P_2}{\rho_0} + \frac{Q_{24}^{LS}}{\rho_0 L}; \\ &-\frac{L_1}{s_2^L}\frac{dM_s^L}{dt} = \sum_{i=1}^{4} Q_{2i}^{LS} - Q_{2s} \qquad \text{for } T_s = T_{SO}, M_s > 0, \end{aligned}\right\} \quad (5.5.32)$$

where Q_{2s} is the heat flux in the snow cover defined by the equation

$$Q_{2s} = (T_{SO} - \bar{T}_s)(h_s/\lambda + h_L/\lambda_L)^{-1}; \quad (5.5.33)$$

$\bar{T}_s$ is the annual mean temperature of the land surface in the northern box; h_L is the thickness of the active land layer in this box; λ_L is the soil thermal conductivity.

It is assumed that the spring high water ends at the moment when the snow mass M_s^L at the land surface in the northern box vanishes.

Summer and autumn. The temperature of the land surface remains higher than T_{SO}; part of the precipitation at $W_2 < W_{02}$ is consumed for run-off formation; another part is used for the change in the soil moisture content; otherwise (as $W_2 \geq W_{02}$) all moisture excess transforms into run-off. Thus,

$$\frac{dW_2}{dt} = \begin{cases} \left(1 - \gamma_{R2}\dfrac{W_2}{W_{02}}\right)\dfrac{P_2}{\rho_0} + \dfrac{Q_{24}^{LS}}{\rho_0 L} & \text{for } W_2/W_{02} < 1, \\ 0 & \text{for } W_2/W_{02} \geq 1, \end{cases}$$

$$R_2 = \begin{cases} \gamma_{R2}\dfrac{W_2}{W_{02}}\dfrac{P_2}{\rho_0} + \dfrac{s_{12}^L}{s_2^L}R_1 & \text{for } W_2/W_{02} < 1, \\ \left(\dfrac{P_2}{\rho_0} + \dfrac{Q_{24}^{LS}}{\rho_0 L}\right) + \dfrac{s_{12}^L}{s_2^L}R_1 & \text{for } W_2/W_{02} \geq 1, \end{cases} \qquad (5.5.34)$$

Q_{24}^{LS} is defined in the same way as Q_{14}^{LS}.

During this period the temperature of the land surface in the northern box, as well as the temperature of the land surface in the southern box throughout the year, is obtained from the heat budget equation for the underlying surface written in the form (the first subscript is omitted)

$$\sum_{i=1}^{4} Q_i^{LS} - Q_s = 0, \tag{5.5.35}$$

where $Q_s = (T_s - \bar{T}_s)(h_L/\lambda_L)^{-1}$ is the heat flux in the active land layer.

It remains to determine the areas of the northern and southern boxes. We assume that the boundary between them coincides with the annual mean isotherm $-4.5\,°C$, which location is defined by the approximation of the meridional distribution of the annual mean surface air temperature by the first two terms of the series in even Legendre polynomials. The last condition, together with (5.5.1)–(5.5.35), forms a closed system to estimate all the climatic characteristics of the Northern Hemisphere that we are interested in.

We now begin discussion of the results of a numerical experiment on the simulation of natural seasonal variability (Figures 5.5–5.7). As can be seen, the qualitative features of the seasonal variability of the climatic system in the Northern Hemisphere are reproduced correctly. In particular, we note a phase shift between air and land surface temperature variations in the northern and southern boxes, and also between water and air temperature variations in both boxes; the absence of marked seasonal variations of water temperature in the area of cold deep water formation, and in the DOL, and their presence in the UML in the upwelling area; a decrease in duration of the period of heating and an abrupt increase in convective heat exchange at the ocean surface during the cold half of the year in the area of cold deep water formation; an increase in amplitudes of seasonal oscillations of the surface air temperature and land surface temperature in the northern box compared with their values in the southern box; an increase in the atmospheric moisture content in summer and its decrease in winter; a phase shift between oscillations of evaporation from the land and ocean surfaces in the southern box that is connected with the delay in timing of maximum and minimum values of the sea surface temperature; a significant seasonal variability in evaporation in the area of cold deep water formation; an increase in precipitation in the southern box accompanying an enhancement of local evaporation; an opposite change in precipitation during the warm half of the year in both boxes and, as a consequence of a summer decrease in evaporation, the appearance of spring and autumn precipitation maxima in the northern box.

We pay attention to the existence of a positive difference between

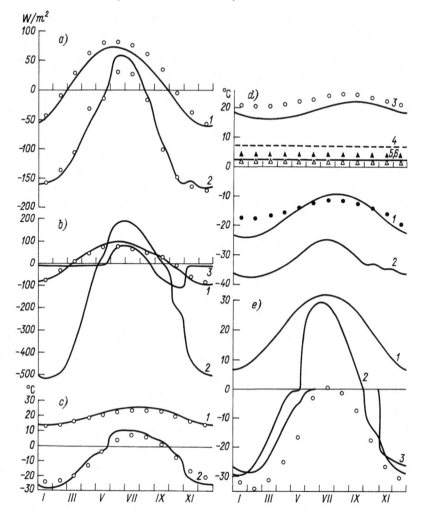

Figure 5.5 Seasonal variability of the thermodynamic cycle characteristics in the 0.5-dimensional box model of the climatic system, according to Kagan *et al.* (1990); (*a*) the net radiation flux at the upper atmospheric boundary in the southern (1) and northern (2) boxes, circles are experimental data from Stephens *et al.* (1981); (*b*) the resulting heat flux at the ocean–atmosphere interface in the upwelling area (1), in the area of cold deep water formation (2), and in the polar ocean (3), circles are data from Strokina (1982); (*c*) the surface air temperature in the southern (1) and northern (2) boxes, circles are data from Warren and Schneider (1979); (*d*) the mass-weighted average air temperature in the southern (1) and northern (2) boxes, the water temperature in the UML (3) and deep layer (4) in the upwelling area, in the area of cold deep water formation (5), and in the deep layer of the polar ocean (6), the solid circles are data from Oort and Rasmusson (1971), the open circles are data from Strokina (1982), the triangles are data from Stepanov (1974), solid triangles are in the deep layer of the upwelling area, open triangles are in a zone of cold deep water formation; (*e*) the land surface temperature in the southern (1) and northern (2) boxes and sea ice surface temperature (3) circles are data from Untersteiner (1961).

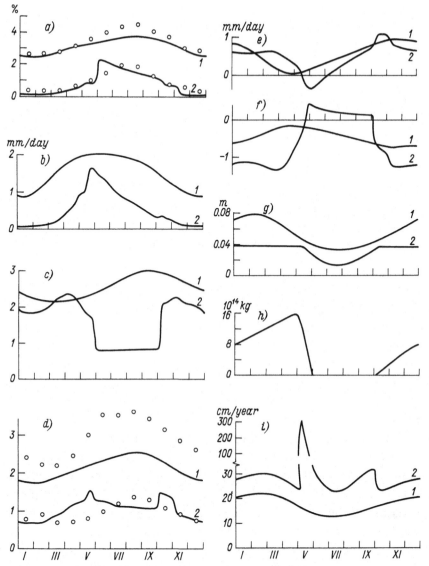

Figure 5.6 Seasonal variability of the hydrological cycle characteristics in the 0.5-dimensional box model, according to Kagan *et al.* (1990); (*a*) the air specific humidity in the southern (1) and northern (2) boxes, circles are data from Oort and Rasmusson (1971); (*b*) evaporation from the land surface in the southern (1) and northern (2) boxes; (*c*) evaporation from the ocean surface in the upwelling area (1), and in the area of cold deep water formation (2); (*d*) precipitation in the southern (1) and northern (2) boxes, circles are data from Jaeger (1983); (*e*) the precipitation–evaporation difference at the land surface in the southern (1) and northern (2) boxes; (*f*) the precipitation–evaporation difference at the ocean surface in the upwelling area (1), and in the area of cold deep water formation (2); (*g*) the soil moisture content in the southern (1) and northern (2) boxes; (*h*) the snow mass at the land surface; (*i*) the run-off in the southern (2) and northern (1) boxes.

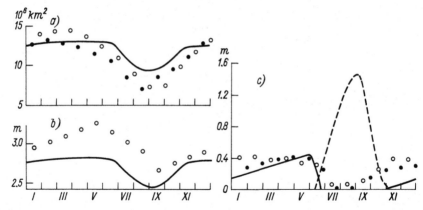

Figure 5.7 Seasonal variability of sea ice characteristics in the 0.5-dimensional box model, according to Kagan *et al.* (1990); (*a*) ice area; (*b*) ice thickness; (*c*) thickness of snow cover (solid curve), and of melted water (dotted line), solid and open circles are data from different sources.

precipitation and evaporation on the land, and a negative difference in the ocean, leading to closure of the hydrological cycle. By the way, the minimum values of this difference on the land of the northern box fall in summer due to decreasing precipitation and increasing evaporation, and on the land of the southern box these minimum values fall in spring, mainly due to increasing evaporation.

We also point to the fact that the soil moisture content in high latitudes, as would be expected, is less than its critical value only during the warm half of the year (after snow melting) and that in the southern box it is subjected to more distinctive seasonal oscillations with minimum in summer and maximum in winter. The run-off in the southern box is subjected to similar changes, with variations of the precipitation–evaporation difference and run-off shifting with respect to each other by two or three months. As a result, maximum run-off in the southern box occurs at the end of February, and minimum run-off occurs in the middle of July. The seasonal change of run-off in the northern box contains two maxima. One (the spring maximum) is caused by intensive snow melting, the other (the autumn maximum) is caused by a precipitation increase. Results of calculations also reproduce the asymmetry of the seasonal change in the area and the thickness of sea-ice, the complete disappearance of snow in summer and the phase shift between variations of the sea ice area and air temperature.

Figures 5.5–5.7 illustrate the degree of agreement between calculated and observed values of the climatic characteristics. As can be seen, satisfactory agreement cannot be reached for all of them. So, the amplitude of seasonal oscillations of the mass-weighted average air temperature T_A in the southern box turns out to be twice as large as the observed value, and the amplitude of seasonal oscillations of specific humidity q in the same box is only two-thirds of the observed value. Overestimation of the amplitude of oscillations of T_A is likely to be connected with prescribing constant cloudiness throughout the year; meanwhile underestimation in the amplitude of oscillations of q is probably due to application of the condition of instantaneous precipitation of moisture excess. By the way, underestimation in the oscillation amplitude of q leads to overestimation of precipitation and latent heat of oscillation amplitudes and, therefore, it is one more cause of overestimation of the oscillation amplitude of T_A. But it should be remembered that the basic variable in the model is the surface air temperature T_a, seasonal oscillations of which are reproduced quite satisfactorily: maximum discrepancies between calculated and observed values of T_a occur in August and amount to about 1.5 °C.

Considerable discrepancies are also inherent characteristics of sea ice during the warm half of the year. This is connected with the limited potential of the model, which does not allow for the appearance and disappearance of leads and the drainage of melted water. At the same time, agreement of calculated and observed annual mean values of the area, thickness and export of sea ice turns out be satisfactory: according to calculated results the annual mean sea ice area amounts to $12.0 \times 10^6 \ km^2$, thickness 2.72 m and export $0.18 \times 10^6 \ m^3/s$, while from the observational data they are equal to $11 \times 10^6 \ km^2$, 2.7 m and $(0.1–0.2) \times 10^6 \ m^3/s$ respectively. A similar conclusion can be made in regard to the run-off into the ocean. According to the calculated results the annual mean run-off into the oceans of the Northern Hemisphere normalized to the unit area of land, and the annual mean run-off in the Arctic Ocean normalized to a unit area of land in the northern box, amount to 20.5 and 42.4 cm/year. The experimental data yield 22.0–22.4 cm/year and 30.2 cm/year respectively.

We should mention in particular the estimate of the heat transfer into the atmosphere in the area of cold deep water formation. In winter, according to calculated results, it can even exceed the value of the absorbed solar radiation flux. Indications in favour of the existence of very large values (up to 500–600 W/m²) of heat transfer into the atmosphere can be found in Killworth (1983).

An ideologically similar box model of the climatic system was proposed by Harvey and Schneider (1985). This differs from the one discussed in the rejection of the description of the hydrological cycle and sea ice, the fixing of temperature in the area of cold deep water formation, the prescribed seasonal change in the UML thickness in the upwelling area, the replacement of the mixed DOL by a layer where the vertical temperature distribution adheres to the condition of local balance between the vertical heat diffusion and the vertical ordered heat transport and, finally, in the approximation of the vertical diffusion coefficient and upwelling velocity by some prescribed functions of the vertical coordinate. These and similar assumptions impoverish the model in the sense of accounting for different feedbacks regulating the behaviour of the climatic system, but they do allow us to carry out a great number of numerical experiments for little cost. Let us agree that each model has its own advantages and a great deal, if not all, depends on the way that we use these models.

One of the shortcomings of the box models, as well as of the one-dimensional models, is the assumption that the meridional heat transport in the atmosphere is realized only by quasi-two-dimensional turbulence of synoptic scale parametrized with the help of the diffusive approximation being constant or variable depending on the local meridional temperature gradient coefficient of horizontal diffusion, and in the ocean by quasi-two-dimensional turbulence and ordered motions comprising the meridional circulation cell. Let us show that there is one more mechanism of the meridional heat transport in the ocean. We define the meridional heat transport MHT_O, as well as the meridional component of the current velocity v_O and water temperature T_O in the ocean, as

$$MHT_O = \int_0^{2\pi} f'_O c_O m_O \widehat{v_O T_O} a \cos \varphi \, d\lambda,$$

$$v_O = [v_O] + v_O^*,$$

$$T_O = [T_O] + T_O^*,$$

where f'_O is the ocean fraction in the zonal belt of unit meridional extent; c_O is the specific heat of sea water; m_O is the mass of a water column of unit cross-section; the symbol $\frown$, the square brackets and the asterisk mean averaging over depth, zonal averaging and departure from zonal averaging respectively.

Substitution of the relations for v_O and T_O into the definition of MHT_O yields

$$MHT_O = \int_0^{2\pi} f'_O c_O m_O ([\widehat{v_O}][\widehat{T_O}] + [\widehat{v_O^* T_O^*}]) a \cos \varphi \, d\lambda,$$

where the first term in parentheses describes overturning in the meridional plane (the heat transport by the meridional circulation cell), the second one, representing the horizontal correlation between velocity and temperature departures, describes the gyre effect (the heat transport in the horizontal plane).

From general considerations it is clear that the meridional heat transport in tropical and subtropical latitudes of the ocean is mainly due to circulation in the meridional plane because of the predominance of the vertical temperature contrast over the horizontal one. The data systematized by Bryan (1982) serve as confirmation. As for temperate and high latitudes of the ocean, the relation between the separate components of MHT_0 cannot be considered as reliably established. In particular, this is demonstrated by very contradictory estimates obtained within the framework of three-dimensional ocean general circulation models. Be that as it may, until this problem is solved finally the justification for an allowance for the meridional heat transport owing to the gyre effect cannot be doubted. This distinguishes the 1.5-dimensional models favourably from the 0.5-dimensional (box) and one-dimensional models.

5.6 1.5-dimensional models

The thermodynamic model of the ocean–atmosphere–continent ice system developed by Verbitskii and Chalikov (1986) is a 1.5-dimensional model. In this model the World Ocean is represented in the form of three meridional oceans of different zonal extent Λ, a southern ring (the analogue of the Southern Ocean), and the North Polar Ocean. The depth of all the oceans is assumed to be constant and equal to $H = 3700$ m. Meridional oceans limited from the West and East by continents are divided into two areas: the western boundary layer of constant angular width λ_0 and the open ocean. Both these areas, in turn, are divided into two layers: the upper (with thickness $h = 600$ m) and a deep layer. Within each of them the temperature is assumed to be equal to its mean (over the vertical and longitude) value. Thus, the thermal regime of the meridional ocean is described by a system of four one-dimensional heat budget equations derived from the initial equations of heat conductivity after averaging over depth and longitude within the limits of the layer in question, and after replacing the average products of the velocity and temperature by the product of averages, with the temperature at the western boundary layer/open ocean interface or at the upper layer/deep layer being equal to half the sum of its average values. When the heat transport across the eastern ocean boundary is equal to zero these equations take the form

$$\frac{\partial T_{11}}{\partial t} - \frac{u_{11}(\lambda_0)}{a(\Lambda - \lambda_0)\cos\varphi}\frac{T_{11} + T_{21}}{2} + \frac{\partial}{a\cos\varphi\,\partial\varphi}\left([v_{11}]T_{11} - k_H\frac{\partial T_{11}}{a\,\partial\varphi}\right)\cos\varphi$$

$$+ \frac{[w_1]}{h}\frac{T_{11} + T_{12}}{2} = \frac{q_1^s - q_1^h}{h}, \quad (5.6.1)$$

$$\frac{\partial T_{12}}{\partial t} - \frac{u_{12}(\lambda_0)}{a(\Lambda - \lambda_0)\cos\varphi}\frac{T_{12} + T_{22}}{2} + \frac{\partial}{a\cos\varphi\,\partial\varphi}\left([v_{12}]T_{12} - k_H\frac{\partial T_{12}}{a\,\partial\varphi}\right)\cos\varphi$$

$$- \frac{[w_1]}{H - h}\frac{T_{11} + T_{12}}{2} = \frac{q_1^h}{H - h}, \quad (5.6.2)$$

$$\frac{\partial T_{21}}{\partial t} + \frac{u_{21}(\lambda_0)}{a\lambda_0\cos\varphi}\frac{T_{11} + T_{21}}{2} + \frac{\partial}{a\cos\varphi\,\partial\varphi}\left([v_{21}]T_{21} - k_H\frac{\partial T_{21}}{a\,\partial\varphi}\right)\cos\varphi$$

$$+ \frac{[w_2]}{h}\frac{T_{21} + T_{22}}{2} = \frac{q_2^s - q_2^h}{h}, \quad (5.6.3)$$

$$\frac{\partial T_{22}}{\partial t} + \frac{u_{22}(\lambda_0)}{a\lambda_0\cos\varphi}\frac{T_{12} + T_{22}}{2} + \frac{\partial}{a\cos\varphi\,\partial\varphi}\left([v_{22}]T_{22} - k_H\frac{\partial T_{22}}{a\,\partial\varphi}\right)\cos\varphi$$

$$- \frac{[w_2]}{H - h}\frac{T_{21} + T_{22}}{2} = \frac{q_2^h}{H - h}, \quad (5.6.4)$$

where T is the average (over depth and longitude) temperature; v and w are the meridional and vertical velocity components defined in a similar fashion; $u(\lambda_0)$ is the zonal velocity component at the western boundary layer/open ocean interface; q^s is the resulting heat flux at the ocean surface normalized to sea water heat capacity; k_H is the coefficient of the horizontal eddy diffusion; the first subscript indicates belonging to the open ocean (1) and the western boundary layer (2); the second subscript indicates belonging to the upper (1) and deep (2) layers.

System (5.6.1)–(5.6.4) contains unknown values of horizontal (u, v) and vertical (w) velocity components and heat fluxes q^h at the interface between the upper and deep ocean layers. To find the velocity components in the upper layer of the open ocean the following relationships can be used:

$$u_{11} = \hat{u}_1 - \frac{g\alpha_T h}{fa}\left(1 - \frac{h}{H}\right)\left(m\frac{\partial T_{11}}{\partial\varphi} + n\frac{\partial T_{12}}{\partial\varphi}\right), \quad (5.6.5)$$

$$v_{11} = \hat{v}_1 - \frac{g\alpha_T h}{fa\cos\varphi}\left(1 - \frac{h}{H}\right)\left(m\frac{\partial T_{11}}{\partial\lambda} + n\frac{\partial T_{12}}{\partial\lambda}\right) - \left(1 - \frac{h}{H}\right)\frac{\tau/\rho_0}{fh}, \quad (5.6.6)$$

$$w_1 = -\frac{h}{a \cos \varphi} \left(\frac{\partial u_{11}}{\partial \lambda} + \frac{\partial}{\partial \varphi} v_{11} \cos \varphi \right), \tag{5.6.7}$$

which are obtained by integration of the initial linearized equations of motion and continuity over depth within the limits of the upper layer with an allowance for assumptions of standard vertical temperature distribution of the type

$$T_1 = \begin{cases} T_{12} + (T_{11} - T_{12})\vartheta(z) & \text{at } z \leq h, \\ T_{12} & \text{at } z > h, \end{cases} \tag{5.6.8}$$

and also those about the absence of momentum fluxes at the interface between upper and deep layers and at the ocean bottom. Here, $\hat{u}_1, \hat{v}_1$ are depth-averaged zonal and meridional velocity components defined by the equalities

$$\hat{u}_1 = (\Lambda - \lambda_0)\frac{\partial}{\partial \varphi} (\hat{v}_1 \cos \varphi), \qquad \hat{v}_1 = \frac{1}{\beta H} \text{curl}_z \tau/\rho_0, \tag{5.6.9}$$

the first one of which is a result of the continuity equation integrated over the ocean depth (note that at the eastern boundary of the ocean, that is, at $\lambda = \Lambda$, the zonal velocity component is considered to be zero); the second equality represents the Sverdrup relation; τ/ρ_0 is the zonal component of the surface wind stress normalized to the mean density ρ_0 of sea water; β is the change in the Coriolis parameter f with latitude; $\vartheta(z)$ is a prescribed empirical function of the vertical coordinate; m and n are numerical factors depending on the choice of h, H and $\vartheta(z)$.

The zonal and meridional velocity components in the deep layer are defined by the condition of vanishing integral (within the limits of the open ocean depth) velocity deviations from their depth-averaged values. This condition, combined with (5.6.5) and (5.6.6), yields

$$u_{12} = \hat{u}_1 + \frac{g\alpha_T h}{fa} \frac{h}{H} \left(m \frac{\partial T_{11}}{\partial \varphi} + n \frac{\partial T_{12}}{\partial \varphi} \right), \tag{5.6.10}$$

$$v_{12} = \hat{v}_1 + \frac{g\alpha_T h}{fa \cos \varphi} \left(m \frac{\partial T_{11}}{\partial \lambda} + n \frac{\partial T_{12}}{\partial \lambda} \right) + \frac{\tau/\rho_0}{fH}, \tag{5.6.11}$$

In the vicinity of the equator where the quasi-geostrophic relations (5.6.5), (5.6.6), (5.6.10) and (5.6.11) are not fulfilled, the baroclinic component of the meridional velocity v_{11} caused by horizontal inhomogeneity of the temperature field (the second term on the right-hand side of (5.6.6)) and the vertical velocity w_1 at the interface between the upper and deep layers are assumed to be equal to zero. In this case to estimate $u_{11}(\lambda)$, $[v_{11}]$, $u_{12}(\lambda_0)$

and $[v_{12}]$, instead of (5.6.5), (5.6.6), (5.6.10) and (5.6.11), Equations (5.6.9) are used.

The velocity components in the western boundary layer are found from the condition of water mass conservation in the zonal ocean section. In other words, it is assumed that the meridional mass transport in the open ocean is compensated by the opposite transport in the western boundary layer, and that this compensation occurs locally within the limits of every selected layer, that is,

$$[v_{21}] = -[v_{11}](\Lambda - \lambda_0)/\lambda_0, \qquad [v_{22}] = -[v_{12}](\Lambda - \lambda_0)/\lambda_0, \Bigg\} \tag{5.6.12}$$
$$[w_2] = -[w_1](\Lambda - \lambda_0)/\lambda_0.$$

The heat fluxes at the interface of two layers are parametrized in the form

$$q_1^h = \frac{k_v}{H/2}(T_{11} - T_{12}), \qquad q_2^h = \frac{k_v}{H/2}(T_{21} - T_{22}), \tag{5.6.13}$$

where k_v is the coefficient of the vertical eddy diffusion, assumed to be equal to $5 \times 10^{-4}\,\mathrm{m^2/s}$ for $T_{11} \geq T_{12}$, $T_{21} \geq T_{22}$ and $5 \times 10^{-2}\,\mathrm{m^2/s}$ for $T_{11} < T_{12}$, $T_{21} < T_{22}$.

Finally, it is assumed that the water temperature in the northern and southern polar oceans remains constant, and to estimate it the integral condition of the heat budget is used. This condition is written in the form

$$\int H\hat{v}\hat{T}a \cos \varphi \, \mathrm{d}\lambda = \int q^s a^2 \cos \varphi \, \mathrm{d}\lambda \, \mathrm{d}\varphi, \tag{5.6.14}$$

where integration on the left-hand side of the equation is performed along the length of the liquid boundary of the polar ocean; integration on the right-hand side is extended over the area of the polar ocean; the symbol $\hat{}$ means averaging over the vertical within the whole depth of the ocean.

Equation (5.6.14) does not take into account the possibility of water phase transitions in the process of ice formation and, hence, requires correction in the case where the temperature in the polar oceans falls below freezing point. In this case it is assumed that the water temperature is maintained at freezing point and that the resulting heat flux q^s at the ocean surface is equal to zero. The last condition means that when water temperature reaches freezing point the ocean is instantly covered with ice. This condition, together with the condition of vanishing total (advective + diffusive) heat transport across the coastal boundary, completes the formulation of the ocean submodel.

The atmospheric submodel includes quasi-stationary equations for the heat and moisture budget (of the type given by Equation (5.5.1)), written with an allowance for the meridional resolution, and with rather simplified (compared with 5.5.24) parametrization of heat sources and sinks. In particular, the coefficients of short-wave and long-wave radiation absorption and the upward and downward long-wave radiation emissivities are assumed to be constant. An analogous simplification is made with regard to the relative humidity at the land surface that excludes the necessity of examining the continental part of the hydrological cycle. The meridional heat and moisture transport are parametrized in terms of a non-linear diffusive approximation.

When constructing the third submodel (unit of continental ice) it is assumed that ice sheets behave as a viscous liquid spread under its own weight, and because of this their height is related to the area, and the latter depends in turn on continental size and ice sheet mass budget originating from solid precipitation, ablation and iceberg discharge when the ice sheet reaches the edge of the continent. The resulting relations obtained for the ice sheet height and area, together with the integral equation for the heat budget, as well as an approximation of the vertical temperature distribution in the ice sheet by a polynomial of second degree and with appropriate boundary conditions, provide all the information required from this submodel: the height and area of the ice sheet and the temperature of its upper surface.

Numerical integration of the model equations is performed using artificial synchronization of separate subsystem states, namely, the equations for the evolution of the ice sheet are integrated with a time step equal to 1000 years. Then at every such step a stationary solution of the atmospheric equations and a time-dependent solution of the ocean equations are found, with the time step for equations of the ocean submodel assumed to be equal to three days and the duration of the integration period assumed to be 10 years. The integration is completed when the climatic system as a whole has achieved an approximate statistical equilibrium.

Tests of the model as applied to modern conditions reveal that it simulates the meridional distribution of the mass-weighted temperature in the atmosphere, as well as ocean and land surface temperatures, the meridional heat transport and characteristics of the Antarctic ice sheet. But perhaps the most remarkable peculiarity is that the model demonstrates indications of intransitivity: the climatic system has four steady states with fixed external parameters and different initial conditions for the ice sheet area in the Northern and Southern Hemisphere and for the ocean temperature. The first state corresponds to the present-day climate characterized by the absence of the ice sheet in the Northern Hemisphere (the Greenland ice sheet is not reproduced on a grid with five-degree angular resolution) and by the presence of the Antarctic ice sheet; the second and third states are characterized by maximum development and the total disappearance of ice sheets in both hemispheres;

and the fourth state is characterized by maximum development of the ice sheet in the Northern Hemisphere and its absence in the Southern Hemisphere. A series of numerical experiments performed by Verbitskii and Chalikov (1986) shows that the distinctions of steady states become apparent only in the temperate and high latitudes of both hemispheres and do not extend to equatorial latitudes. The scales of climatic changes associated with them can be judged from the following figures. The appearance of the continental ice sheet is accompanied by a decrease in the mass-averaged temperature of the atmosphere and the temperature of the underlying surface in temperate and high latitudes of the Atlantic and Pacific Oceans of almost 5 °C. The temperature of the land surface changes most of all in the following areas: where land is covered by ice its surface temperature falls by 20 °C. But at small distances from the edge of the ice sheet the decrease in surface temperature does not exceed 5 °C.

One more important result of numerical experiments is the high sensitivity of climatic characteristics and of the number of stationary solutions to mutual locations of land and ocean. This result is easier to understand when recollecting that the area of the southern ice sheet is controlled by Antarctica, and the northern ice sheet is controlled by ocean temperature. Thus, according to Verbitskii and Chalikov (1986) in the case where at the initial time that the southern boundary of the northern ice sheet was located to the south from parallel 40 °N, and the temperature in northern parts of the Atlantic and Pacific Oceans was higher than at present, the ice sheet receded immediately. If the water temperature was lower than at present then at first heat transfer from the Southern Hemisphere to the Northern Hemisphere, together with an increase in water temperature, occurred and only after the ice sheet receded.

5.7 Two-dimensional (zonal) models

Imagine that the underlying surface is homogeneous along a latitude circle, and non-zonal motions are determined only by the instability of the zonal circulation. We average the initial hydrothermodynamic equations along the latitude circle. The resulting equations will contain statistical moments. To close the equations it is necessary to find the connection between these moments and the characteristics of the zonal circulation whose existence at the homogeneous (in a zonal direction) Earth is ensured by the fact that all statistical moments generated by zonal averaging have to be invariant to a rotation relative to the Earth's axis. Thus, in the case discussed it is possible in principle to construct a zonal atmospheric model.

When taking into account inhomogeneities of the underlying surface (say, continents and oceans) the property of invariance of statistical moments is not valid. Moreover, owing to the change in the thermophysical properties of the underlying surface in a zonal direction, steady areas of high pressure over the land and depressions over the ocean are formed. These areas are not described by zonal models. But if for the land and ocean such longitudinal distinctions are decisive due to the boundedness of their extensions in a zonal direction, for the atmosphere they do not play such an important role, and because of this zonal models of the atmosphere turn out to be representative, from the viewpoint of the reproduction of the meridional structure of the atmospheric circulation, in this case as well.

The distinctive property of the atmosphere forming the basis for the derivation of zonal model equations is the approximate symmetry of the thermodynamic regime relative to the axis of the Earth's rotation. A high degree of zonality in the distribution of many characteristics of the atmospheric circulation due to the zonality of the diurnal insolation points to this fact. We represent any dependent variable appearing in the initial hydrothermo-dynamic equations as $a = [\bar{a}] + \bar{a}^* + a'$, where $[\bar{a}]$ is the zonal average stationary component; $\bar{a}^*$ and a' are stationary and time-dependent deviations from the zonal average defined by the equalities $[\bar{a}^*] = 0$, $\bar{a}' = 0$; the overbar means time averaging. Then after averaging over time (with the period of averaging being *a fortiori* less than the annual cycle in order to take seasonal variability into account) and over longitude the initial system of atmospheric hydrothermodynamic equations reduces to the form

$$
\frac{\partial}{\partial t}[\overline{p_s}][\bar{u}] + \frac{\partial}{a\cos\varphi\,\partial\varphi}\{[\overline{p_s}]([\bar{u}][\bar{v}] + [\bar{u}^*\bar{v}^*] + [\overline{u'v'}])\cos\varphi\}
$$

$$
+ \frac{\partial}{\partial\sigma}\{[\overline{p_s}]([\bar{u}][\bar{\omega}] + [\bar{u}^*\bar{\omega}^*] + [\overline{u'\omega'}])\}
$$

$$
= f[\overline{p_s}][\bar{v}] + [\overline{p_s}]\{[\bar{u}][\bar{v}] + [\bar{u}^*\bar{v}^*] + [\overline{u'v'}]\}\frac{\tan\varphi}{a} + [\overline{F_\lambda}];
$$

$$
\frac{\partial}{\partial t}[\overline{p_s}][\bar{v}] + \frac{\partial}{a\cos\varphi\,\partial\varphi}\{[\overline{p_s}]([\bar{v}][\bar{v}] + [\bar{v}^*\bar{v}^*] + [\overline{v'v'}])\cos\varphi\} \qquad (5.7.1)
$$

$$
+ \frac{\partial}{\partial\sigma}\{[\overline{p_s}]([\bar{v}][\bar{\omega}] + [\bar{v}^*\bar{\omega}^*] + [\overline{v'\omega'}])\}
$$

$$
= -f[\overline{p_s}][\bar{u}] + [\overline{p_s}]\{[\bar{u}][\bar{u}] + [\bar{u}^*\bar{u}^*] + [\overline{u'u'}]\}\frac{\tan\varphi}{a}
$$

$$
- \frac{[\overline{p_s}]}{a}\left(\frac{\partial[\Phi]}{\partial\varphi} + R[\bar{T}]\frac{\partial[\overline{p_s}]}{\partial\varphi}\right) + [\overline{F_\lambda}];
$$

(contd)

$$\frac{\partial [\overline{p_s}]}{\partial t} = -\frac{1}{a \cos \varphi} \int_0^1 \frac{\partial}{\partial \varphi} [\overline{p_s}][\overline{v}] \cos \varphi \, d\sigma;$$

$$\frac{\partial [\overline{\omega}]}{\partial \sigma} + \frac{1}{a \cos \varphi} \frac{\partial}{\partial \varphi} [\overline{v}] \cos \varphi = 0;$$

$$\frac{\partial}{\partial t} [\overline{p_s}][\overline{\theta}] + \frac{\partial}{a \cos \varphi \, \partial \varphi} \{[\overline{p_s}]([\overline{\theta}][\overline{v}] + [\overline{\theta^* v^*}] + [\overline{\theta' v'}]) \cos \varphi\}$$

$$+ \frac{\partial}{\partial \sigma} \{[\overline{p_s}]([\overline{\theta}][\overline{\omega}] + [\overline{\theta^* \omega^*}] + [\overline{\theta' \omega'}])\} = [\overline{F_T}];$$

$$\frac{\partial}{\partial t} [\overline{p_s}][\overline{q}] + \frac{\partial}{a \cos \varphi \, \partial \varphi} \{[\overline{p_s}]([\overline{q}][\overline{v}] + [\overline{q^* v^*}] + [\overline{q' v'}]) \cos \varphi\}$$

$$+ \frac{\partial}{\partial \sigma} \{[\overline{p_s}]([\overline{q}][\overline{\omega}] + [\overline{q^* \omega^*}] + [\overline{q' \omega'}])\} = [\overline{F_q}].$$

$$(5.7.1)$$

Here $\Phi = gz$; $\omega = dp_s/dt$ is the isobaric vertical velocity; $\sigma = p/p_s$, p_s is the surface pressure; R is the gas constant for air; F_λ, F_φ, F_T and F_q are sources and sinks of momentum, heat and moisture, respectively; other symbols are the same.

System (5.7.1) contains 20 unknown second moments (additional unknowns appear in zonal averaging of the terms F_λ, F_φ, F_T and F_q if they have non-linear structure and when formulating boundary conditions at the underlying surface if its topography is taken into consideration). Among the 20 unknown second moments some (with primes) owe their origins to transient synoptic disturbances, others (with asterisks) owe their origins to steady motions. The common method of description of second moments with the help of the hypothesis of the semiempirical theory of turbulence (second moments are considered to be proportional to gradients of appropriate averages with proportionality factors having the sense of coefficients of virtual diffusion of momentum, heat and moisture) does not work here because these factors turn out to be complicated functions of spatial coordinates and some of them (say, those appearing by the parametrization of $[\overline{v'u'}]$), even become negative.

A more acceptable method of parametrization of the meridional flux $[\overline{u'v'}]$ of zonal momentum was proposed by Williams and Davis (1965). They proceeded from the fact that synoptic disturbances draw their energy from the available potential energy of zonal motions, a measure of which is the meridional temperature gradient. Subsequently, the momentum flux

$[\overline{u'v'}]$ is presented in the form

$$[\overline{u'v'}] = -kL_0 \frac{R}{a} \frac{\partial T}{\partial \varphi}, \qquad (5.7.2)$$

providing the possibility of the existence of negative viscosity in jet streams. Here $L_0 = c/\Omega$ is the horizontal scale of barotropic synoptic disturbances; c is the sound velocity; Ω is the angular velocity of the Earth's rotation; k is the non-dimensional factor assumed to be proportional to $(z/H_0)^2$; z is the vertical coordinate; $H_0 = c^2/\kappa g$ is the height of the homogeneous atmosphere; $\kappa = c_p/v_v$ is the ratio of heat capacities at constant pressure and constant volume.

The remaining second moments created by transient synoptic disturbances were described by ordinary diffusion formulae with positive coefficients of virtual viscosity, heat conductivity and diffusion. Such a parametrization of second moments of the $[\overline{a'b'}]$ type was used by Dymnikov *et al.* (1979), and by Williams and Davis (1965).

A different method of parametrization of the second moments $[\overline{a'b'}]$ was developed by Green (1970). The essence of this is to connect zonal average fluxes of conservative characteristics created by transient synoptic disturbances with the average gradients of these characteristics using the matrix of transport coefficients, the dependence of which on spatial coordinates can be established from an analysis of zonal flow instability. Considering this, the intensity of the potential temperature transport is expressed in terms of horizontal and vertical gradients of potential temperature (anisotropic diffusion), and the intensity of quasi-geostrophic potential vorticity transport is expressed only in terms of the horizontal gradient of potential vorticity.

Following Green (1970) we obtain the expression for second moments. We start with the definition of the quasi-geostrophic potential vorticity. In Cartesian coordinates it has the form

$$\Pi = \zeta + f + \beta y + \frac{f}{\rho} \frac{\partial}{\partial z} \left(\frac{\rho \delta \eta}{B} \right), \qquad (5.7.3)$$

where Π and ζ are the potential and relative vorticities; η is the logarithm of potential temperature; $B = \partial \ln \eta / \partial z$ is the parameter of static stability in the environment; ρ is the density; β is the change in the Coriolis parameter f with latitude; y is the meridional coordinate; the symbol δ signifies the departure from an undisturbed state which is a function of the vertical coordinate only.

On the basis of (5.7.3) the meridional potential vorticity flux will be equal to

$$[\overline{\Pi'v'}] = [\overline{\zeta'v'}] + \frac{f}{\rho}\left[\overline{v'\frac{\partial}{\partial z}\left(\frac{\rho'\delta\eta'}{B}\right)}\right],$$

or, using the thermal wind relation $\partial v'/\partial z = (g/f)(\partial\delta\eta'/\partial x)$ and periodicity of all the variables with longitude,

$$[\overline{\Pi'v'}] = [\overline{\zeta'v'}] + \frac{f}{\rho}\frac{\partial}{\partial z}\left(\rho\,\frac{[\overline{v'\delta\eta'}]}{B}\right), \tag{5.7.4}$$

But

$$[\overline{\xi'v'}] = -\frac{\partial}{\partial y}[\overline{u'v'}], \tag{5.7.5}$$

so that the first term on the right-hand side of (5.7.4) describes the meridional momentum flux and the second term describes the intensity of excitation of the zonal motion.

Represent the fluxes of heat and potential vorticity created by transient synoptic disturbances as

$$\left.\begin{aligned}
[\overline{v'\delta\eta'}] &= -k_{vy}\frac{\partial[\bar\eta]}{\partial y} - k_{vz}\frac{\partial[\bar\eta]}{\partial z}, \\[2mm]
[\overline{w'\delta\eta'}] &= -k_{wy}\frac{\partial[\bar\eta]}{\partial y} - k_{wz}\frac{\partial[\bar\eta]}{\partial z}, \\[2mm]
[\overline{\Pi'v'}] &= -k_{vy}\frac{\partial[\bar\Pi]}{\partial y},
\end{aligned}\right\} \tag{5.7.6}$$

where k_{vy}, k_{vz}, k_{wy}, k_{wz} are transport coefficients assumed proportional to the local temperature gradients.

Substitution of (5.7.6) into (5.7.4) yields for $B = const$

$$[\overline{v'\zeta'}] = \frac{f}{B}\frac{\partial}{\partial z}\left(k_{vy}\frac{\partial[\bar\eta]}{\partial y} + k_{vz}\frac{\partial[\bar\eta]}{\partial z}\right) - k_{vy}\frac{\partial[\bar\Pi]}{\partial y},$$

from which, with an allowance for (5.7.3), it follows that

$$[\overline{v'\zeta'}] = -k_{vy}\left(\beta + \frac{\partial[\bar\zeta]}{\partial y}\right) + \frac{f}{B}\frac{\partial[\bar\eta]}{\partial y}\frac{\partial k_{vy}}{\partial z} + f\frac{\partial k_{vz}}{\partial z}. \tag{5.7.7}$$

Thus, to determine $[\overline{v'\zeta'}]$ and, hence, $[\overline{u'v'}]$ it is necessary to know only the spatial distribution of transport coefficients k_{vy} and k_{vz}. For this purpose the theory of baroclinic instability of the zonal flow and additional energetic considerations (say, estimates of the ratio between time of growth and degeneration of baroclinic disturbances) are used. The case is aggravated by the existence of the integral condition which the adopted method of parametrization has to meet. Indeed, let the range of determination of the solution be restricted in the meridional direction by walls where the meridional flux $[\overline{u'v'}]$ vanishes. Then, according to (5.7.5)

$$\int_0^Y [\overline{\zeta'v'}]\,dy = 0, \tag{5.7.8}$$

where Y is the meridional extension of the domain examined. It is clear that Equation (5.7.8) has to be fulfilled at any instant of time and at any height in the atmosphere, and its rejection is equivalent to the introduction of false sources and sinks of momentum.

One more method for the parametrization of second moments on the basis of the theory of baroclinic instability was proposed by Stone (1972). He defined second moments by formulae of the form

$$\left. \begin{aligned}
[\overline{u'v'}] &= -k_1 \mathcal{U}^2 (1 + \mathrm{Ri})^{3/2} \left[\left(\frac{z}{H_0} - \frac{1}{2} \right)^2 + \frac{1}{12} \right], \\[2mm]
[\overline{u'w'}] &= -k_2 f H_0 \mathcal{U} (1 + \mathrm{Ri})^{1/2} \left[\left(\frac{z}{H_0} - \frac{1}{2} \right)^2 - \frac{1}{4} \right], \\[2mm]
[\overline{v'w'}] &= -k_3 f H_0 \mathcal{U} (1 + \mathrm{Ri}) \left[\left(\frac{z}{H_0} - \frac{1}{2} \right)^2 - \frac{1}{16} \right], \\[2mm]
[\overline{v'\theta'}] &= k_4 H_0 \mathcal{U} \frac{\partial \theta}{\partial z} \frac{(1 + \mathrm{Ri})^{1/2}}{\mathrm{Ri}} \frac{4}{Y} \left(1 - \frac{4}{Y} \right), \\[2mm]
[\overline{w'\theta'}] &= k_5 H_0^2 \frac{\partial \theta}{\partial z} \frac{(1 + \mathrm{Ri})^{1/2}}{\mathrm{Ri}} \frac{z}{H_0} \left(1 - \frac{z}{H_0} \right),
\end{aligned} \right\} \tag{5.7.9}$$

where $\mathcal{U} = -(gH_0/fT)(\partial\theta/\partial y)$ is the zonal component of the thermal wind; $\mathrm{Ri} = (gH_0^2/\mathcal{U}^2 T)(\partial\theta/\partial z)$ is the Richardson number; $k_1, \ldots, k_5$ are numerical factors.

It has been shown that the relations (5.7.9) connecting momentum and heat fluxes with the temperature field are valid for forced disturbances and invalid for free disturbances. According to Lorenz (1979) the forced disturbances are disturbances for which changes in external heat sources and sinks determine the changes in the temperature gradient, and they, in turn, determine changes in the heat flux. In such disturbances the variations of the heat flux and temperature gradient turn out to be positively correlated. As for the case of free disturbances, the opposite situation occurs: the changes in heat flux arising from the internal instability of the atmospheric circulation and changes in the temperature gradient generated by them turn out to be negatively correlated. The disturbances in the heat flux with spatial scales less than the planetary scale, and also planetary oscillations with time scales more or less than the seasonal time scale are free; the seasonal oscillations of the planetary scale are forced.

A similar method of parametrization of second moments was developed by Petuhov (1989). This method, as well as all the others mentioned above, contains many hypotheses and empirical constants obtained as applied to the present-day conditions and this, naturally, restricts the implementation of such parametrizations. An approach based on closing the atmospheric hydro-thermodynamic equations with the help of equations for second moments (so-called Friedman–Keller equations), and on simplifying the latter by exclusion of third moments, has higher universality. Such simplification also represents rather restrictive hypotheses but it has the advantage that it has much less influence on the characteristics of the zonal average circulation. This approach, proposed by Monin (1965), has still not gained wide acceptance and was realized only for the case of the barotropic atmosphere approximated in the form of an incompressible, non-dissipative, two-dimensional spherical film.

Finally, one more, still not accepted, method of parametrization of second moments reduces to the application of the following relations (see Chalikov, 1982):

$$
\left.
\begin{aligned}
[\overline{u'v'}] &\sim R^{1/2}\mathcal{U}\left(\frac{\partial\theta}{\partial\sigma}\right)^{1/2} F_1(\sigma, \text{Ri}, \text{Ki}, \text{Ma}, \text{Re}), \\[2mm]
[\overline{v'v'}] &\sim R^{1/2}\mathcal{U}\left(\frac{\partial\theta}{\partial\sigma}\right)^{1/2} F_2(\sigma, \text{Ri}, \text{Ki}, \text{Ma}, \text{Re}), \\[2mm]
[\overline{\theta'v'}] &\sim R^{3/2}f^{-2}\left(\frac{\partial\theta}{\partial y}\right)\left|\frac{\partial\theta}{\partial y}\right|\left(\frac{\partial\theta}{\partial\sigma}\right)^{1/2} F_3(\sigma, \text{Ri}, \text{Ki}, \text{Ma}, \text{Re}), \\[2mm]
[\overline{q'v'}] &\sim R^{3/2}f^{-2}\left(\frac{\partial q}{\partial y}\right)\left|\frac{\partial\theta}{\partial y}\right|\left(\frac{\partial\theta}{\partial\sigma}\right)^{1/2} F_4(\sigma, \text{Ri}, \text{Ki}, \text{Ma}, \text{Re}),
\end{aligned}
\right\} \quad (5.7.10)
$$

which were obtained from experimental data handling processing with an allowance for dimensional considerations. Here $\text{Ri} = \sigma f^2(\partial\theta/\partial\sigma)/R(\partial\theta/\partial y)^2$ is the Richardson number; $\text{Ki} = f^{-1}(\partial\mathcal{U}/\partial y)$ is the Kybel number; $\text{Ma} = \mathcal{U}(gR(\partial\theta/\partial\sigma))^{-1/2}$ is the Mach number; $\text{Re} = \mathcal{U}a/K$ is the Reynolds number; K is the macroviscosity coefficient describing motions of a subgrid scale; other designations are the same.

The functions $F_1, \ldots, F_4$ appearing in (5.7.10) contain too many dimensionless arguments to determine them from empirical data. But the number Re for $K = 10^5 \, \text{m}^2/\text{s}$ is of order 10^3, and the numbers Ma and Ki are of order 10^{-2} and 10^{-1} respectively. Therefore, it is expected that Equations (5.7.10) will have the property of self-similarity on Re, Ma and Ki and, hence, the question of their use reduces to finding functions $F_1, \ldots, F_4$ of two arguments σ and Ri.

As for the second moments created by the stationary non-zonal motions, there are no justifiable parametrization schemes for these at present. But, as has been shown by Stone and Miller (1980), the net meridional heat transport corresponding to stationary and transient disturbances of the zonal circulation is correlated with the meridional temperature gradient more than with any separate constituent of this transport. This circumstance, as well as the fact that both these components complement each other, point to the existence of negative feedback between them and to one source of their origin – the baroclinic instability of the zonal circulation. This lifts some of the burden from researchers, allowing them to use (for the meridional heat transport resulting from stationary disturbances at least) the same parametrization as for the meridional heat transport initiated by transient disturbances, or even to consider the joint effect of non-zonal motions without separating them into stationary and transient motions.

We make some brief remarks here about the potential of zonal models by reference to one of them, developed by Dymnikov *et al.* (1979). This model successfully simulates the observed three-cell structure of the meridional circulation in the troposphere of both hemispheres. It also provides good reproduction of the easterly winds in tropical and polar latitudes and the jet flow in the vicinity of the 200 hPa isobaric surface in temperate latitudes. Meanwhile, the velocity of the zonal winds in temperate latitudes turns out to be slightly excessive due to rejecting the effects of the Earth's surface topography and overestimating of the meridional temperature gradient. This last circumstance is caused by underestimation (by about 10 K) of the air temperature in polar latitudes, and this underestimation is caused, in its turn, by inaccurate specification of the underlying surface temperature in polar regions, by underestimation of the meridional heat transport due to eddy disturbances and by the coarse vertical resolution.

5.8 Three-dimensional models

In this section we will discuss three-dimensional models of the ocean–atmosphere system occupying the highest position in terms of their complexity in the hierarchy of climatic models. As before, much thought will be given to the principles of construction of these models and to their application to simulation of the present-day climate.

An obvious question arises at this point: why use models to simulate the present-day climate if it is already known from observational data? In other words, why make efforts to solve problems that are already solved (even in the first approximation)? It is both easy and difficult to answer this question. Indeed, no climatic model will completely comply with the actual climatic system so that aspirations to understanding the distinctions and similarities between them is an indispensable condition, if we wish climatic models to have, according to Monin (1982), convincing and predicting strength. On the other hand, what one dreams of and what one can actually do might differ for a number of reasons (say, technical or economic ones). It is essential to understand whether what can be done is useful. As for this question the point is that an answer to it can be given only with precise knowledge about the capability of the models to reproduce the modern climate.

At present, seven global models of the ocean–atmosphere system have been tested and verified. These are the models developed at the Laboratory of Geophysical Fluid Dynamics, Princeton University (GFDL model), at the P. P. Shirshov Institute of Oceanology, the Russian Academy of Sciences (IOAS model), at the Computer Centre of the Siberian Branch of the Russian Academy of Science (CC SBAS model), at the National Centre for Atmospheric Research in the United States (NCAR model), at the Oregon State University (OSU model), at the UK Meteorological Office (UKMO model), and at the Max-Planck-Institut für Meteorologie (MPI model). This list could also include models that either fix the sea surface temperature or present the ocean in the form of a 'swamp' with zero or finite heat capacity. In the latter case, the ocean serves as an infinite source of moisture for the atmosphere but does not provide heat transport in the meridional direction. Many such models exist but here we discuss only those listed, i.e. coupled ocean–atmosphere global circulation models, in the order in which they are mentioned.

GFDL model

The earlier version of this model and the annual mean state of the ocean–atmosphere system obtained with its help are described in an article by Manabe and Bryan (1969). A generalization of this model version for the case of seasonal variability of the climatic system is presented by Manabe *et al.* (1979).

The GFDL model has the following structure. It consists of atmospheric and ocean submodels and a general library where data necessary to calculate the characteristics of the ocean– atmosphere interaction are continuously updated. The atmospheric submodel includes hydrothermodynamics equations in which the independent variables are time t, longitude λ, latitude φ and non-dimensional pressure $\sigma = p/p_s$. The state of the atmosphere is described by zonal u, meridional v and vertical ω (in the σ-system of coordinates) components of wind velocity, as well as by temperature T, specific humidity q and pressure p_s at the underlying surface. These variables are described by equations of motion:

$$\left.\begin{aligned}\frac{d}{dt}p_s u &= \left(f + u\frac{\tan\varphi}{a}\right)p_s v - \frac{p_s}{a\cos\varphi}\frac{\partial\Phi}{\partial\lambda} + F_\lambda + g\frac{\partial\tau_\lambda}{\partial\sigma}, \\ \frac{d}{dt}p_s v &= -\left(f + u\frac{\tan\varphi}{a}\right)p_s u - \frac{p_s}{a}\frac{\partial\Phi}{\partial\varphi} + F_\varphi + g\frac{\partial\tau_\varphi}{\partial\sigma},\end{aligned}\right\} \tag{5.8.1}$$

by the hydrostatic equation

$$\Phi = \Phi_s + R\int_0^1 (T/\sigma)\,d\sigma, \tag{5.8.2}$$

by the equation for surface pressure

$$\frac{\partial p_s}{\partial t} + \frac{1}{a\cos\varphi}\int_0^1\left(\frac{\partial p_s u}{\partial\lambda} + \frac{\partial}{\partial\varphi}p_s v\cos\varphi\right)\partial\sigma = 0, \tag{5.8.3}$$

and by the equations for the heat and moisture budget

$$\frac{d}{dt}p_s T = \frac{pT}{c_p\sigma}\dot{p} + \frac{g}{c_p}\frac{\partial}{\partial\sigma}(F_R + H) + p_s(Q_C + Q_P + Q_D), \tag{5.8.4}$$

$$\frac{d}{dt}p_s q = g\frac{\partial E}{\partial\sigma} + p_s(P_C - C + P_D). \tag{5.8.5}$$

Here F_λ, F_φ, Q_D and P_D are terms describing the horizontal diffusion of momentum, heat and moisture respectively; τ_λ and τ_φ are components of the vertical eddy momentum flux; H and E are vertical eddy fluxes of heat and moisture; F_R is the radiative heat flux; Q_C and P_C are normalized heat and moisture influxes of convective origin; Q_P is the rate of temperature change due to water vapour phase transitions; C is the rate of the moisture content change in the atmosphere due to precipitation:

$$\frac{d}{dt}p_s(\ ) = \frac{\partial}{\partial t}p_s(\ ) + \frac{\partial}{a\cos\varphi\,\partial\lambda}p_s u(\ ) + \frac{\partial}{a\cos\varphi\,\partial\varphi}p_s v(\ )\cos\varphi + \frac{\partial}{\partial\sigma}p_s\omega(\ )$$

is the operator of the total time derivative; ω and $\dot{p}$ are analogues of the vertical velocity in σ- and p-systems of coordinates defined by the diagnostic relations

$$\omega = -p_s^{-1}\left[\sigma\frac{\partial p_s}{\partial t} + \int_0^\sigma\left(\frac{\partial p_s u}{a\cos\varphi\,\partial\lambda} + \frac{\partial p_s v\cos\varphi}{a\cos\varphi\,\partial\varphi}\right)d\sigma\right], \qquad (5.8.6)$$

$$\dot{p} = p_s\omega + \sigma\left(\frac{\partial p_s}{\partial t} + \frac{u\,\partial p_s}{a\cos\varphi\,\partial\lambda} + \frac{v\,\partial p_s}{a\,\partial\varphi}\right). \qquad (5.8.7)$$

The calculation scheme for the radiative fluxes includes a description of the absorption by ozone, carbon dioxide and water vapour, and also a description of cloud effects and of their albedo dependence on the height of clouds. The annual mean fields of three-level cloudiness and ozone distribution are prescribed from the climatological data. The CO_2 concentration is considered to be constant. The albedo of the underlying surface varies depending on the type of surface (water, land surface, snow, sea and continental ice).

The evaporation from land and the moisture content in the soil active layer are found by integration by the prognostic equation for soil moisture, taking into account the precipitation, snow melting, evaporation and run-off. In so doing, it is assumed that the land moisture capacity (that is, maximum amount of moisture which can be retained in the land) does not exceed 15 cm, and the effect of land moisture on evaporation manifests itself only in the case where the land moisture content is less than a certain critical value amounting to 75% of the moisture capacity. In this case the evaporation decreases by a value which is proportional to the ratio of current and critical values of the land moisture content. The thickness of the snow cover is determined from the snow mass budget equation, taking into account accumulation, sublimation and melting of snow.

To calculate the convection and water vapour phase transitions, a convective adjustment scheme is used which provides the vertical redistribution of moisture and heat in the presence of hydrostatic instability. The redistribution is performed with an allowance for the integral conservation of the heat and moisture content, after which the condensed moisture is assumed to be precipitated in the form of rain or snow.

The effects of horizontal mixing are parametrized in the following way. The terms describing the horizontal momentum diffusion are represented in the form of the product of appropriate components of the velocity deformation tensor and the horizontal eddy viscosity coefficient. The latter is obtained from the condition of equality between local generation and dissipation of the turbulent energy, and the mixing length (the turbulence scale) is assumed to be equal to the grid step. The same coefficient is used to determine the horizontal heat and moisture diffusion effect.

The vertical eddy fluxes of momentum, heat and moisture are calculated with the help of a semiempirical model of turbulence where the turbulence scale is prescribed as linear, increasing up to a height of 75 m, and then as linearly decreasing to zero at a height of 2.5 km (the upper boundary of the atmospheric planetary boundary layer). To estimate friction stresses and sensible and latent heat fluxes at the underlying surface, the bulk formulae with fixed values of resistance and heat and moisture exchange coefficients are used. The soil temperature is derived from the heat budget equation on the assumption that the heat flux into soil is equal to zero; the sea surface temperature is found from the ocean submodel.

Equations of the atmospheric submodel are approximated by their difference analogues on the latitude–longitude grid with 64 nodal points along the latitude circle and with 19 nodal points along the meridian (between the equator and the pole). Along the vertical the atmosphere is divided into nine layers with a concentration of levels in the planetary boundary layer and in the low stratosphere. The real topography of the Earth's surface is replaced by a smoothed relief, taking into account the major mountain systems, and also the Antarctic and Greenland ice sheets.

The ocean submodel unites the primitive equations of motion in the Boussinesq approximation

$$
\left.\begin{aligned}
\frac{du}{dt} &= \left(f + u\,\frac{\tan\varphi}{a}\right)v - \frac{\partial p}{\rho_0 a \cos\varphi\,\partial\lambda} + F_\lambda + k_M\frac{\partial^2 u}{\partial z^2}, \\
\frac{dv}{dt} &= -\left(f + u\,\frac{\tan\varphi}{a}\right)u - \frac{\partial p}{\rho_0 a\,\partial\lambda} + F_\varphi + k_M\frac{\partial^2 v}{\partial z^2},
\end{aligned}\right\}
\tag{5.8.8}
$$

the hydrostatic equation

$$
\partial p/\partial z = -g\rho
\tag{5.8.9}
$$

and the continuity equation

$$
\frac{\partial u}{a\cos\varphi\,\partial\lambda} + \frac{\partial v\cos\varphi}{a\cos\varphi\,\partial\varphi} + \frac{\partial w}{\partial z} = 0,
\tag{5.8.10}
$$

as well as the evolution equations for the potential temperature θ and salinity S:

$$
\frac{d}{dt}(\theta, S) = k_H\nabla^2(\theta, S) + k_M\frac{\partial^2}{\partial z^2}(\theta, S),
\tag{5.8.11}
$$

where ρ_0 is the reference density of sea water; k_M and k_H are coefficients of the vertical and horizontal eddy exchange; F_λ and F_φ are the zonal and meridional components of the frictional force (they are calculated using a linearized version of appropriate formulae for the atmospheric submodel);

$$\frac{d}{dt}(\ ) = \frac{\partial}{\partial t}(\ ) + \frac{\partial u(\ )}{a \cos \varphi\, \partial \lambda} + \frac{\partial v(\ ) \cos \varphi}{a \cos \varphi\, \partial \varphi} + \frac{\partial}{\partial z} w(\ ),$$

$$\nabla^2(\ ) = \frac{1}{a^2 \cos^2 \varphi}\left[\frac{\partial^2}{\partial \lambda^2}(\ ) + \frac{\partial}{\partial \varphi}\left(\cos \varphi \frac{\partial}{\partial \varphi}(\ )\right)\right]$$

are the operator of the total time derivative and the Laplace operator respectively.

To find the sea water density the linear equation of state in Eckart's form is used:

$$\rho = \frac{p_0 + p_s}{1.000\,027[a + b(p_0 + p_s)]}, \tag{5.8.12}$$

where a, b and p_0 are parameters defined as $a = 1779.5 + 11.25\theta - 0.074\,5\theta^2 - (3.8 + 0.01\theta)S$; $b = 0.698$; $p_0 = 5890 - 38\theta - 0.375\theta^2 + 3S$; the density is measured in g/cm^3; the temperature θ in degrees Celsius, the salinity in ‰.

As in the atmospheric submodel the appearance of the hydrostatic instability is considered to be possible. It is resolved by means of convective adjustment, that is, by instantaneous equalization of the temperature and salinity profiles with an allowance being made for the integral conservation of the latter.

The vertical structure of the upper ocean layer is determined, apart from the convection, by wind mixing. Its effect is described within the framework of the local upper mixed layer model proposed by Kraus and Turner (1967). At the ocean surface the vertical velocity is assumed to be equal to zero, and components of the wind stress and vertical eddy fluxes of heat and salt determined in the atmospheric submodel are considered to be prescribed. The no-flow condition and the condition of vanishing of the vertical heat and salt fluxes are specified at the bottom. On the coastline the normal component of the velocity and the horizontal fluxes of heat and salt are assumed to be zero. The system of equations and boundary conditions is supplemented by the evolution equation for sea ice thickness, taking into account the possibility of change in sea ice thickness due to local factors (accretion, melting, sublimation, etc.) and the influence of the thickness on the sea ice drift velocity.

When solving the ocean submodel equations the standard staggered grid with finer resolution in the vicinity of the west coast is used. At high latitudes where the grid step decreases, the procedure of spatial Fourier smoothing is applied. The ocean is divided along the vertical into 12 layers whose thickness increases unevenly with distance from the free surface to the bottom.

The first numerical experiments within the framework of the coupled model were carried out for the domain representing the truncated spherical sector. From west to east it was limited by meridians separated at 120°, and from south to north by parallels ±81.7°. The zonal extension of the atmosphere and ocean was assumed to be equal everywhere excluding regions of high latitudes. Such a configuration of the domain in question resembles the Atlantic Ocean adjacent to North and South America. In the atmosphere the periodicity condition was prescribed at the boundary meridians and the no-flow conditions were specified at the boundary parallels. The land and ocean surfaces were considered to be flat, the ocean depth was taken to be 4 km. To overcome the difficulties connected with the different heat inertia of the atmosphere and the ocean, an artificial synchronization, that is, combining the ocean state at each time step with the time-averaged atmospheric state, was realized. This was performed using a procedure equivalent to an exponential filter with the weighted function being equal to $\lambda^{-1} \exp(t/\lambda)$ as $t \leq 0$ and to zero as $t > 0$, where λ is the time constant equal to one week.

Analysis of the results of integration of the model equations showed that an allowance for the heat transport in the ocean leads to the cooling of the atmosphere in low latitudes and its heating in mid latitudes. In its turn, a decrease in temperature difference between the equator and temperate latitudes causes a weakening of the zonal circulation, and, as a result of the baroclinic atmospheric instability, a decrease in the kinetic energy of macro-turbulence in temperate latitudes and in intensity of the Ferrel meridional cell. The heat advection in the ocean also favours deepening of the subtropical and polar atmospheric pressure maxima and baric depression in temperate latitudes providing the transport of air heated over the ocean to the east coast of a model continent. The existence of the cold equatorial upwelling effects a decrease in the precipitation rate in the tropical zone of the oceans and its increase over the continents. In subtropical and high latitudes the warm water transport by the subtropical gyre increases the sensible and latent transfer from the ocean to the atmosphere, as well as the precipitation along the eastern coast of the continent; in northern latitudes this transport causes the appearance of similar effects along the western coast of the continent. In total, the results of calculations using the coupled ocean–atmosphere model turn out to be more realistic than those using the model where the ocean is represented as a 'swamp' of zero heat capacity.

Subsequently, the main results of the above-mentioned numerical experiment were confirmed by more detailed calculations carried out as applied to the real geometry of continents and oceans. The calculations were performed on a grid containing nine levels in the atmosphere and 12 levels in the ocean. The spatial step was taken as equal to 500 km; synchronization of the atmospheric and ocean states was realized by equating 1.3 atmospheric years to 430 ocean years.

Let us dwell on the results of the calculation of the ocean fields. Though the equilibrium of the thermal regime has not been attained, the ocean temperature distribution turned out to be very similar to the observed temperature distribution. It has also been possible to simulate major features of the global salinity distribution: in the Atlantic the salinity was higher than in the Pacific Ocean, and in the Pacific Ocean it was found to be higher in the south than in the north. Also, the model properly reproduced the meridional circulation in the ocean, forming two cells with upwelling at the equator and downwelling in high latitudes. As it turned out, the meridional heat transport was found to reach a maximum in the subtropics where thermohaline and drift components of the current velocity are oriented in one direction. On the other hand, in temperate latitudes where drift and thermohaline components are oriented in opposite directions the meridional heat transport decreases.

Along with the above the inevitable descrepancies between calculated and observed fields of ocean characteristics can be mentioned. They primarily concern overestimation of temperature in high and polar latitudes, underestimation of salinity, clearly expressed in the North Atlantic, and a weakening of the intensity of the circulation as a whole. The last circumstance, as well as underestimation of the rate of cold deep water formation in high latitudes, is explained, according to Manabe *et al.* (1979), by the absence of the seasonal change of insolation in the model.

The inclusion of the seasonal variations of insolation with small changes in the model (coefficients of the vertical and horizontal eddy diffusion of momentum, heat and salinity in the ocean were prescribed by functions of depth; a synchronization was performed by equating two atmospheric years to 1200 ocean years) improved the reproduction of a number of climatic characteristics of the ocean–atmosphere system such as the distribution of pack ice in the Arctic and snow cover on the continents of the Northern Hemisphere. Consideration of the seasonal cycle of insolation resulted in the elimination of the unrealistically large thickness of sea ice (now it amounts to about 3 m instead of 35 m as it was before) and of the continuous accumulation of snow.

The time–space distribution of the zonal average surface air temperature over the ocean and land turned out to be very satisfactory. Specifically, it has been possible to simulate a phase shift (lag) and decrease in amplitudes of the seasonal oscillations of the surface air temperature over the ocean compared with their values over land in the temperate and low latitudes and an increasingly earlier occurrence of maximum nearer the pole at high latitudes. The last feature is caused by the influence of sea ice preventing heat exchange between the ocean and the atmosphere, and thereby contributing to the transformation of a maritime climate into a continental one.

A similar conclusion, in terms of accuracy of simulation of the time–space distribution of climatic characteristics, can be made with respect to the average (in the Northern and Southern Hemispheres) values of the air temperature at various heights in the atmosphere, and the water temperature at different depths in the upper ocean layer. It can be seen from Figure 5.8 that in the Northern Hemisphere where the land occupies about 40% of the hemisphere area the amplitude of the seasonal air temperature oscillations increases when approaching the underlying surface, basically because of the high variability of the albedo that relates to the appearance and disappearance of snow on continents. This feature is not found in the Southern Hemisphere. The diagram also demonstrates the asymmetry of the seasonal cycle of the water temperature in the upper ocean layer: maximum positive departures of temperature from its annual mean value in summer are higher than maximum negative departures in winter by almost 0.5 °C. The cause of this is the existence of winter convection limiting the drop in sea surface temperature, and of the seasonal thermocline shielding the warm upper layer from the cold deep ocean. This feature is well reproduced by the model.

The same can be said with reference to seasonal variations of the heat content and the meridional heat transport in the atmosphere and ocean. It was established by Manabe *et al.* (1975) that in temperate latitudes the meridional heat transport in the atmosphere takes maximum values at the beginning of winter and minimum values at the beginning of summer, for the most part due to appropriate changes in intensity of the baroclinic instability. At low latitudes the meridional heat transport even becomes negative, that is, directed from north to south. In the equatorial area this is performed by the Hadley cell transporting heat from the summer hemisphere to the winter hemisphere.

Data obtained indicate a close correlation between the meridional heat transport in the ocean and the intensity of the zonal atmospheric circulation: the heat transport to the equator is due to the action of westerly winds, and poleward due to the action of easterly winds. In winter when the westerly

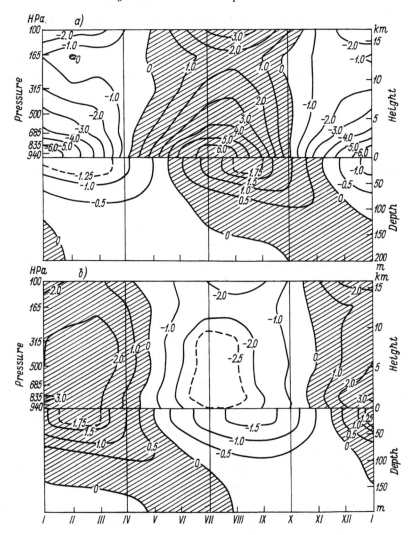

Figure 5.8 Seasonal variability of the hemispheric average air temperature (°C) at different heights in the atmosphere, and of the water temperature at different depths in the ocean, according to Manabe *et al.* (1979): (*a*) the Northern Hemisphere; (*b*) the Southern Hemisphere.

and easterly winds become simultaneously stronger, the heat transport to the pole increases in tropical latitudes of the ocean and decreases in mid latitudes. This feature reproduced by the model is in good agreement with actual data. The amplitudes of seasonal oscillations of the meridional heat transport in the ocean are in worse agreement (they turn out to be several times underestimated). But the worst of this is that the underestimation cannot be explained.

The amplitudes of seasonal oscillations of the sea surface temperature in mid latitudes of the Northern Hemisphere, and of the surface air temperature in high latitudes of the Southern Hemisphere also turned out to be under-estimated. On the other hand, the amplitudes of seasonal oscillations of the atmospheric heat content at high latitudes and also the amplitudes of seasonal oscillations of latent heat in snow and sea ice were overestimated. The discrepancies listed are explained by Manabe *et al.* (1979) to be the result of inaccuracy in prescribing cloudiness (particularly since observational data are unavailable, the field of cloudiness in the Southern Hemisphere is assumed to be the same as in the Northern Hemisphere), and of too low values of the mass transport in the major ocean currents.

IOAS model

The main distinction of this model (see Zilitinkevich *et al.*, 1978) from the GFDL model is the rejection of the artificial synchronization of the oceanic and atmospheric states, and the subdivision of the ocean into the upper active layer naturally synchronized with the atmosphere, and the deep ocean, the state of which is calculated separately. Such an approach was designed for a description of the relatively short-range processes (i.e., seasonal oscillations and interannual variability) in which the state of the deep ocean can be considered as prescribed to a first approximation.

The model consists of four submodels: free atmosphere, the atmospheric planetary boundary, the active ocean layer, and the deep ocean. To simulate the climatic state of the free atmosphere the hydrothermodynamic equations are used in exactly the same form as in the GFDL model; they are distinguished only by the methods of parametrization of the physical processes. Parametrization of small-scale interaction between the atmosphere and the underlying surface accepted in the IOAS model is based on the similarity theory for the Ekman boundary layer, with the latter being considered to be submerged into the low grid layer containing 15% of the total atmospheric mass. The space–time variability of this layer is calculated with the help of equations, and the vertical structure is assumed to be universal, that is, as the similarity theory for the Ekman boundary layer predicts.

The calculation of the characteristics of interaction between the atmosphere and the underlying surface is realized within the framework of laws of resistance, heat and moisture exchange considering the influence of the density stratification and change in the underlying surface roughness on land. The underlying surface temperature is determined from the heat budget equation; in the ocean this equation serves to determine the resulting heat flux at the ocean–atmosphere interface. In the presence of ice cover at the ocean surface the temperature of its upper surface is found in the same fashion as that on land, that is, it is assumed that the ice completely isolates the ocean from the atmosphere. For the permanent ice cover, including pack ice, a restriction is introduced which does not allow the surface temperature to be higher than 0 °C. It is assumed that in this case the resulting heat flux is expended for ice melting.

The IOAS model uses coarse vertical resolution of the atmosphere: the latter is divided into layers with interfaces at heights 1.5, 4.5 and 11 km. But, unlike the GFDL model, cloudiness is not prescribed from climatic data but is determined from the empirical formulae at every time step. A scheme of calculation of radiative fluxes is also simplified in accordance with the coarse vertical resolution. The effects of three-level cloudiness and of absorption by water vapour are taken into account. The mesoscale convection and water vapour phase transitions are calculated in a similar way as in the GFDL model but instead of equalization of the relative humidity with height the condition of similarity for the convective transfer of specific humidity and equivalent potential temperature is used.

Due to limitations in available computer resources the effects of salinity in the active ocean layer are not taken into account. The influence of vertical motions is not taken into account either, and the depth-averaged horizontal components of the current velocity in the active layer are decomposed into climatic constituents borrowed from the deep ocean submodel and drift constituents determined by the wind stress. It is also supposed that the thermal structure of the ocean active layer possesses universality: there is a distinctive upper mixed layer and its underlying thermocline with self-similar temperature distribution. The temperature at the low boundary of the active layer is taken from the deep layer submodel, and the temperature and thickness of the upper mixed layer are determined from the heat and turbulent energy budget equations integrated over the depth of the active layer.

These equations are complemented by the following algorithm, roughly describing the influence of ice cover: if the temperature found from the heat budget equation for the ocean surface becomes less than the sea water freezing point (-1.8 °C), it is assumed that ice appears, completely isolating the active layer from the atmosphere. Accordingly, the resulting heat flux at the water–air interface and the kinetic energy flux from the atmosphere into the ocean are assumed to be equal to zero, and the temperature of the subice upper mixed layer is prescribed as equal to the sea water freezing temperature. In so doing, the temperature of the upper ice surface is determined from the heat budget equation for the ocean surface on the assumption that the resulting heat flux is equal to zero. As soon as the temperature of the upper ice surface becomes higher than the sea water freezing point the ice is believed to disappear.

The two-dimensional hydrothermodynamic model of the ocean global circulation, proposed by Kagan *et al.* (1974), is used as a submodel of the deep ocean. It is based on the following assumptions: the sea water density depends only on temperature; variations in the latter take place only in the baroclinic layer with 2 km thickness below which the temperature is fixed, and in the layer itself it is presented in the form of the product of some standard function of the vertical coordinates and the required function of horizontal coordinates and time, which is determined from the heat transport equation integrated within the limits of the whole ocean thickness. The barotropic components of velocity appearing in this equation are found from the equation for the integral stream function; the vertically averaged velocity components of the drift current are determined from the Ekman equations by the prescribed wind stress at the ocean surface. Finally, the vertically averaged baroclinic velocity components are estimated with the help of quasi-geostrophic relations. The boundary conditions are specified in such a way as to provide realization of heat and mass conservation laws in the World Ocean.

The equations of the deep ocean submodel are integrated for a latitude-longitude grid with 5° angular resolution from the initial state complying with the state of the rest for the horizontally homogeneous (but stratified along the vertical) ocean, up to establishing an equilibrium regime. The wind stress and the resulting heat flux at the ocean surface are calculated by the annual mean fields of atmospheric pressure, air temperature and radiation budget at the ocean surface with fixed values of the Bowen ratio equal to 0.5. Note that only fields of the current velocity and temperature at the low boundary of the active layer need to be provided for the coupled ocean–atmosphere model.

The equations for the atmospheric submodel and the active ocean layer are integrated on Kurihara's spherical grid with a horizontal step of about 1000 km. The annual mean meridional distributions of temperature in the atmosphere and the upper mixed layer of the ocean, the adiabatic vertical distribution of air temperature, the absence of wind as well as constant (in a horizontal plane) values of the surface atmospheric pressure, the relative air humidity and the upper mixed layer thickness are used as initial conditions.

A calculation using the natural synchronization of atmospheric and ocean states was carried out over a period of 1000 days, with an allowance for the seasonal cycle of insolation. An equilibrium quasi-periodical regime of the atmosphere–active ocean layer system was reached after about a year. By that time the mass-averaged wind velocity amounted to 17 m/s (up to 40 m/s in the upper layer); the mass-averaged temperature of the atmosphere amounted to 244 K; the underlying surface temperature amounted to 282 K (minimum monthly averaged values in the Antarctic and maximum in northern Africa turned out to be equal to 234 K and 308 K respectively); the zonal average humidity amounted to 1.6 g/kg; the total fraction of cloudiness amounted to 0.47; evaporation and precipitation amounted to about 3.1 mm/day; the net radiative flux at the underlying surface amounted to 470 W/m^2 (80% of this flux is spent as long-wave emission of the Earth's surface, the remaining 20% is used for evaporation and sensible heat exchange). The atmospheric pressure over the ocean turned out to be less than that over land (994 as against 1040 hPa), and, on the other hand, the air temperature, specific humidity, cloudiness, evaporation and precipitation turned out to be more (by 5 °C, 0.9 g/kg, 0.4 and 2.9 mm/day respectively).

The model has been shown to simulate the main features of time–space variability of the atmospheric temperature, precipitation, evaporation, cloudiness, components of the heat budget and the vertical mass flux at the ocean surface, as well as characteristics of the upper mixed ocean layer: its temperature and thickness. In order to confirm the above we mention, in particular, the model detected asymmetry of seasonal changes in air temperature with respect to the equator connected with the different relationship between land and ocean areas in both hemispheres; winter maximum of precipitation in the equatorial region over oceans and summer maximum in tropical latitudes on land; enhancement of evaporation in temperate latitudes of the ocean in winter and on land in summer caused by an increase in temperature difference between the underlying surface and the atmosphere; predominance of cloudiness over the ocean rather than over the land; a six-month phase shift between seasonal oscillations of the resulting heat flux at the ocean surface in the Northern and Southern Hemispheres; and, finally, an increase in the amplitudes of seasonal oscillations of the mass flux in temperate latitudes of the ocean and the retention of its minimum values at the equator throughout a year. The last feature is the result of the small water–air temperature difference and the shielding influence of cloudiness.

The global distribution of the upper mixed layer thickness obtained using the IOAS model demonstrates that from the beginning of summer heating it decreases to several tens of metres in temperate latitudes and to 120–150 m in the subtropics. In summer its distribution is characterized by strong spatial variability. In the winter period the upper mixed layer extends practically over the whole active ocean layer.

CC SBAS model

This model (its detailed description can be found in Marchuk *et al.*, 1984) differs from the models discussed above by its more economic numerical algorithm. Its basis is the use of a combination of two types of splitting (for the time derivative operator and the advective transport and diffusive operators) leading to the fulfilment of the integral laws of conservation of mass, angular momentum, energy (on the adiabatic approximation), moisture, etc. It is appropriate to emphasize here that this algorithm allows us to perform integrations with much larger time steps than is accepted in traditional difference schemes.

An atmospheric submodel of the CC SBAS model contains the evolution equations for the zonal and meridional components of the wind velocity, specific humidity and surface pressure. With a view to designing an absolutely stable difference scheme the first four equations are 'symmetrized' by transition from u, v, T and q to the variables $p_s^{1/2}u$, $p_s^{1/2}v$, $p_s^{1/2}T$ and $p_s^{1/2}q$. Parametrization of the horizontal eddy viscosity is carried out with an allowance for two conditions. That is, the term describing the horizontal momentum diffusion is required to be dissipative, and the angular momentum is to be retained. These conditions are met by the following expressions:

$$F_\lambda = \frac{1}{a^2 \cos^2 \varphi} \left[\frac{\partial}{\partial \lambda} k_H \sigma \frac{\partial u}{\partial \lambda} + \frac{\partial}{\partial \varphi} k_H \sigma \cos^3 \varphi \frac{\partial}{\partial \varphi} \left(\frac{u}{\cos \varphi} \right) \right],$$

$$F_\varphi = \frac{1}{a^2 \cos^2 \varphi} \left[\frac{\partial}{\partial \lambda} k_H \sigma \frac{\partial v}{\partial \lambda} + \frac{\partial}{\partial \varphi} k_H \sigma \cos^3 \varphi \frac{\partial}{\partial \varphi} \left(\frac{v}{\cos \varphi} \right) \right],$$

where, as before, $\sigma = p/p_s$, k_H is the horizontal eddy viscosity coefficient. Indeed, we multiply F_λ by u, and F_φ by v and integrate over the whole area of the atmosphere. Then, for example, for uF_λ we obtain

$$\int uF_\lambda a^2 \cos \varphi \, d\lambda \, d\varphi = \int \frac{u}{\cos \varphi} \frac{\partial}{\partial \varphi} k_H \sigma \cos^3 \varphi \frac{\partial}{\partial \varphi} \left(\frac{u}{\cos \varphi} \right) d\lambda \, d\varphi$$

$$= - \int k_H \sigma \left[\frac{\partial}{\partial \varphi} \left(\frac{u}{\cos \varphi} \right) \right]^2 \cos^3 \varphi \, d\lambda \, d\varphi \le 0,$$

that is, there is dissipation. Further, we multiply F_λ by $\cos \varphi$ and integrate over the whole area of the atmosphere. As a result we obtain

$$\int F_\lambda a^2 \cos^2 \varphi \, d\lambda \, d\varphi = 0,$$

whence the conservation of angular momentum follows. The horizontal eddy diffusion of heat and moisture is parametrized in the traditional way:

$$F_{T,q} = \frac{1}{a^2 \cos^2 \varphi} \left[\frac{\partial}{\partial \lambda} \left(k_H \sigma \frac{\partial}{\partial \lambda} T, q \right) + \cos \varphi \frac{\partial}{\partial \varphi} \left(k_H \sigma \cos \varphi \frac{\partial}{\partial \varphi} T, q \right) \right].$$

The parametrization scheme for the atmospheric planetary boundary layer reduces to singling out the logarithmic layer and the overlying well mixed layers within the limits of which wind velocity, temperature and humidity are considered to be constant with height. At the same time, it is assumed that the position of the upper boundary of the planetary boundary layer coincides with the grid level closest to the underlying surface, and the angle between the wind velocity vector in the mixed layer and wind stress at the underlying surface remains constant and equal to 30° in an extra-tropical area over land, 20° over ocean, and 10° over ice. In the tropics this angle is assumed to be equal to zero. To estimate the eddy fluxes of momentum, heat and moisture at the ocean surface the well-known bulk formulae with resistance and heat exchange coefficients depending on wind velocity and stratification are applied.

The parametrization of the processes of convective adjustment and large-scale condensation is performed similarly to that in the GFDL model. The fraction of cloudiness is calculated using empirical relations. The thickness of clouds, their albedo and absorption capacity, as well as the albedo of the ocean surface and ice, ozone and carbon dioxide concentrations, are assumed to be fixed. The albedo of the land surface is determined as a function of the snow cover thickness in water equivalent.

The ocean submodel of the CC SBAS model includes the complete system of the ocean hydrothermodynamic equations, which does not differ much from that adopted in the GFDL model. But the method of solution differs. For this purpose a splitting technique is used where the extended spatial operator is decomposed into a number of simpler ones. As in the GFDL model, artificial synchronization of the atmosphere and ocean states is introduced, whereby one atmospheric year is equated to about 100 ocean years. The time sampling of the information exchange between the atmospheric and ocean submodels is chosen as appropriate to the characteristic time τ of mixing in the ocean upper mixed layer with 100 m thickness. For $\tau = 14$ days and time step equal to 40 minutes in the atmosphere and two days in the ocean the information exchange is performed every six atmospheric and every seven ocean steps. All atmospheric data transmitted to the ocean submodel are smoothed in time with the difference analogue of the exponential filter.

The model described was tested on January average conditions. The calculation was carried out in two stages. First, the equilibrium regime of the atmospheric circulation with fixed sea surface temperature and the prescribed distribution of sea and continental ice were calculated. Then the ocean submodel with four levels in the vertical was run. These levels were located at 100, 500, 1500 and 3000 m. The total duration of the calculation amounted to 11 ocean years, that is, equivalent to two atmospheric months.

An analysis of the results obtained detects a general decrease in temperature in the ocean of the Northern Hemisphere and its increase in temperate and high latitudes of the Southern Hemisphere. This, naturally, affects the atmospheric circulation. Specifically, there is a shift in the tropical belt of precipitation by about 10° to the south, a diminution (especially in the Southern Hemisphere) of the meridional temperature gradient, a decrease (of approximately 25%) of the available potential energy and its transformation into kinetic energy, an enhancement of the direct Hadley cell in the Northern Hemisphere and a weakening of the opposite Ferrel cells in both hemispheres, a decrease of wind stress, and a deepening of the centres of low pressure over the continents of the Southern Hemisphere.

Consideration of the effects of ocean–atmosphere interaction implies the following sequence of events. At first, on the western coasts of the Pacific Ocean and in the north-west part of the Indian Ocean, negative anomalies of water temperature appear which cause the appearance of strong western boundary currents and the enhancement of upwelling. This, in turn, leads to intensification of the meridional circulation, smoothing of horizontal temperature gradients in the ocean and weakening of the interaction between the ocean and atmosphere.

NCAR model

This model differs from the GFDL model mainly by finite difference approximation of equations of the atmospheric submodel in the vertical direction; by parametrization of the physical processes of the subgrid scale; and by the method of synchronization of the atmosphere and ocean states. In the NCAR model the atmosphere is divided into eight layers, each three kilometres thick. Accordingly, the methods calculation of radiative fluxes, cloudiness and eddy fluxes of momentum, heat and moisture at the underlying surface, as well as the temperature of the latter, are modified. Specifically, when calculating radiative influxes, more accurate functions of absorption of short-wave radiation and of transmission of long-wave radiation are used, and the absorptive properties of various types of underlying surface are taken into account. The diurnal change in insolation is described explicitly. The fractional cover of low and middle level clouds is determined from empirical relations. The base of low-level clouds is located at a height of 1.5 km and the top of the clouds is fixed. The middle-level clouds are assumed to be infinitely thin. The fraction of upper-level clouds is prescribed from the climatological data and their base is placed at a height of 10.5 km over the equator and 7.5 km over the poles; the thickness is taken as 1.5 km.

Eddy fluxes of momentum, heat and moisture are calculated, taking into account the dependence of resistance and heat exchange coefficients on stratification of the surface atmospheric layer, and the temperature of the underlying surface – with an allowance for the type of surface. Three types of underlying surface are singled out: the ocean surface, the sea ice surface and the land surface both covered and not covered with vegetation. In turn, the sea ice surface and the land surface not covered with ice are differentiated depending on whether they are covered with snow or not, and depending on the soil moisture. To each type or subtype of the underlying surface is assigned a certain value of albedo, assumed to be a function of snow thickness and of the solar zenith angle. In the presence of vegetation, the temperature distinctions of vegetation cover and the underlying surface, transpiration and evaporation, as well as the transformation of the vertical structure of wind in vegetation cover, are taken into account. A detailed description of the atmospheric submodel of the NCAR model is presented by Washington and Williamson (1977) and Washington *et al.* (1980).

The ocean submodel of the NCAR model does not differ much from that used in the GFDL model. Compared with the latter, the NCAR model excludes the effects of the horizontal transport and friction in the equation describing the evolution of sea ice thickness (thus, the sea ice model becomes a purely thermodynamic model), and employs other values of coefficients of horizontal and vertical eddy diffusion of heat and momentum. Discretization in the vertical direction is produced differently: the ocean is divided into four layers with thickness (from above) of 50, 450, 1500 and 2000 m.

The method of synchronization of atmospheric and ocean states reduces to the following. The equations of the ocean submodel are integrated with fixed (for January, April, July and October) values of atmospheric parameters for a period of five years. The temperature of the underlying surface and sea ice area obtained in the last year of this period are used as initial information when integrating the equations of the atmospheric submodel separately for each of the four months mentioned above. The values of the atmospheric parameters obtained are assumed to be the initial values when integrating the equations of the ocean submodel during the second five-year period. Again, the atmospheric parameters corresponding to each of the four months of the annual cycle are found and this is continued until a quasi-equilibrium regime is established in all links of the ocean–atmosphere system. According to estimates by Washington *et al.* (1980), this method of synchronization, from a computing viewpoint, is ten times more effective than that approved for the GFDL model.

The zonal air temperature, longitudinally averaged zonal component of wind velocity, ocean surface temperature, current velocity in the upper 50 metre layer and its underlying 450 metre ocean layer, and, finally, the vertical velocity at the interface between these layers obtained at the end of the fourth five-year period turned out to be close to those observed in terms of quality but not quantity. For example, in January the zero isotherm is at latitude 38 °N according to calculated results, and at latitude 50°N according to observational data. This points to an underestimate of the calculated values of air temperature in the Northern Hemisphere in winter. Next, according to calculated results the summer air temperature at all levels in the troposphere has a maximum in the region 40–50 °N and this is not confirmed by observational data. We note also an overestimation of the meridional air temperature gradient in the Northern and Southern Hemispheres, a shift of the equatorial zone of westerly winds into the Southern Hemisphere in winter and into the Northern Hemisphere in summer and large discrepancies (both in height and in direction) between calculated and observed wind velocities in high latitudes of the Southern Hemisphere.

The model simulated the main features of seasonal variability of the ocean surface temperature and location of the thermal equator correctly. But the horizontal temperature gradients in the regions of the Gulf Stream and Kurosio turned out to be underestimated, and temperature values in the subtropics and in midlatitudes turned out to be overestimated compared with those observed. The first circumstance is caused by the coarse (5°) grid resolution, the second by underestimation of the wind velocity and its associated decrease in the heat exchange between the ocean and atmosphere, and in the heat transport by drift currents. It is emphasized that overestimation of the surface temperature in the Southern Ocean has led to a decrease in the sea ice around the Antarctic. According to the authors of the model this is due to the same causes and to the prescription of a too large value of the horizontal eddy heat diffusion coefficient.

The qualitative agreement of calculated and observed fields is satisfactory. Suffice it to say that in the first grid layer (Figure 5.9) the Antarctic Circumpolar Current, Gulf Stream, Kurosio, West-Australian, Californian and Bengwale currents are apparent. The westward equatorial currents in the Atlantic and Pacific Oceans are found to be in the second grid layer, and the narrow Pacific equatorial countercurrent is found to be in the first layer. Some currents (such as Agulhas and Labrador) are not presented at all

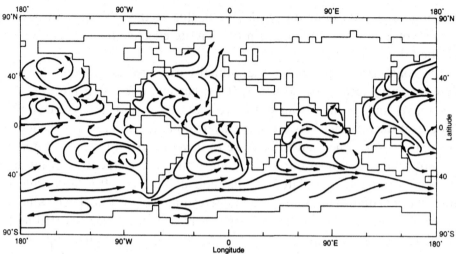

Figure 5.9 Field of currents in the upper 50-metre layer of the World Ocean in January, according to Washington *et al.* (1980).

because of the coarse grid resolution. The velocities of all currents turned out to be about one-third of those observed.

Literally everything mentioned above can be applied to the vertical motion velocities. The model simulated the intensive equatorial upwelling and regions of downwelling located 15–20° to the north and south of the equator. In both cases extreme velocities of vertical motions (according to calculated results and observational data) take place in the eastern part of the Pacific Ocean. But calculated values of the vertical velocity in coastal upwelling regions turn out to be less than those observed. This cause is the same: the coarse spatial resolution of the grid.

OSU model

The OSU model (see Gates *et al.*, 1985, and Han *et al.*, 1985) differs from the GFDL model mainly by the coarser horizontal and vertical resolution, by the inclusion of cloudiness in a number of unknowns to be determined, and by the rejection of artificial synchronization of ocean and atmospheric states.

The model has the following structure. It consists of two atmospheric and six ocean layers. The upper boundary ($\sigma = 0$) of the atmosphere is coincident with the isobaric surface $p_T = 200$ hPa; the lower boundary is coincident with the isobaric surface where the pressure p is equal to the surface pressure p_s; the interface ($\sigma = 1/2$) between layers is coincident with the surface $p = (p_T + p_s)/2$. Each chosen atmospheric layer is divided in turn into two sublayers with interfaces $\sigma = 1/4$ ($p \approx 369$ hPa) and $\sigma = 3/4$ ($p \approx 788$ hPa).

The horizontal components of wind velocity, air temperature and specific humidity are determined at these levels. The isobaric vertical velocity is calculated at the level $\sigma = 1/2$ and is assumed to be equal to zero at levels $\sigma = 0$ and $\sigma = 1$. The interfaces between ocean layers are located at depths of 50, 250, 750, 1550, 2750 and 4350 metres. All layers are divided in half by intermediate depths where horizontal components of the current velocity, sea water temperature and salinity are calculated. The vertical velocity is determined at the interfaces between layers. At the ocean surface it is taken equal to zero, at the bottom it is assumed equal to the orographic vertical velocity defined by the kinematic relation. The upper 50-metre layer is assumed to be well mixed, and the temperature at the intermediate depth $z = 25$ m is assumed to coincide with the sea surface temperature. The horizontal resolution (equal in the ocean and the atmosphere) is preset equal to 4° in latitude and 5° in longitude.

As has already been noted the model provides the determination of cloudiness. Depending on the mechanism of its formation four types of cloud are distinguished, corresponding to penetrative and cumulus convection at the middle level $\sigma = 1/2$ (type 1), large-scale condensation at levels $\sigma = 3/4$ (type 2) and $\sigma = 1/2$ (type 4) and to cumulus convection at the level $\sigma = 3/4$ (type 3). Cumulonimbus clouds with base at $\sigma = 1/4$ and top at $\sigma = 0$ are clouds of the first type; stratus and cirrus clouds of the lower level with base at $\sigma = 3/4$; and top at the level $\sigma = 1/2$ are clouds of the second type; thin cumulus and cirrus clouds of the middle level at $\sigma = 3/4$ are clouds of the third type; and stratus and cirrus clouds of the upper level with base at $\sigma = 1/2$ and top at $\sigma = 1/4$ are clouds of the fourth type. It is considered that cumulus clouds are formed in the case where the vertical gradient of the equivalent potential temperature becomes less than zero, and the local relative humidity becomes more than 80%. Meanwhile, stratus and cirrus clouds are formed if the local relative humidity is more than 80% and the vertical gradient of equivalent potential temperature takes either positive or zero values (the atmosphere is hydrostatically stable). It is also assumed that clouds of the second, third and fourth types cannot coexist simultaneously with clouds of the first type, and that clouds of the third type cannot coexist simultaneously with clouds of the second type. However, the coexistence of clouds of the second and third types with clouds of the fourth type is possible. In the last case all clouds are ascribed to the lower level.

Calculation of atmospheric characteristics is carried out in two stages: first (every ten minutes) changes are found that are determined by horizontal and vertical advection, then their values are corrected at the expense of internal and external sources and sinks of heat and momentum. Non-adiabatic factors are addressed every hour. The time step for integrating the ocean submodel equations is one hour, which makes it possible to synchronize the heat and momentum change between the ocean and atmosphere without the need for any additional procedures of time smoothing or averaging.

Preliminary results concerning the evolution of the ocean–atmosphere system during the first 16 years of the simulation are discussed by Han *et al.* (1985). We dwell only on some global characteristics here. The model correctly simulates maximum positive values of the net radiative flux at the upper atmospheric boundary in January and underestimates its minimum negative values in June. This last circumstance is connected with under-estimation of the outward long-wave radiation flux, and the latter in turn with errors when determining cloudiness in winter in the Southern Hemisphere, and in summer in the Northern Hemisphere. The annual mean global net radiation flux at the upper atmospheric boundary amounts to 4–5 W/m^2 instead of 0, as it needs to be in the steady state. The authors explain this feature by the influence of numerical viscosity and by disregarding the trans-formation of mechanical energy into heat due to viscous forces.

The model underestimates the extremal values of the resulting heat flux at the underlying surface in June and February caused by overestimation of sensible and latent heat fluxes in temperate and high latitudes of the ocean in winter, and as a result does not reproduce observed variations of the sea surface temperature: after ten years from the start of integration the calculated and observed seasonal oscillations of the sea surface temperature turn out to be out of phase with each other. The cause is the progressive decrease of the sea ice area in the Southern Ocean leading to an increase in the contribution of the oceans of the Southern Hemisphere to the formation of seasonal oscillations of the global average sea surface temperature (in reality these oscillations are mainly controlled by the oceans of the Northern Hemisphere).

We note a systematic underestimation (by approximately 1.5 °C) of the average temperature of the troposphere and a gradual increase in seasonal oscillations of the surface air temperature. Underestimation of the average tropospheric temperature relates to the coarse vertical resolution, and an increase in seasonal oscillations of the surface temperature relates to a decrease in the sea ice area in the Southern Ocean. By the way, the disadvantages detected (underestimation of the sea ice area and, accordingly, overestimation of the surface temperature in the Southern Ocean) are inherent not only in this model but also in other coupled ocean–atmosphere general circulation models. But it is interpreted differently in other models: the authors of the GFDL model explain it by overestimation of the absorbed solar radiation flux at the ocean surface and by underestimation of the heat transport to the equator by drift currents; the authors of the NCAR model explain it by overestimation of the horizontal eddy heat diffusion coefficient; and the authors of the OSU model explain it by overestimation of the absorbed solar radiation flux and by the inadequate reproduction of the Antarctic upwelling when using models with a coarse horizontal resolution. It is obvious that the first, second and third explanations have some grounds but it is not clear whether these are sufficient or not.

UKMO model

This model consists of atmospheric, ocean and sea ice submodels. The atmospheric submodel is an 11-level version of the climatic model of the United Kingdom Meteorological Office which is more advanced, as compared with that presented by Corby *et al.* (1977) and that presented by Slingo and Pearson (1987), in the following respect. First, it contains a parametrization of the orographic gravity waves drag determined by the interaction between gravity waves and elements of the Earth's surface orography. Second, it is assumed that the planetary boundary layer of the atmosphere includes the surface layer and the overlying mixed layer with the inversion above it. The height of the planetary boundary layer identified with the inversion base is assumed to be variable. This is found with the help of the evolution equation incorporating air entrainment in a regime of convective instability. Provision is made for the fact that under stable stratification the thickness of the planetary boundary layer can decrease down to its minimum value determined by mechanical mixing. Third, the model describes the vertical distribution of temperature in the active soil layer. For this purpose the soil temperature is calculated on four levels. Finally, the cloudiness is assumed to be interactive, that is, generated by the model, and radiative properties of clouds are assumed to be constant and equal to their characteristic values for the Earth as a whole.

The ocean submodel is represented by the 17-level version of the model taking account of isopycnic diffusion and the dependence of the vertical diffusion coefficients on stratification. Sea ice is described within the framework of the thermodynamic model where the prognostic variables are the thickness of ice and snow, surface temperature and ice concentration, the latter being included in the number of unknowns allows for the presence of leads.

The model equations are integrated on a latitude–longitude $2.5° \times 3.75°$ grid. The exchange of information between ocean and atmospheric submodels is performed discretely: the fluxes of heat, fresh water and momentum which are necessary for renewal of the boundary conditions in the ocean submodel, as well as the sea surface temperature, area and concentration of sea ice which are necessary for renewal of the boundary conditions in the atmospheric submodel, are transmitted from one submodel to another every five days.

Data of calculations presented by Foreman *et al.* (1988) were obtained as the result of a four-year integration of the coupled atmospheric–ocean–sea ice model. Such an integration period is not sufficient to reach an equilibrium state of the system but it is acceptable for detecting tendencies to changes in the solution. In particular, it has been found that the model overestimates (sometimes by several degrees) the sea surface temperature in the tropics. One of the reasons for this is an underestimation of sensible and latent heat fluxes in regions with high sea surface temperature and low wind velocity. Another reason is the existence of local feedback between the temperature and salinity of the ocean surface: overestimation of sea surface temperature contributes to intensification of convection in the atmosphere, and the latter leads to an increase in precipitation, the appearance of a surface layer with relatively fresh water, an increase in static stability, weakening of the vertical mixing, and, finally, to a rise in the sea surface temperature. One more possible reason for the overestimation of sea surface temperature in the tropics can be underestimation of cloudiness, causing overestimation of the absorbed solar radiative flux. It follows from the data presented that the model under-estimates mass transports of the western boundary currents. For example, the maximum mass transport of the Gulf Stream turned out to be equal to 35×10^6 m^3/s, while from observations it is of the order of 100×10^6 m^3/s. The cause of this is clear: the coarse spatial resolution of the grid does not allow simulation of the real western boundary currents with scales less than the grid size.

The inadequate reproduction of the meridional heat transport in the ocean of the Northern Hemisphere is closely connected with underestimation of the mass transport in the western boundary currents. This explains the under-estimation of calculated values of sea surface temperature to the north of the 30 °N parallel. The model also underestimates seasonal variations of the sea ice area in the Southern Ocean that, according to the authors, is caused by neglecting to take the dynamics of sea ice into account. But the worst of it is that the annual mean area of the Antarctic sea ice tends to decrease with time. On the other hand, the ice area in the Arctic turns out to be overestimated throughout the year and is almost unaffected by seasonal variations. It is also remarkable that there is an underestimation, by 10–15 °C, of summer surface temperatures over the continental areas to the north of 55°, connected with overestimation of the ice cover area.

The amount of low-level cloud over midlatitudes of land in the Northern Hemisphere turns out to be underestimated, especially in winter and in spring.

This results not only in an overestimation of the active surface temperature but in a delay (almost by a month) of the beginning of snow melting. As a result the summer period becomes shorter, and the summer temperatures become less than those observed by 4–5 °C. This circumstance has one more consequence: a decrease in the advective heat transport from the continents to the Arctic basin. Accordingly, the sea ice area in the Arctic increases, and the temperature of its active surface decreases.

The above effect is intensified further due to the significant decrease in summer values of the absorbed solar radiation flux in the Arctic Ocean. This is determined not only by the existence of feedback between the albedo and temperature of the underlying surface but also by underestimation by more than twice the incoming solar radiation. The last feature is caused by fixing the radiative properties of clouds: the amount of low-level clouds in the Arctic turns out to be realistic but, owing to the fact that their radiative properties are assigned as being uniform everywhere and complying with global average values, that is, more appropriate for clouds in low and temperate latitudes, the absorption of solar radiation by clouds turns out to be too large, and the solar radiation flux to the underlying surface is too small. All this is aggravated by underestimation of the meridional heat transport in the polar zone of the atmosphere and ocean.

MPI model

At first this model was intended not so much for simulation of the present-day climate as for a demonstration of the procedure of elimination of one objectionable feature appearing during the process of solving the problem. We mean the so-called *solution drift*, that is, the slow transition of the solution from one steady state complying with autonomous (non-interactive) models of the ocean and atmosphere to a new steady state corresponding to interactive models of the two media. This phenomenon is caused by inadequacy in the separate submodels and a mismatch between them, a consequence of which is that the new steady state turns out to be far from the real state despite the fact that both autonomous models have been thoroughly calibrated and tested for their agreement with observational data prior to their coupling.

The essence of the proposed procedure reduces to the following (Sausen *et al.*, 1988). If the vector of the atmospheric state is designated as Φ and the vector of the ocean state as Ψ, and if it is considered that the evolution of the model climatic system differs from the evolution of the real climatic system by an error equal to E_A for the atmosphere and E_O for the ocean, then the equations describing the evolution of the interacting atmosphere and ocean can be presented in the form

$$\begin{rcases} \partial\Phi/\partial t = G_A(\Phi, t) + F(\Phi^b, \Psi^b, t) + E_A(\Phi, \Psi, t), \\ \partial\Psi/\partial t = G_O(\Phi, t) - F(\Phi^b, \Psi^b, t) + E_O(\Phi, \Psi, t), \end{rcases} \quad (5.8.13)$$

where G_A, G_O and F are sources and sinks of substances in the atmosphere and ocean and the exchange between them respectively; superscript 'b' designates boundary values of the functions Φ and Ψ (say wind velocity, temperature and moisture in the surface atmospheric layer, the sea surface temperature, etc.) which are necessary to calculate the exchange.

Similarly, the equations describing the evolution of uncoupled atmosphere and ocean take the form

$$\begin{rcases} \partial\Phi_u/\partial t = G_A(\Phi_u, t) + F(\Phi^b_u, \Psi^b_m, t) + E_A, \\ \partial\Psi_u/\partial t = G_O(\Psi_u, t) - F(\Phi^b_m, \Psi^b_u, t) + E_O. \end{rcases} \quad (5.8.14)$$

Here, as is common practice, the boundary values Φ^b_u, Ψ^b_u of the variables Φ_u, Ψ_u are replaced by empirical data Φ^b_m, Ψ^b_m.

In (5.8.13) we add the components ΔF_A, ΔF_O and define them in such a way that the solutions obtained within the framework of interactive and autonomous models should coincide with each other. Then instead of (5.8.13) we arrive at the following equations:

$$\begin{rcases} \partial\Phi/\partial t = G_A(\Phi, t) + F(\Phi^b, \Psi^b, t) + E_A + \Delta F_A, \\ \partial\Psi/\partial t = G_O(\Psi, t) - F(\Phi^b, \Psi^b, t) + E_O - \Delta F_O. \end{rcases} \quad (5.8.15)$$

In (5.8.15) we replace Φ and Ψ by Φ_u and Ψ_u and then subtract Equations (5.8.14). As a result we obtain

$$\begin{rcases} \Delta F_A = F(\Phi^b_u, \Psi^b_m, t) - F(\Phi^b_u, \Psi^b_u, t), \\ \Delta F_O = F(\Phi^b_m, \Psi^b_u, t) - F(\Phi^b_u, \Psi^b_u, t), \end{rcases} \quad (5.8.16)$$

from which it follows that ΔF_A, ΔF_O are differences of the exchange found using empirical information and information from the autonomous models of the ocean and atmosphere, without using empirical data. Thus, the problem reduces to the determination of correcting the components ΔF_A, ΔF_O and the subsequent integration of Equations (5.8.15).

To illustrate the procedure, which is called the *procedure of exchange correction*, we examine, in accordance with Saussen *et al.* (1988), a simple box model of the ocean–atmosphere system, the type we discussed in Section 1.1. Let the evolution of the heat budget in the atmosphere and ocean be described by the system of equations

$$\begin{rcases} c_A \, dT_A/dt = R_A - \lambda_A T_A + \lambda_{AO}(T_O - T_A), \\ c_O \, dT_O/dt = R_O - \lambda_O T_O - \lambda_{AO}(T_O - T_A), \end{rcases} \quad (5.8.17)$$

where T_A and T_O are temperatures of the atmosphere and ocean; R_A and R_O are heat sources of radiative origin; λ_A and λ_O are the parameters of the feedback between radiative heat sinks and the temperature of the respective medium; λ_{AO} is the heat exchange coefficient at the ocean–atmosphere interface; c_A and c_O are atmospheric and ocean heat capacities.

We take R_A as constant and R_O as varying, depending on the ocean temperature, as

$$R_O(T_O) = \begin{cases} R_O^{(1)} & \text{at } T_O \le T_O^{(1)}, \\ R_O^{(1)} + \dfrac{T_O - T_O^{(1)}}{T_O^{(2)} - T_O^{(1)}}(R_O^{(2)} - R_O^{(1)}) & \text{at } T_O^{(1)} < T_O < T_O^{(2)}, \\ R_O^{(2)} & \text{at } T_O \ge T_O^{(2)}, \end{cases} \quad (5.8.18)$$

where $R_O^{(2)} > R_O^{(1)}$. Such a prescription of radiative sources of heat in the ocean allows for the existence of feedback between the albedo of the snow–ice cover and temperature: for $T_O \le T_O^{(1)}$ the ocean surface is assumed to be covered with ice; for $T_O \ge T_O^{(2)}$ it is assumed to be free from ice.

In the case where autonomous models of the atmosphere and ocean are used, and the evolution of the heat budget in the atmosphere is controlled, apart from everything else, by a change in the observed ocean temperature $T_O^{(m)}$, and the evolution of the heat budget in the ocean is controlled by a change in the observed atmospheric temperature $T_A^{(m)}$, Equations (5.8.17) are rewritten in the form

$$c_A \, dT_A^{(u)}/dt = R_A - \lambda_A T_A^{(u)} + \lambda_{AO}(T_O^{(m)} - T_A^{(u)}), \quad (5.8.19)$$

$$c_O \, dT_O^{(u)}/dt = R_O - \lambda_O T_O^{(u)} - \lambda_{AO}(T_O^{(u)} - T_A^{(m)}). \quad (5.8.20)$$

Equation (5.8.19) has the unique steady state solution

$$T_A^{(u)} = (R_A + \lambda_{AO} T_O^{(m)})/(\lambda_A + \lambda_{AO}), \quad (5.8.21)$$

while Equation (5.8.20) can have either one or three steady state solutions, with two of these three solutions being stable and one being unstable. Stable steady solutions comply with the top and bottom lines of Equation (5.8.18) and are defined by the formula

$$T_O^{(u)} = (R_O^{(i)} + \lambda_{AO} T_A^{(m)})/(\lambda_O + \lambda_{AO}), \quad (5.8.22)$$

where the index i can take values 1 or 2 depending on whether $T_O^{(u)}$ belongs to one or another range of temperature change in (5.8.18).

Let $c_A = 10^7$ J/m^2 K, $c_O = 10^8$ J/m^2 K, $R_A = 130$ W/m^2, $R_O^{(1)} = 120$ W/m^2, $R_O^{(2)} = 125$ W/m^2, $\lambda_{AO} = 10$ W/m^2 K, $\lambda_A = 0.5242$ W/m^2 K and $\lambda_O = 0.3472$ W/m^2 K. These model parameters and observed values of temperature in the atmosphere ($T_A^{(m)} = 286$ K) and ocean ($T_O^{(m)} = 288$ K) are met by the unique stable steady solution $T_A^{(u)} = 286.12$ K, $T_O^{(u)} = 288.20$ K which differs only slightly from the empirical estimates. Respective heat flux values for autonomous models of the atmosphere and ocean are equal to $F_A = F(T_A^{(u)}, T_O^{(m)}) = \lambda_{AO}(T_O^{(m)} - T_A^{(u)}) = 18.78$ W/m^2 and $F_O = F(T_A^{(m)}, T_O^{(u)}) = \lambda_{AO}(T_O^{(u)} - T_A^{(m)}) = 22.01$ W/m^2

In the interactive model one or two stable steady state solutions:

$$\left. \begin{aligned} T_A &= (R_A + \lambda_{AO} T_O)/(\lambda_A + \lambda_{AO}), \\ T_O &= \frac{(\lambda_{AO} + \lambda_A)R_O^{(i)} + \lambda_{AO} R_A}{\lambda_A \lambda_O + (\lambda_O + \lambda_A)\lambda_{AO}}, \quad i = 1, 2, \end{aligned} \right\} \quad (5.8.23)$$

are also obtained and for chosen model parameters the desired variables T_A, T_O and $F(T_A, T_O) = \lambda_{AO}(T_O - T_A)$ turn out to be equal to 295.57 K, 297.94 K and 23.7 W/m². This solution differs markedly from the initial solution found with the help of empirical data. Hence, solution drift in the interactive model of the ocean– atmosphere system is inevitable.

We take advantage of the procedure described above. With this purpose in mind we turn to (5.8.16) and find $\Delta F_A = F(T_A^{(u)}, T_O^{(m)}) - F(T_A^{(u)}, T_O^{(u)}) = -\lambda_{AO}(T_O^{(u)} - T_O^{(m)}) = -2.01$ W/m², $\Delta F_O = F(T_A^{(m)}, T_O^{(u)}) - F(T_A^{(u)}, T_O^{(u)}) = \lambda_{AO}(T_A^{(u)} - T_A^{(m)}) = 1.22$ W/m². Then we rewrite Equations (5.8.17) in the form

$$\left. \begin{array}{l} c_A \, dT_A/dt = R_A - \lambda_A T_A + \lambda_{AO}(T_O - T_A) - \lambda_{AO}(T_O^{(u)} - T_O^{(m)}), \\ c_O \, dT_O/dt = R_O - \lambda_O T_O - \lambda_{AO}(T_O - T_A) - \lambda_{AO}(T_A^{(u)} - T_A^{(m)}), \end{array} \right\} \quad (5.8.24)$$

and integrate them. As a result it turns out that the temperature of the atmosphere and ocean at different instants of time will coincide exactly with the initial values $T_A^{(u)} = 286.12$ K, $T_O^{(u)} = 288.20$ K, and, therefore, the procedure of exchange correction completely excludes the solution drift.

We now turn to testing the procedure of exchange correction as applied to the MPI global model. We first indicate its features. The atmospheric submodel is represented by a low resolving spectral model from the European Centre for Medium-Range Weather Forecasts (ECMWF), where the fields of prognostic variables (relative vorticity, velocity divergence, temperature, specific humidity, geopotential and surface pressure) are approximated in the form of finite series in spherical functions. As a result, the initial equations reduce to the appropriate equations for expansion coefficients depending only on time and the vertical coordinate. The vertical structure of the expansion coefficients is found with the help of discrete presentation. The version of the model adopted uses 21 spherical harmonics and 16 levels in the vertical.

The distinguishing features of the model are the introduction of a so-called hybrid vertical coordinate (a combination of the σ-coordinate with the isobaric coordinate) tracking the topography of the underlying surface in the low layer and coinciding with the isobaric coordinate in the upper atmospheric layers, as well as an incorporation of the 'enveloping' orography which approximately takes into account undulations of the underlying surface of the subgrid scale. A number of features which parametrization of the physical processes should include are, first, a detailed description of cloudiness determined as a function of relative humidity and height; second, an allowance for the dependence of radiative fluxes on cloudiness, temperature, specific humidity and concentration of carbon dioxide, ozone and aerosols; third, the inclusion of the influence of stratification on eddy fluxes of momentum, heat and moisture within the framework of the Monin–Obukhov similarity theory in the surface atmospheric layer and K-theory beyond the limits of this layer, and, finally, the determination of the time when the moisture convection appears depending not only on establishing a superadiabatic temperature gradient but on the difference in the divergence of the large-scale moisture transport in the upper layers, and the divergence of the eddy moisture flux in the surface layer of the atmosphere.

To simulate the ocean climate a quasi-geostrophic model is used. Its basis is the principle of decomposition of current velocity into two components: barotropic and baroclinic. It is assumed that the barotropic component is immediately adapted to changes in wind stress and sea water density, and the baroclinic component obeys the geostrophic relation. Thus, temperature and salinity are only prognostic variables. In the vicinity of the equator and in the coastal regions of the ocean, where the geostrophic relation is not valid, the equations of motion are complemented by terms providing for the existence of viscous boundary layers. It is also assumed that heat and salt transport in the ocean is performed only by ordered motions. Because of this the terms describing diffusive transfer are omitted in the respective budget equations. Provision is made for the appearance of the hydrostatic instability at high latitudes, realized through the procedure of convective adjustment. The sea ice effect is taken into account indirectly: when the temperature reaches freezing point the sea water density remains constant. The equations of the ocean submodel are integrated numerically with the help of an implicit scheme on a uniform latitude–longitude grid with a step of 4.0°. The time step is one month. The horizontal components of velocity, temperature and salinity are determined at the depths 75, 150, 300, 1000 and 3000 m, the vertical components of velocity are calculated at the intermediate depths 112, 225, 650, 2000 and 6000 m.

The solution is found by the following sequence (Cubasch, 1989). First the annual mean fields of resulting heat and salt fluxes at the ocean surface and of wind stress are calculated within the framework of the autonomous atmospheric submodel, and the annual mean fields of the sea surface temperature and of sea ice thickness are calculated within the framework of the autonomous ocean submodel. Initial information for the atmospheric submodel is the annual mean field of sea surface temperature and the distribution of sea ice obtained from climatological data; for the ocean model the fields of wind stress, temperature and salinity at the ocean surface are found in the same way. Calculation within the framework of autonomous submodels is complete when the solution has achieved an equilibrium state (in the atmosphere this occurs after a year, the ocean it occurs after 10 000 years). Then the calculation is continued for four years within the framework of the coupled ocean–atmosphere model with and without

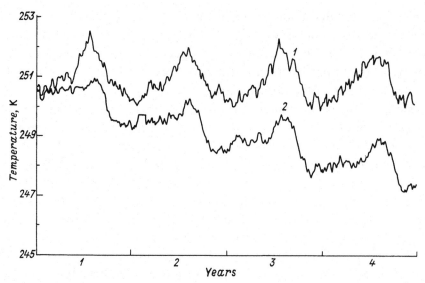

Figure 5.10 Evolution of the global average temperature in the atmosphere with (1) and without (2) an allowance for a heat flux correction at the ocean–atmosphere interface, according to Cubasch (1989).

consideration of correction of the resulting heat flux at the ocean–atmosphere interface.

Results of calculations presented in Figure 5.10 point to the fact that the introduction of a correction in fact eliminates the inter-annual trend of the global average atmospheric temperature. Unfortunately, one cannot say the same with respect to the global average sea surface temperature: it decreases at a rate of 0.37 °C/year when considering correction and 0.35 °C/year when not considering correction. According to Cubasch (1989), this circumstance can be caused by violation of the global budget of fresh water in the ocean and by the short duration of the integration period.

The principal systematic errors common to these models, as *a propos* to all current atmospheric GCMs (see IPCC, 1990), are the general coldness of the lower stratosphere and upper troposphere in polar latitudes and much of the lower troposphere in the tropics and middle latitudes; too low surface air temperature over eastern Asia in winter and too high surface air temperature over the Antarctic ice sheet; the generally inadequate reproduction of the southeast Asia summer monsoon rainfall, the zonal rainfall gradient across the tropical Pacific, the extent of the aridity over Australia and seasonal variations of soil moisture, especially in the tropics; a substantial underestimation of the ocean heat transport and intensity of western boundary currents; warmer-than-observed temperatures in the Southern Ocean, and colder-than-observed temperatures in tropical and high latitudes of the ocean. Finally, none of the models produces the proper deep-water circulation, especially in the Southern Hemisphere. The existence of such common deficiencies despite the considerable differences in the spatial resolution, numerical treatment and physical parametrization, means that in all of these models some physical mechanisms responsible for formation of the climate system and its variability are either not described at all or are represented in a very simplified way.

In completing the discussion about coupled ocean–atmosphere GCMs we emphasize once more that any model reproduces a real picture of the world only within the framework defined by the limitation of our knowledge about processes which occur in it. Three-dimensional models of the ocean–atmosphere system are no exception, and therefore, in a number of cases, unsatisfactory agreement with empirical data was predetermined. But despite all the disadvantages the three-dimensional models are irreplaceable from the viewpoint of the perception of the causes of natural and anthropogenic climate changes. This argument refers in the same degree to simpler models, providing qualitatively correct simulation of the time–space variability of climatic characteristics. Use of these models is attractive in two respects. First, it can save overexpenditure of scarce computer time, and, second, it allows one to single out and understand the nature of the internal mechanisms governing the behaviour of the climatic system. But, as was subtly noticed by Lorenz (1975): 'yet it can be very dangerous to place too much confidence in models whose behavior depends too strongly upon the details of the parametrizations'. Moreover, the base of any parametrization is formed by empirical data and, even if these parametrizations are applicable for the modern epoch, the possibility of extending them to other epochs is strictly limited.

5.9 ENSO as a manifestation of the inter-annual variability of the ocean–atmosphere system

The existence of El Niño (an anomalous rise in water temperature to the west of the equatorial zone of the Pacific Ocean accompanied by suppression of the upwelling of cold deep water, rich with biogenes, and therefore having disastrous consequences for fisheries in the coastal regions of Ecuador and Peru) and Southern Oscillation (fluctuations of surface pressure, dominating winds as well as air temperature and precipitation in the tropics and neighbouring regions of the Indian and Pacific Oceans) has been known for a long time. But the fact that El Niño and Southern Oscillations represent two sides of one and the same phenomenon became clear only in the late 1960s, due mainly to the work of Bjerknes. Since that time this phenomenon has been called El Niño/Southern Oscillations or in abbreviated form ENSO. We reproduce Bjerknes' chain of arguments, taking subsequent refinements into account (see Philander and Rasmusson, 1985, in particular).

For time scales of several weeks (the relaxation time of the statistically steady state of the atmosphere) and longer, the large-scale atmospheric circulation in the tropics is controlled by a system of *Walker cells*, every one of which is characterized by the ascent of air over a heated continent and a neighbouring area with high sea surface temperatures, by air transport in the upper layers of the troposphere towards cold water areas, by descending air over this area, and by trade winds closing a circulation cell in the lower layers of the troposphere (Figure 5.11). In the Pacific Ocean, ascending branches of the Walker cell are located over the Indonesian archipelago and the northern part of Australia, along the inter-tropical convergence zone to the north of the equator and along the Pacific convergence zone south-west of the Pacific Ocean; the descending branch of the Walker cell is located over the south-east part of the tropical zone where the sea surface temperature is less than in the western part by 8 °C in March and by 13 °C in September. The Walker cell is closed by the south-eastern trade winds.

The south-eastern trade wind is are subject to marked seasonal oscillations. It is strengthened and penetrates far to the north in August and September when the inter-tropical convergence zone is situated near the parallel 12 °N, and, on the other hand, it is weakened and displaced to the south in February and March when the inter-tropical convergence zone is in the vicinity of the equator. The westward component of the south-eastern trade winds leads to the accumulation of warm water in the western part of the Pacific Ocean and to upwelling of cold deep water in the east. Accordingly, the thickness of the upper mixed layer changes: it exceeds 150 m in the western part of the tropical zone and decreases, down to zero, in fact, in the vicinity of the South American coast. It is clear that the strengthening of the south-eastern trade winds in

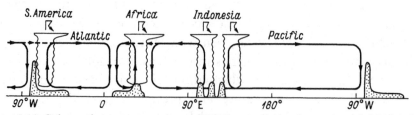

Figure 5.11 Schematic representation of Walker cells in the equatorial belt of the Earth.

August and September must contribute to a reduction in the atmospheric temperature, a rise in the surface pressure, a weakening of convection and a decrease in precipitation in the eastern part of the tropical zone, and to opposite changes in the west.

Now, imagine that the surface temperature rise has occurred in the south-eastern part of the tropical zone of the Pacific Ocean, that is, El Niño has appeared. It should lead to a drop in the zonal gradient of surface pressure, to a weakening of the south-eastern trade winds, to a rise in the mean ocean level in the eastern part and a drop in the western part of the tropical zone, to a decrease in inclination of the free ocean surface and the thermocline underlying the upper mixed layer, to a weakening of the South Equatorial Current and an amplification of the easterly propagating subsurface water mass transport, to a westward displacement of the boundary of the intensive convection area in the atmosphere (this boundary is easily identified from satellite measurements of outgoing long-wave radiation and is assumed to coincide with the 240 W/m^2 contour or with the 27.5 °C surface isotherm), and to an increase in precipitation in the central part of the equatorial zone of the Pacific Ocean. These are the consequences of El Niño's compliance with the warm phase of ENSO.

Next comes the cold phase of ENSO, with its inherent anomalous reduction in surface temperature in the south-eastern part of the tropical zone of the Pacific Ocean. This phase is identified with the opposite of El Niño which was given the name La Niña. The consequences of La Niña are opposite to those described above: the drop in water temperature in the south-eastern part of the tropical zone determines a local rise in surface pressure, strengthening of the south-eastern trade winds, a rise in the mean ocean level in the west and a decrease in the east, an increase in the zonal inclination of the free surface and thermocline, intensification of the Southern Equatorial Current, and weakening of easterly propagating subsurface water mass transport, a shift to the west of the convective zone in the atmosphere and a decrease in precipitation in eastern and central parts of the equatorial zone of the Pacific Ocean.

Certainly, the sole fact of the appearance of the local anomaly of the surface temperature does not yet mean that its influence will be felt at large distances from it: it is necessary to have a combination of appropriate conditions. For example, a local rise in surface temperature in the region of downward atmospheric motions can turn out to be insufficient to establish intense convection. On the other hand, a local rise in surface temperature in a region of upward atmospheric motions favours intensification of convection, an increase in precipitation, enhancement of heat release due to the phase transitions of water vapour and subsequent intensification of upward vertical motions. Because of this the influence of the surface temperature anomaly is determined not only by its magnitude, sign and location but also by the season of the year. It is no mere chance that the mature stage of El Niño development falls in a period of maximum sea surface temperature, that is, in December–February.

The short-term rise of water temperature in the eastern part of the equatorial zone, as well as a weakening of the south-east trade winds and even the appearance of anomalous westerly winds in the central part of the equatorial zone in September–November of the preceding year serve to herald El Niño. In March–May a rise in temperature is felt over eastern and central parts of the equatorial zone of the Pacific Ocean. It is also maintained here in June–August. The first signs of the end of El Niño and of the restoration of normal conditions are found in September–November. This is demonstrated by a decrease in the *South Oscillation index* (the difference of surface pressure between points located in the south-eastern part and in the Indonesian–Australian sector of the Pacific Ocean), weakening of westerly winds in the central part of the equatorial zone, and suppression of atmospheric convection in the east. In December and during January and February of the following year the degeneration of El Niño continues. Positive anomalies of surface temperature in the equatorial zone disappear (warm phase of ENSO terminates) in June–August. The cold phase of ENSO starts in September–October. It is characterized by sharp strengthening of the south-eastern trade winds, by a westward shift from the 180° meridian of the intense convection zone, by a rapid increase in the difference in the surface pressure between Tahiti and Darwin, and by the appearance of negative anomalies of surface temperature in the equatorial zone of the Pacific Ocean. In June– August of the second year after the occurrence of a mature phase of El Niño the negative anomalies of surface temperature spread over the whole equatorial zone and stay there until the end of the year. Thus, the duration of the complete cycle of ENSO is about three years.

The sequence of events described above is typical of ENSO, although it does not exclude individual pecularities in certain years. For example, the El Niño of 1982–3 differed from all the others by an extremely large rise in temperature in the eastern part of the equatorial zone of the Pacific Ocean. Suffice it to say that for the preceding 30 years the ocean surface temperature east of the 140 °W meridian had never been higher than 29 °C. In 1982 the 29 °C isotherm reached the coast of South America. This led to a reduction, practically down to zero, of the difference in surface temperatures between the eastern and western parts of the equatorial zone of the Pacific

Ocean, to a sharp strengthening of westerly winds, to an increase in frequency of typhoons and thunderstorms in the central part, to a degeneration of the easterly subsurface water mass transport, and to the disappearance of thermocline inclination along the equator.

No two cycles of ENSO ever repeat each other. The same can be said with respect to the interval between two subsequent cycles of ENSO, as attested by the chronology of El Niño events. Indeed, since 1935 El Niño has appeared 13 times: in 1940–2, 1946–7, 1951–2, 1953–4, 1957–8, 1963–4, 1965–7, 1969–70, 1973–4, 1977–8, 1982–3, 1986–7 and 1991–2. In two cases (in 1940–2 and in 1982–3) it was extreme: the respective South Oscillation index exceeded the amplitude of seasonal oscillations of surface pressure by 2.5 times. Therefore, ENSO represents a distinct aperiodic phenomenon.

All models of ENSO can be divided into two types. The basis of the models of the first type is the assumption that the ocean–atmosphere system in the tropics has two quasi-equilibrium states complying with warm and cold phases of ENSO. It is also assumed that the transition from one state to another is determined by an internal instability inherent in the system which acts like a trigger and is excited by high-frequency (compared to the ENSO frequency) stochastic disturbances. But even in the presence of such disturbances the instability occurs only when the oscillations of the system become critical. Because of this the appearance of the instability and, hence, the tendency towards the development of events and the duration of the existence of one or another state depend not only on high-frequency stochastic disturbances but on low-frequency deterministic oscillations of the system.

In models of the second type ENSO is determined by an instability of low-frequency equatorial disturbances generated by the ocean–atmosphere system. The appearance of the instability can be explained with the help of the following qualitative considerations (see Philander, 1985). Let us assume that in the western part of the equatorial zone a local positive anomaly of surface temperature has occurred for some reason, as a result of which a convective zone with upward vertical motions and wind velocity convergence in the low layers of the atmosphere will have formed immediately over it, that is, with westerly winds towards the west and easterly winds to the east of the anomaly. The wind velocity convergence in the low layers of the atmosphere will lead to ocean current velocity convergence in the upper mixed ocean layer and, hence, to a downwelling and deepening of the thermocline. As is known, the sea surface temperature decreases from west to east. Therefore, an advective transport to the west of the anomaly will contribute to a further temperature rise and to its eastward displacement.

A different situation arises to the east of the anomaly. Here a change in temperature is determined by the competition of two factors: by upwelling created by local easterly winds, and by downwelling formed by baroclinic Kelvin waves propagating towards the east (these are induced by westerly winds at the western periphery of the anomaly). The first factor causes a decrease in surface temperature; the second leads to its increase. Domination of one factor or another depends on the relative intensity and the zonal extent of bands of westerly and easterly winds. Since, all other things being equal, the intensity and zonal extent of a band of westerly winds on the equatorial beta plane are greater than those of easterly winds, the downwelling will dominate over the upwelling and in the process of its evolution the positive anomaly will increase and be displaced to the east.

An increase in a negative surface temperature anomaly can be explained in a similar manner: a temperature drop results in suppression of convection, decrease in westerly winds, domination of upwelling over downwelling and, eventually, further temperature reduction to the east of the anomaly. If this temperature drop is accompanied by a decrease in the dimensions of the convective zone in the atmosphere and by strengthening of easterly winds then intensification of upwelling, a further temperature drop and eastward displacement of the negative anomaly are inevitable.

We illustrate the above by calculations carried out by McCreary (1986). The model used by McCreary (1986) describes the instantaneous response of the first baroclinic mode in the atmosphere to a heat inflow from the ocean, and the evolution of the ocean upper mixed layer forced by changes in momentum and heat fluxes at the ocean–atmosphere interface. In a model of the upper mixed layer the effects of mechanical and convective mixing, entrainment at the thermocline boundary and momentum and heat advection are taken into account.

The results of numerical experiments presented in Figure 5.12 are in respect of three different distributions of land and ocean. In the first experiment the ocean is approximated in the form of a zonally symmetric (with respect to the equator) band surrounding the whole Earth, in the second experiment the zonal ocean is divided into two disconnected basins with the same angular distance between boundary meridians. In the third experiment the western ocean basin decreases by half, and the remaining part of the zonal band is occupied by land adjoining a large ocean basin to the east and a small ocean basin to the west. Such a configuration is reminiscent of the system of the Indian and Pacific Oceans. In all three cases the initial distribution of temperature and upper mixed layer thickness is considered to be homogeneous, and motions in the ocean and the atmosphere are considered to be absent. The

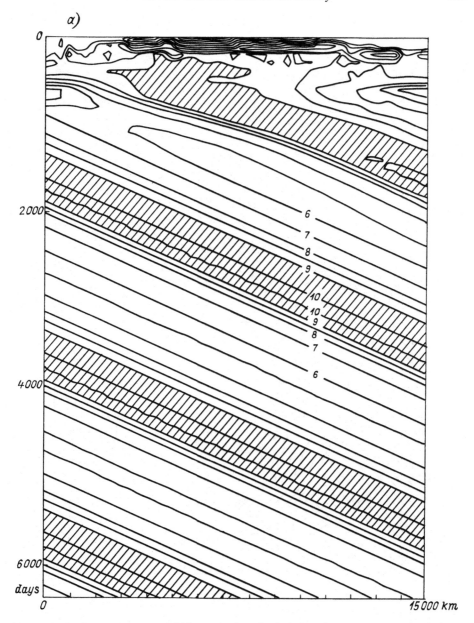

Figure 5.12 Time–space variability of the deviation of the upper mixed layer temperature from the thermocline temperature at the equator for different ocean configurations (*a*, *b*, *c*), according to McCreary (1986). Zones in which temperature deviations exceed 9 °C are shaded. Further explanation may be found in the text.

(*continued*)

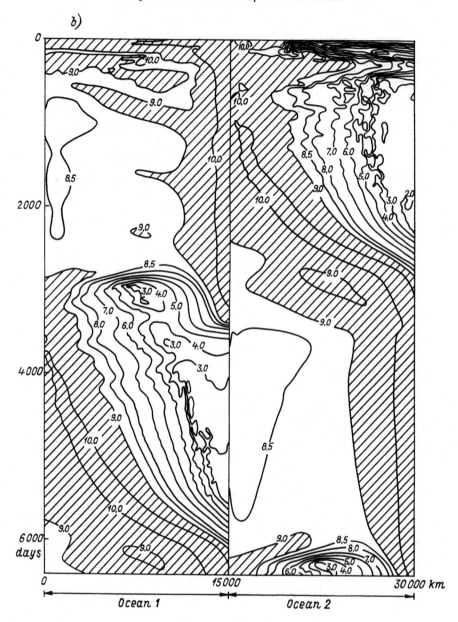

Figure 5.12 (*continued*).

perturbations of the initial distributions of the desired characteristics are introduced by changes in the wind stress field during the first 100 days of integration. These changes are then excluded and perturbations generated by the model are reproduced without the superposition of any external forcings.

Figure 5.12 shows the separate fragments of the time–space variability of

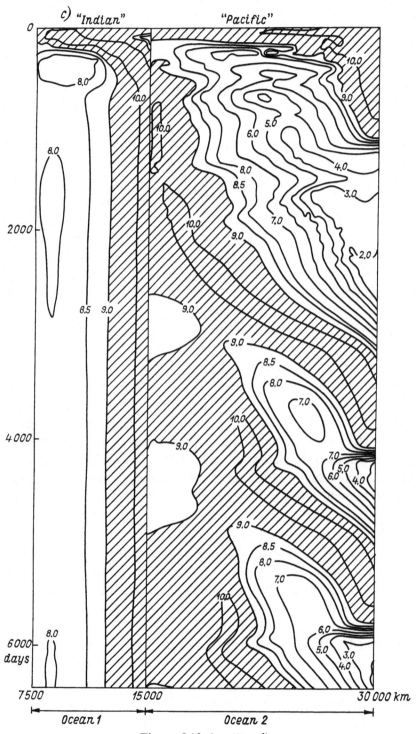

Figure 5.12 *(continued)*.

temperature deviations in the upper mixed layer from the thermocline temperature at the equator for the three above-mentioned ocean configurations. Initially, temperature deviations are assumed to be identical and equal to 10°. It can be seen that in the first case (*a*) the formation of positive and negative deviations is terminated after 1000 days, and then in fact they do not change practically but are displaced slowly to the east. In the second case (*b*) the situation is different: positive deviations only appear at the western boundary of the eastern ocean. Then they slowly propagate to the east and in 2800 days reach the eastern boundary. As this takes place the negative deviations are formed next to the western boundary of the eastern ocean, and, at the western boundary of another ocean, positive temperature deviations arise. These do not remain but, rather, are slowly displaced to the east. Upon reaching the eastern boundary of the western ocean the positive deviations act in the vicinities of the western boundary of the neighbouring (eastern) ocean through the westerly transport in the atmosphere. As a result, positive deviations are formed here. Then the cycle of temperature oscillations in the system of the two oceans is repeated.

Case (*c*) is crucially different from the first two cases. Now positive deviations after the termination of a period of relaxation are localized at the eastern boundary, and negative deviations are localized at the western boundary, of the small ocean basin. Such a state turns out to be stable in the sense that it does not change over time. In contrast to this, low-frequency oscillations with periods of about five years occur in the large ocean basin. This last peculiarity of the solution, as well as the existence of a constant temperature rise (and therefore intense convection) in the vicinity of the land dividing both oceans, and also the slow displacement of temperature deviations to the east, have much in common with observational data for the equatorial zone of the Indian and Pacific Oceans.

The calculated results presented in Figure 5.12 reproduce the successive interchange of warm and cold phases of ENSO but leave the question as to the causes of their alternation unanswered. An answer to this question is given by Graham and White's (1988) conceptual model. We describe it briefly starting from the time of the origin of a positive surface temperature anomaly in the central part of the equatorial zone of the Pacific Ocean.

It has already been mentioned that an anomalous increase in the surface temperature leads to the appearance of anomalous westerly winds in the equatorial zone and to a weakening of easterly winds beyond its limits. This, in turn, must lead to an enhancement of cyclonic vorticity in the wind velocity field, intensification of upwelling, and decrease in surface temperature in the central part of the extra-equatorial zone. The negative surface temperature anomaly formed here is transferred by baroclinic Rossby waves to the western ocean boundary where, as a result of reflection, these waves are transformed into baroclinic Kelvin waves propagating towards the equator and then along it to the east. The negative anomaly of surface temperature is carried along, together with the baroclinic Kelvin waves.

But since the phase velocity of the baroclinic Rossby waves is much less than the phase velocity of baroclinic Kelvin waves, then a temperature rise in the eastern part of the ocean, created by the transport of the anomaly from the central part of the equatorial zone, is followed by a temperature decrease caused by the transport of the negative anomaly from the central part of the extra-equatorial zone. As this takes place the lag between times of increase and decrease in temperature will be determined by the time of displacement of the negative anomaly from the central part of the extra-equatorial zone into the central part of the equatorial zone, amounting to one or two years. Similar (in terms of sequence but not sign) changes must occur when a negative anomaly of surface temperature appears in the central part of the equatorial zone.

Thus, in accordance with the concepts stated, a change in ENSO phases is explained by the influence of baroclinic Rossby waves whose domain of existence is restricted by extra-equatorial latitudes. In particular, from this it follows that any attempt to predict the complete ENSO cycle without taking into account the interacting processes in equatorial and extra-equatorial latitudes of the ocean will be condemned to failure.

6

Response of the ocean–atmosphere system to external forcing

After listing mathematical methods of analysis of sensitivity we shall discuss the results of numerical experiments on the equilibrium (steady-state) response of the climatic system to external forcing. In reality, the response of the climatic system is time-dependent in nature because of the large heat capacity of the ocean, hence the results obtained cannot be used directly to predict the climate evolution for a short period (on the ocean relaxation time scale). The above numerical experiments have another purpose: to give an idea of the possible limiting changes of the climatic system. That discussion will be followed by a brief description of the transient response of the ocean–atmosphere system to an increase of atmospheric CO_2 emission resulting from human activities.

6.1 Sensitivity of the climatic system: mathematical methods of analysis

We consider the available mathematical methods of analysis of sensitivity, using as an example a dynamic system governed by the equation

$$dy/dt = v(y, \alpha). \tag{6.1.1}$$

In the simplest situations the state y of the system and the parameter α are finite-dimensional vectors. We shall also consider the more general case where, for any α, the vector y is a function of the spatial variables x changing in a domain Ω. In the general case the phase space Y of the system (6.1.1) is infinite dimensional. If the field v is specified by a differential operator, its description has to include the information about boundary conditions ensuring single-valued solvability of the Cauchy problem for Equation (6.1.1). Let us denote the resolving operators by v_α^t so that the solution to the Cauchy problem with an initial condition $y|_{t=0} = y_0$ will be defined by the formula $y = v_\alpha^t(y_0)$.

Two types of stability of the solution of Equation (6.1.1) are distinguished: stability to the initial condition perturbations for the fixed field v and stability to perturbations of the field v including perturbations of the parameter α. In climate theory it is the second type of stability that is associated with the concept of sensitivity. The term '*robustness*' is applied when the same problems are considered in the theory of dynamical systems. The dynamical system is identified as robust (lattice-stable) if small perturbations of the field v do not change its topological properties. The simplest example of the robust system is the system (6.1.1) having only an equilibrium or time-periodic solution whose stability spectrum is located in the left-hand side of the complex plane.

Systems whose limiting states are exhausted by time-periodic and equilibrium regimes have been a principal subject of interest in mathematics and applications. The reasons for such a constraint are well known: for a small dimension of the phase space ($n = 1, 2$), the field of any smooth system may be transformed by small deformations into the field of the Morse–Smale system, i.e. the system in which all motions at $t \to \infty$ tend to periodic trajectories (including equilibrium states) with a number of time-periodic solutions being finite and no spectra crossing the imaginary axis. However, for $n \geq 3$ the situation is changed drastically: there are robust dynamical systems which have different and generally more complicated forms of asymptotes at large t. Furthermore, for large dimensions the robust systems become, in a sense, non-typical.

Direct and adjoint equations of sensitivity. Let us consider three problems associated with Equation (6.1.1): the Cauchy problem, the problem of the construction of equilibrium solutions and the problem of the construction of time-dependent solutions. Each of these problems may be formulated as the problem of the determination of solutions of the functional equation

$$G(y, \alpha) = 0, \tag{6.1.2}$$

with a suitable choice of the operator G. An analysis of the sensitivity of solutions of the problems mentioned above is performed by using an identical scheme. We shall describe the essence of this in brief.

There is a broad class of non-linear systems whose response to small perturbations of parameters is almost linear: the world is linear in the small if it is not pathological. According to Arnold, the class of pathological systems forms a 'thin set' within a set of all systems. The class of non-pathological systems is described by the implicit function theorem (Sattinger, 1980). If conditions of the theorem are fulfilled the determination of the solution for one value $\bar{\alpha}$ of the parameter allows us to describe system states for all α quite close to $\bar{\alpha}$ without solving Equation (6.1.2) once again. In the case where G is a scalar function of the scalar variables y, α the theorem guarantees correctness of the formula:

$$y(\alpha) - y(\bar{\alpha}) = \zeta(\alpha - \bar{\alpha}) + O(\|\alpha - \bar{\alpha}\|), \qquad (6.1.3)$$

where ζ is defined by the relation

$$\zeta = \left(\frac{\partial G}{\partial y}\bigg|_{y=\bar{y}, \alpha=\bar{\alpha}}\right)^{-1} \frac{\partial G}{\partial \alpha}\bigg|_{y=\bar{y}, \alpha=\bar{\alpha}} \qquad (6.1.4)$$

Having defined the derivatives in a reasonable way we can extend the relations (6.1.3) and (6.1.4) to the infinite-dimensional functional space.

Let us denote a vector with components y, α by z and assume that the operator G is Fréchet differentiable. The operator G is said to be Fréchet differentiable at a point $\bar{z} = (\bar{y}, \bar{\alpha})$ provided there is a bounded linear operator A from a set V to Y such that the quantity $R(\bar{z}, h) = G(\bar{z} + h) - G(\bar{z}) - Ah$ is $o(h)$ as $\|h\| \to 0$, that is,

$$\lim_{\|h\| \to 0} \|h\|^{-1} \|R(\bar{z}, h)\| \to 0.$$

In the case being investigated,

$$Ah = \frac{\partial G}{\partial y}\bigg|_{y=\bar{y}, \alpha=\bar{\alpha}} (y - \bar{y}) + \frac{\partial G}{\partial \alpha}\bigg|_{y=\bar{y}, \alpha=\bar{\alpha}} (\alpha - \bar{\alpha}),$$

where $\partial G/\partial y$ and $\partial G/\partial \alpha$ are linear operators from V to Y. Certainly, these operators depend on $\bar{z} = (\bar{y}, \bar{\alpha})$, with the dependence being non-linear.

We now write down the formulation of the implicit function theorem. *Let G be a Fréchet differentiable operator. Suppose $G(\bar{y}, \bar{\alpha}) = 0$ and $\partial G(\bar{y}, \bar{\alpha})/\partial y$ is an invertible operator. Then for sufficiently small $\|\alpha - \bar{\alpha}\|$ there is a Fréchet differentiable function $y(\alpha)$ satisfying the relations (6.1.3) and (6.1.4). Furthermore, in a sufficiently small neighbourhood $V' \subset V$ there is the only solution of Equation (6.1.2).*

If G has k derivatives the function $y(\alpha)$ has k continuous derivatives, too. Equations for derivatives are found by formal differentiation of the Equation (6.1.2) with respect to α. Basic facts of finite-dimensional analysis (the rule of differentiation of a superposition, Taylor's theorem and the Lagrange formula) have their analogues in the non-linear functional analysis. In particular, as applied to the theorem conditions, the function $y(\alpha)$ can be approximated up to quantities of the order of $\|\alpha - \bar{\alpha}\|^k$ by the appropriate Taylor polynomial. In accordance with (6.1.4), evaluation of the operator

ζ is reduced to solving the *direct equations of sensitivity*:

$$\left(\frac{\partial G}{\partial y}(\bar{y}, \bar{\alpha})\right)\zeta = -\frac{\partial G}{\partial y}(\bar{y}, \bar{\alpha}). \tag{6.1.5}$$

If y is a function of spatial variables and time it is important to know averaged values of this function determined by the functionals $R(y, \alpha)$. Let R be a differentiable functional. Its response to perturbations of the parameter α may be computed by the formula

$$R(y, \alpha) - R(\bar{y}, \bar{\alpha}) = \left(\frac{\partial R}{\partial y}(\bar{y}, \bar{\alpha})\right)(y - \bar{y}) + \left(\frac{\partial R}{\partial \alpha}(\bar{y}, \bar{\alpha})\right)(\alpha - \bar{\alpha})$$

$$+ o(\|y - \bar{y}\| + \|\alpha - \bar{\alpha}\|). \tag{6.1.6}$$

In Hilbert space, a typical example of which is a set of scalar quadratically summed functions, the derivatives $\partial R/\partial y$ and $\partial R/\partial \alpha$ (*functionals of sensitivity*) admit the following representation

$$\left.\begin{aligned}\left(\frac{\partial R}{\partial y}(\bar{y}, \bar{\alpha})\right)(y - \bar{y}) &= (r_y, y - \bar{y}), \\ \left(\frac{\partial R}{\partial \alpha}(\bar{y}, \bar{\alpha})\right)(\alpha - \bar{\alpha}) &= (r_\alpha, \alpha - \bar{\alpha}),\end{aligned}\right\} \tag{6.1.7}$$

where r_y and r_α are elements of spaces Y and Λ respectively; Λ is a set of admissible values of α.

From here and from (6.1.3)–(6.1.6) it follows that

$$R(y, \alpha) - R(\bar{y}, \bar{\alpha}) = VR + o(\|\alpha - \bar{\alpha}\|), \tag{6.1.8}$$

where

$$VR = (r_y, \zeta(\alpha - \bar{\alpha})) + (r_\alpha, \alpha - \bar{\alpha}). \tag{6.1.9}$$

In geophysical applications another formula for VR being a consequence of the relation (6.1.9) is generally used, namely,

$$VR = \left(\zeta^*, \frac{\partial G}{\partial \alpha}(\bar{y}, \bar{\alpha})(\alpha - \bar{\alpha})\right) + (r_\alpha, \alpha - \bar{\alpha}), \tag{6.1.9'}$$

where ζ^* is a solution of the problem

$$\left(\frac{\partial G}{\partial y}(\bar{y}, \bar{\alpha})\right)^* \zeta^* = -r_y, \tag{6.1.5'}$$

adjoint to (6.1.5). Equation (6.1.5') is named *the adjoint equation of sensitivity* and, together with (6.1.5), it can be utilized to evaluate the variance of the functional R in terms of the variance of the parameter α.

Detailed information about sensitivity can be obtained if the Green's functions $g(x, x')$ and $g^*(x, x')$ of operators $\partial G/\partial y$ and $(\partial G/\partial y)^*$ are known. The solutions to the problems are connected with the Green's functions by relations

$$\zeta(x) = -\int_\Omega g(x, x')r_\alpha(x')\,dx',$$

$$\zeta^*(x) = -\int_\Omega g^*(x, x')r_y(x')\,dx'.$$

Applying these formulae we find for $r_\alpha = 0$

$$y(x, \alpha) - y(\bar{x}, \bar{\alpha}) = \int_\Omega [\alpha(x) - \bar{\alpha}(x)]Z(x, x')\,dx + o(\|\alpha - \bar{\alpha}\|),$$

$$R(y, \alpha) - R(\bar{y}, \bar{\alpha}) = \int_\Omega \int_\Omega [\alpha(x) - \bar{\alpha}(x)]r_y(x')Z^*(x, x')\,dx\,dx',$$

where the functions

$$Z(x, x') = -g(x, x')g_\alpha(x),$$

$$Z^*(x, x') = -g^*(x, x')g_\alpha(x),$$

are referred to as *densities of sensitivity*. They determine uniquely the functional derivatives of y with respect to α.

Estimations of feedback influence. In climate models it is common practice to accept simplifications based on adopting empirical values of the parameter α depending on a system state. The use of these models for prediction entails inevitable errors due to variations of α as y changes.

An allowance for a dependence of α on y in such cases is described as an allowance for a feedback influence. One possible method of obtaining *a priori* estimates of the feedback influence is the application of the above formulae (6.1.9) and (6.1.9'). A feedback can be taken into account in Equation (6.1.2) if the latter is supplemented by a relation describing the dependence of α on y. Then, instead of Equation (6.1.2), we have the system

$$G(y, \alpha) = 0, \qquad \alpha = \bar{\alpha} + \lambda f(y), \tag{6.1.10}$$

where f is a prescribed operator and λ is a new parameter describing the intensity of the feedback.

The state y of the system without an allowance for feedback is derived from solution of the equation

$$G(\bar{y}, \bar{\alpha}) = 0. \tag{6.1.11}$$

The problem is to estimate the quantities $y - \bar{y}$ and $R(y, \alpha) - R(\bar{y}, \bar{\alpha})$, where R is a non-linear functional satisfying the conditions (6.1.6) and (6.1.8). In this case the formulae (6.1.9') and (6.1.3) cannot be used directly because the parameter variations depend on system state variations. But use could be made of the old formulae representing λ as a new parameter. Then, instead of the relations (6.1.3) and (6.1.9') we have

$$
\left.
\begin{aligned}
y(\alpha) - y(\bar{\alpha}) &= \zeta\lambda + o(|\lambda|), \\
\zeta &= -\left(\frac{\partial G}{\partial y}\right)^{-1}\frac{\partial G}{\partial \alpha}f(\bar{y}),
\end{aligned}
\right\}
\tag{6.1.12}
$$

$$
\left.
\begin{aligned}
R(y, \alpha) - R(\bar{y}, \bar{\alpha}) &= VR + o(|\lambda|), \\
VR &= \lambda\left(\zeta^*, \frac{\partial G}{\partial \lambda}f(\bar{y})\right) + \lambda(r_\alpha, f(\bar{y})),
\end{aligned}
\right\}
\tag{6.1.13}
$$

where, as before, the function ζ^* is determined by Equation (6.1.5'), and derivatives $\partial G/\partial y$ and $\partial G/\partial \alpha$ are evaluated at the point $\bar{y}, \bar{\alpha}$.

Following Cacuci and Hall (1984) we consider the problem of determination of temperature T complying with the conditions

$$
\frac{dT}{dt} = -\alpha T^4 + \beta, \qquad T|_{t=a} = T_a, \qquad a < t < b.
\tag{6.1.14}
$$

As a measure of sensitivity we use the functional $R(T, \alpha) = \int_a^b \alpha T^4\, dt$ assuming that the feedback is prescribed by the relation

$$
\alpha = \bar{\alpha} + \lambda(T - T_a)^4
\tag{6.1.15}
$$

and find the estimates of the quantity $R(T, \alpha) - R(\bar{T}, \bar{\alpha})$, where $\bar{T}$ is a solution to the problem (6.1.14) and (6.1.15) for $\alpha = \bar{\alpha}$.

Let us denote $\Theta = T - T_a$, $\bar{\Theta} = \bar{T} - T_a$, $G(\Theta, \alpha) = d\Theta/dt + \alpha(\Theta + T_a)^4 - \beta$, and take as a domain of definition of the operator G a set of pairs (Θ, λ) where the function Θ satisfies the conditions

$$
\int_a^b \Theta^2\, dt + \int_a^b \left(\frac{d\Theta}{dt}\right)^2 dt < \infty, \qquad \Theta|_{t=a} = 0,
$$

λ is a numerical parameter from the interval $[-1, 1]$.

Then, the functional R is given by the relation

$$
R(\Theta, \lambda) = \int_a^b (\bar{\alpha} + \lambda\Theta)(\Theta + T_a)^4\, dt,
$$

the difference $\Delta = R(T, \alpha) - R(\bar{T}, \bar{\alpha})$ to be estimated is represented in the form

$$\Delta = VR + o(\|\Theta - \bar{\Theta}\| + |\lambda|),$$

$$VR = \int_a^b r_\Theta (\Theta - \bar{\Theta}) \, dt + \lambda \int_a^b r_\lambda \, dt,$$

$$r_\Theta = 4\bar{\alpha}(\bar{\Theta} + T_a)^3, \qquad r_\lambda = \Theta(\Theta + T_a)^4,$$

and Equation (6.1.5′) is reduced to the following problem relative to ζ^*:

$$-\frac{d\zeta^*}{dt} + 4\alpha(\bar{\Theta} + T_a)^3 \zeta^* = -4\bar{\alpha}(\bar{\Theta} + T_a)^3,$$

$$a < t < b, \qquad \zeta^*|_{t=b} = 0.$$

A solution to this problem has the form

$$\zeta^* = \int_a^b \left(-4\bar{\alpha}(\bar{\Theta}(\tau) + T_a)^3 \exp\left\{ -4\bar{\alpha} \int_t^\tau [T_a + \bar{\Theta}(t')]^3 \, dt' \right\} \right) d\tau.$$

By virtue of (6.1.13) we obtain

$$VR = \lambda \int_a^b [\zeta^*(t)(\bar{\Theta} + T_a)^4 + (\bar{\Theta} + T_a)^4]\bar{\Theta} \, dt.$$

Sensitivity of solutions to time-dependent problems. Let us consider the sensitivity of a solution to the Cauchy problem as applied to Equation (6.1.1). In this case the direct equations of sensitivity coincide with the equations for variations. For simplicity we write down these equations assuming that initial data are independent of the parameter α.

Let the solution $\bar{y}(t)$ being investigated correspond to initial data $\bar{y}(0)$ and the value $\bar{\alpha}$ of the parameter α. Let

$$\left. \begin{aligned} \xi &= \frac{\partial v_\alpha^t(y(0))}{\partial y(0)} \bigg|_{y(0) = \bar{y}(0), \, \alpha = \bar{\alpha}}, \\ \zeta &= \frac{\partial v_\alpha^t(y(0))}{\partial \alpha} \bigg|_{y(0) = \bar{y}(0), \, \alpha = \bar{\alpha}}, \end{aligned} \right\} \tag{6.1.16}$$

where v_α^t is the resolving operator of Equation (6.1.1).

Formal differentiation of Equation (6.1.1) yields

$$\frac{d\xi}{dt} = A(t)\xi; \qquad \xi|_{t=0} = I_Y, \tag{6.1.17}$$

$$\frac{d\zeta}{dt} = A(t)\zeta + B(t); \qquad \zeta|_{t=0} = 0, \tag{6.1.18}$$

where I_Y is the identical operator in Y; the operators $A(t)$ and $B(t)$ are defined as

$$A(t) = \frac{\partial v}{\partial y}\bigg|_{y=\bar{y}(t),\,\alpha=\bar{\alpha}} \quad ; \qquad B(t) = \frac{\partial v}{\partial\alpha}\bigg|_{y=\bar{y}(t),\,\alpha=\bar{\alpha}}. \qquad (6.1.19)$$

Equations for higher derivatives with respect to the parameter α and to initial data are found in an analogous fashion. Each of these derivatives is a solution to the problem

$$\frac{dh}{dt} = A(t)h + f(h), \qquad h|_{t=0} = h(0), \qquad (6.1.20)$$

with the functions $h(0)$ and f defined in a proper way. It should be noted that the equations appearing in (6.1.17), (6.1.18) and (6.1.10) are linear but non-autonomous in the case where the field v in (6.1.1) is independent of time explicitly.

The problem under consideration is equivalent to the functional equation

$$G(y, \alpha) = 0,$$

$$G(y, \alpha) = \frac{dy}{dt} - v(y, \alpha),$$

where the desired function y has to satisfy the condition

$$\int_0^T \|y(t)\|^2 \, dt < \infty \quad \text{as } T < \infty.$$

Let $R(y, \alpha)$ be a functional of a solution to the Cauchy problem complying with the conditions

$$\left.\begin{aligned}
R(y, \alpha) - R(\bar{y}, \bar{\alpha}) &= VR + o\left(\|\alpha - \bar{\alpha}\| + \left(\int_0^T \|y - \bar{y}\|^2 \, dt\right)^{1/2}\right), \\
VR &= \int_0^T (r_y, y - \bar{y}) \, dt + (r_\alpha, \alpha - \bar{\alpha}), \\
\int_0^T \|r_y\|^2 \, dt &+ \|r_\alpha\|^2 < \infty.
\end{aligned}\right\} \qquad (6.1.21)$$

In this case the basic relations of the theory of sensitivity are written as follows:

$$\left.\begin{aligned}
y(t, \alpha) - y(t, \bar{\alpha}) &= \zeta(\alpha - \bar{\alpha}) + o(\|\alpha - \bar{\alpha}\|), \\
R(y, \alpha) - R(\bar{y}, \bar{\alpha}) &= VR + o(\|\alpha - \bar{\alpha}\|), \\
VR &= \int_0^T \left(\zeta^*, \frac{\partial v}{\partial\alpha}(\alpha - \bar{\alpha})\right) dt + (r_\alpha, \alpha - \bar{\alpha}),
\end{aligned}\right\} \qquad (6.1.22)$$

where the function ζ^* is a solution to the problem

$$-\frac{d\zeta^*}{dt} = A^*(t)\zeta^* - r_y, \qquad 0 \le t \le T,$$

$$\zeta^*|_{t=T} = 0,$$

$$(6.1.23)$$

$A^*(t)$ is the operator adjoint to $A(t)$.

The relations (6.1.22) and (6.1.23) are fulfilled *a fortiori* for smoothly bounded fields v. If a field v and an operator $A(t)$ of linearization are unbounded (as is the case for partial differential equations), the problems (6.1.18) and (6.1.23) are ill-posed in general, and the formulae (6.1.20)–(6.1.22) are invalid. A sufficient condition of correctness for the problems (6.1.18) and (6.1.23) is, for example, the Hille–Iosida condition

$$\|(\lambda - A)^{-1}\| \le (\operatorname{Re}\lambda - \lambda_0)^{-1}$$

for all complex λ and a real constant λ_0. The solutions to the problems (6.1.18) and (6.1.23) can be written under this condition in the form

$$\xi(t) = T(t)\xi(0),$$

$$\zeta(t) = \int_0^T T(t - t')B(t')\,dt',$$

where $\{T(t)\}_{t\ge 0}$ is a family of continuous (over t) operators satisfying the condition

$$\|T(t)\| \le \mu\exp(\lambda_0 t);$$

μ is a constant.

In the general case, the operators A and B depend on t, and the solutions to the problems (6.1.17), (6.1.18) and (6.1.23) are determined by the evolution operator V:

$$\xi(t) = V(t, 0)\xi(0), \tag{6.1.24}$$

$$\zeta(t) = \int_0^t V(t, \tau)B(\tau)\,d\tau, \tag{6.1.25}$$

$$\xi^*(t) = -\int_0^T V^*(t, \tau)z_y\,d\tau. \tag{6.1.26}$$

The solution of Equation (6.1.1) is reduced, with the help of the evolution operator, to the solution of an integral equation for the function $u(t) = y(t, \alpha) - y(t, \bar{\alpha})$, i.e.,

$$u(t) = V(t, 0)u(0) + \int_0^t V(t, t')f(u, t')\,dt',$$

$$f(u, \alpha) = v(y, \alpha) - v(\bar{y}, \bar{\alpha}) - A(t, u).$$

$$(6.1.27)$$

If the operator A is unbounded, it is common for the operator V to have smoothing properties. This allows us to justify the relations (6.1.20)–(6.1.22) for a broad class of problems.

Sensitivity of steady-state and time-periodic solutions. The steady-state solution of Equation (6.1.1) complies with Equation (6.1.2) for $G = v$. As applied to this case, Equations (6.1.3), (6.1.5), (6.1.8), (6.1.5′) and (6.1.9′) are written as

$$\left.\begin{aligned} y(\alpha) - y(\bar{\alpha}) = \zeta(\alpha - \bar{\alpha}) + o(\|\alpha - \bar{\alpha}\|), \\ A\zeta + B = 0, \end{aligned}\right\} \tag{6.1.28a}$$

$$\left.\begin{aligned} R(y, \alpha) - R(\bar{y}, \bar{\alpha}) = VR + o(\|\alpha - \bar{\alpha}\|), \\ VR = (\zeta^*, B(\alpha - \bar{\alpha})), \qquad A^*\zeta^* = -r_y, \end{aligned}\right\} \tag{6.1.28b}$$

where the operators A and B are defined by the formulae (6.1.19), and the functional R satisfies the condition (6.1.6).

The implicit function theorem guarantees correctness of the formulae (6.1.28) if the spectrum of the operator A has no eigenvalues at the origin of the imaginary axis. Violation of this condition is usually accompanied by the loss of stability of equilibrium solutions to small perturbations in the initial data. In general, if the spectrum of the operator A lies strictly in the left-half plane, the formulae (6.1.28) determine a linear change in sensitivity to parameter variations.

We consider the equations with time-periodic coefficients

$$\frac{dy}{dt} = v(y, z(t), \alpha), \qquad z(t + \tau) = z(t). \tag{6.1.29}$$

Let $\bar{y}(t)$ be a time-periodic solution of Equation (6.1.2) for a fixed value $\bar{\alpha}$ of the parameter α, i.e.

$$\frac{d\bar{y}}{dt} = v(\bar{y}, z(t), \bar{\alpha}), \qquad \bar{y}(t + \tau) = \bar{y}(t), \tag{6.1.30}$$

Equations for variations referred to the problem (6.1.29) are given, as before, by the relations (6.1.17) and (6.1.18) but now the operators A and B are τ-periodic functions, and the evolution operator of Equation (6.1.18) satisfies the condition

$$V(t + \tau, 0) = V(t, 0)V(\tau, 0). \tag{6.1.31}$$

The operator $V(\tau) \equiv V(\tau, 0)$ is known as the *monodromy operator*, and its eigenvalues are called *multiplicators*. Suppose that $\xi = \exp(\lambda t)W(t)$ in Equation

(6.1.17). If W has a period τ, Equation (6.1.17) guarantees the fulfilment of the relations

$$\frac{dW}{dt} = A(t)W(t) + \lambda W(t); \qquad W(t + \tau) = W(t). \qquad (6.1.32)$$

A set of λ for which the problem (6.1.32) has a solution is named as the *stability spectrum of the time-periodic solution $\bar{y}(t)$*. In regular cases, multiplicators coincide with points $\exp\{\lambda t\}$. If the monodromy operator has a logarithmic singularity, the standard form of the implicit function theorem can no longer be applied and the Floquet representation $V(t, 0) = Q(t)\exp(t, \Gamma)$ is valid. Here Q is a τ-periodic operator: $\Gamma = \tau^{-1}\ln V(\tau)$. The Floquet representation occurs particularly if $\tau^{-1}\int_0^\tau \|A(t)\|\,dt$ does not exceed a constant c depending on the structure of the phase space Y. For Hilbert space this constant is equal to π. The Floquet representation allows us to reduce Equations (6.1.17) and (6.1.18) to analogous equations with constant coefficients.

Let us consider the problem of evaluation of the sensitivity of a time-periodic solution assuming that the spectrum of the monodromy operator is strictly inside the unit circle (at a fixed distance from the unit circle). In this case the stability spectrum lies in the left-half plane and the estimate $\|V(t, t')\| < \mu\exp\{\chi(t - t')\}$ is fulfilled for $\chi < 0$.

As applied to time-periodic solutions, the basic equation of sensitivity takes the form

$$y(t, \alpha) - y(t, \bar{\alpha}) = \zeta(\alpha - \bar{\alpha}) + o(\|\alpha - \bar{\alpha}\|), \qquad (6.1.33a)$$

$$d\zeta/dt = A(t)\zeta + B(t); \qquad \zeta(t + \tau) = \zeta(t), \qquad (6.1.33b)$$

$$R(y, \alpha) - R(\bar{y}, \bar{\alpha}) = \int_0^\tau \left(\zeta^*, \frac{\partial v}{\partial \alpha}(\alpha - \bar{\alpha})\right) dt + o(\|\alpha - \bar{\alpha}\|), \qquad (6.1.33c)$$

$$-d\zeta^*/dt = A^*(t)\zeta^* - r_y; \qquad \zeta^*(t + \tau) = \zeta^*(t), \qquad (6.1.33d)$$

with operators A and B being defined by the formulae (6.1.19) and the functional R satisfying the relations (6.1.21) at $T = \tau$.

We derive formulae to evaluate the functions ζ and ζ^*. Any solution of Equation (6.1.18) admits the representation

$$\zeta(t) = V(t, 0)\zeta(0) + \int_0^t V(t, t')B(t')\,dt'. \qquad (6.1.34)$$

Considering that this solution has to be time-periodic it is necessary and sufficient to have the following equality fulfilled:

$$[I - V(\tau)]\zeta(\tau) = \int_0^\tau V(\tau, t')B(t') \, dt'.$$

Here I is the identical operator.

With the indicated assumptions the operator $I - V(\tau)$ is invertible. Hence,

$$\zeta(\tau) = [I - V(\tau)]^{-1} \int_0^\tau V(\tau, t')B(t') \, dt'.$$

On the basis of this relation and the formula (6.1.34) we obtain

$$\zeta(t) = V(t, 0)[I - V(\tau)]^{-1} \int_0^\tau V(\tau, t')B(t') \, dt' + \int_0^t V(t, t')B(t') \, dt'. \quad (6.1.35)$$

The function ζ^* is constructed in a similar way with the help of the operator $V^*(t, t')$.

The above results are valid *a fortiori* if the inequality

$$\|V(t, t')\| \le \mu \exp\{\chi(t - t')\}$$

with $\chi < 0$ is fulfilled. The results can be extended to a broad class of unbounded non-linear problems. However, the assumptions used are violated for auto-oscillations (time-periodic solutions of the autonomous equation (6.1.1)). In such a situation $V(\tau)v(\bar{y}, \bar{\alpha}) = v(\bar{y}, \bar{\alpha})$ so that one multiplicator (equal to 1) is necessarily sited on the unit circle. If the remaining part of the spectrum of the monodromy operator lies strictly inside the unit circle, auto-oscillations are stable and their sensitivities are linear in nature.

Sensitivity at large time intervals. The basic relations (6.1.20)–(6.1.22) of the theory of sensitivity of time-dependent problems are valid under sufficiently general assumptions at any finite time interval $[0, T]$, although the accuracy of these relations can diminish exponentially as T increases. It may be possible to ensure that the exponential growth will be absent when a solution under investigation converges to an equilibrium or time-periodic state, the stability spectrum of which lies strictly in the left-half plane. In such a situation the evolution operator of equations of sensitivity satisfies the condition $\|V(t, t')\| < \mu \exp\{\chi(t - t')\}$ with $\chi < 0$. One may prove using this condition that solutions of direct and adjoint equations of sensitivity converge exponentially to their limiting values. Accordingly, sensitivity of a limiting state may be calculated by the stabilization method.

The systems, where small perturbations increase exponentially with time, can be examined by probability methods. See, e.g., Kagan *et al.* (1990) for a more detailed discussion of this aspect of the subject.

6.2 Equilibrium response to a change in ocean–land area ratio

The idea that the ratio between ocean and land areas and their mutual locations play a role in climate formation that is no less than that played by the latitudinal zonality of insolation and the composition of atmospheric gases is repeated so often that it has taken on the appearance of a belief. But no idea (even those constantly repeated) can be considered as being correct until it has been corroborated by proof. Such proof can be obtained when using either empirical data or the results of specially planned numerical experiments. In this discussion of the results of numerical experiments we start with an answer to the question as to what climate of the Earth would have had if it had been covered solely by ocean or by land.

At first sight this question may seem senseless. But let us not jump to conclusions and recall instead that in the course of the Earth's geological history the ratio between ocean and land areas in separate hemispheres has been subjected to repeated and quite significant changes generated by the drift of continents. These changes, on a time scale of the order of 10^8 years, must have been accompanied by climate fluctuations characterized by the weakening or enhancement of continentality. Thus, the simulation of climate with the two above-mentioned limiting situations allows us to estimate the probable range of climatic fluctuations caused by the drift of continents and, thereby, helps us to solve the problem of the reconstruction of the paleoclimate. The purpose of the numerical experiments carried out by Kagan *et al.* (1990) can be interpreted in this way.

The 0.5-dimensional (box) model of the climatic system of the Northern Hemisphere, described in detail in Section 5.5, was taken as the basis for the above-mentioned work. It was initially modified in two respects. First, in the case of the ocean-covered Earth, in the equations of conservation of heat for temperate and low latitude areas the correlative term was excluded, i.e. the influence of spatial inhomogeneity of the fields of temperature and thickness of the UML being determined, among other things, by the blocking effect of continents was not taken into account. Second, in the case of the land-covered Earth the soil moisture content was fixed, W/W_c was assumed equal to 1, and the southern boundary of the snow cover was identified with the annual mean isotherm $-1\,^\circ$C, which position, as in the initial model, was found by approximation of the meridional distribution of the annual mean surface air temperature by the first two terms of the series in even Legendre polynomials.

The results of the numerical experiments presented by Kagan *et al.* (1990) demonstrate that the ocean-covered Earth had a warmer climate, and the land-covered Earth had a colder climate (compared to the present climate), with the main differences of climatic characteristics being confined primarily to high latitudes: the difference of the annual mean values of surface air temperature in the two cases mentioned accounted for more than 26 K in the northern box and only 3 K in the southern box. The causes of such climate transformations are obvious. They are: changes in the radiative and thermophysical properties of the underlying surface and the existence of positive feedback between the albedo of the underlying surface and surface air temperature, from which two most remarkable features of the solutions obtained take their origin: the disappearance of sea ice in the first case and a considerable displacement of the snow cover towards the south in the second case.

But the matter is not restricted to the above. In the case of the ocean-covered Earth the temperature of the deep layer in the temperate and low latitude areas is to by more than 12 °C with respect to present-day values. This feature is caused by a weakening of the transport of cold deep water from high latitudes to temperate and low latitudes of the ocean. In turn, a weakening of the transport of cold deep water is caused by a rise in water temperature at high latitudes of the ocean, and this rise is caused by the local decrease in the heat transfer from the ocean into the atmosphere. In the case of the land-covered Earth, the most remarkable feature of the solution, apart from that mentioned above, is a considerable strengthening of the meridional transport of sensible and latent heat in the atmosphere. This feature is explained, first, by the concentration of the meridional heat transport within the atmosphere (in two other cases examined it is redistributed between the ocean and the atmosphere) and, second, by the intensification of disturbances of synoptic scale in the atmosphere determined by an increase in temperature contrast between the northern and southern atmospheric boxes.

The numerical experiments detect one, at first sight unexpected, result: in the case of the land-covered Earth the moisture content of the atmosphere in the southern box not only does not decrease as would be expected with decreasing temperature and evaporation from the underlying surface, but increases compared with the present Earth and the ocean-covered Earth. To interpret this result we turn to Figure 6.1, which illustrates the seasonal oscillations of the main climatic characteristics: temperature, humidity and precipitation. It can be seen that the basic contribution to the formation of annual mean air humidity comprises values in the summer season when the

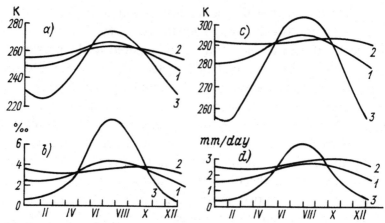

Figure 6.1 Seasonal oscillations of average (over the Northern Hemisphere) climatic characteristics for the present-day ocean–land area ratio (1), and for the ocean-covered (2) or ocean-free Earth (3), according to Kagan *et al.* (1990): (*a*) the mass-weighted average air temperature; (*b*) the mass-weighted average air specific humidity; (*c*) the surface air temperature; (*d*) precipitation.

surface air temperature in the case of the land-covered Earth is higher by 5 or 6 K than for the other two cases. If we now consider that the air humidity and temperature are related to each other by a non-linear dependence then the above-mentioned increase in atmospheric moisture content becomes clear. But then the climate of temperate and low latitudes in the case of the land-covered Earth has to be more humid than the climate complying with present-day distribution of land and ocean. This result, a contradiction of common sense, is a consequence of fixing of soil moisture content. It serves as direct evidence in favour of the importance of a proper description of the hydrological cycle when modelling the paleoclimate.

Let us examine another aspect of the subject in question: the mutual location of ocean and land, and turn again to the analysis of limiting cases. Imagine that the Earth has either one continent occupying a zonal belt between 17 °N and 17 °S (the case of the equatorial continent), or two continents with centres on the poles and with boundaries passing along the parallels 45 °N and 45 °S (the case of the polar continents). For such an approximation the area of the equatorial continent in the first case, and of the two polar continents in the second case, will be equal to the total area of all the present-day continents. In other words, it is necessary to estimate climate sensitivity to variations in the mutual location of ocean and land with the constant ratio of their areas. The problem formulated in such a way was solved by Barron *et al.* (1984) within the framework of a simplified version of the NCAR model which is different from that presented in Section 5.8 in

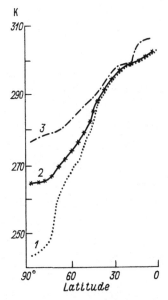

Figure 6.2 The meridional distribution of the annual mean zonally averaged temperature of the underlying surface, according to Barron *et al.* (1984): (1) the case for polar continents with snow cover in high latitudes of land; (2) the same but without the land snow cover; (3) the case for the equatorial continent.

that it represents the ocean in the form of a 'swamp' with zero heat capacity. The derived dependence on latitude of the annual mean zonally averaged temperature of the underlying surface is shown in Figure 6.2.

As can be seen, in the case of the equatorial continent the annual mean zonally averaged temperature of the underlying surface in a region of the pole remains higher than the water freezing temperature; meanwhile in the case of the polar continents it decreases (compared with the previous case) by 12 K if the existence of snow cover at high latitudes is not taken into account, and by 34 K if it is taken into account. These estimations point to the fact that the maintenance of low temperatures of the underlying surface and, hence, ice and snow in polar latitudes, is determined exclusively by nature or, more precisely, by the reflectance of the underlying surface. Another explanation is the temperature decrease in the equatorial and tropical latitudes where it is connected with the enhancement of evaporation when land is replaced by ocean. For the Earth as a whole the annual mean temperature of the underlying surface in the case of the polar continents turns out to be lower by 4.6 K than in the case of the equatorial continent when the ice cover of land is not taken into account, and by 7.4 K when it is taken into account. Thus it is obvious that the climate is quite sensitive to variations of the mutual locations and to the ratio between the ocean and land areas.

Let us direct our attention to one more circumstance. What is in question here is the dependence of climate sensitivity to variations of one climate-forming factor on a change in another, that is, what has come to be called the *sensitivity of sensitivity* (or *cross-sensitivity*) of climate, which we will come across in the next section in the discussion of climate sensitivity to variations of atmospheric CO_2 concentration with and without consideration of the meridional heat transport in the ocean. The same situation arises here. Indeed, it is clear from general reasoning that climate sensitivity to variations in the solar constant or in the concentration of atmospheric CO_2 must be larger if the continents are grouped in the vicinity of the pole, and less if they are in the region of the equator. The cause of this is an enhancement of the positive feedback between albedo and temperature of the underlying surface due to the appearance of sea ice and snow cover of land in the first case, and their absence in the second case. This is confirmed by estimates of climate sensitivity to variations in the solar constant for Late Paleozoic and Late Mesozoic paleogeographical reconstructions of the Earth's surface (see Verbitskii and Chalikov, 1986). From this it follows that the transfer of respective paleoclimatic estimates from one geological epoch to another should be done with caution.

6.3 Equilibrium response to a change in the concentration of atmospheric CO_2

The main climatic role of carbon dioxide, as well as of a number of other gases, is to create the so-called *greenhouse effect* in the atmosphere. The latter consists of the following. On the one hand, CO_2 absorbs the downward flux of short-wave solar radiation weakly; on the other hand, it is a strong absorber of the outgoing long-wave emission in the range of wavelengths 10–18 μm. The long-wave emission absorbed by atmospheric CO_2 is reradiated in all directions. Downward long-wave radiation is absorbed by the Earth's surface and as a result its temperature becomes higher than it would have been in the absence of CO_2 and other 'greenhouse' gases.

A simple estimate of the greenhouse effect for the contemporary composition of the atmosphere follows from the condition of equality between absorbed short-wave solar radiation and long-wave emission, that is,

$$\pi a^2 S_0 (1 - \alpha_p) = 4\pi a^2 \sigma T_e^4,$$

where a and T_e are the radius and effective radiative temperature of the planet, S_0 is the solar constant, α_p is the planetary albedo and σ is the Stefan–Boltzmann constant.

Assuming that $S_0 = 1370 \text{ W/m}^2$ and $\alpha_p = 0.3$ we have $T_e \approx 255 \text{ K}$. The temperature T_e can be interpreted as the temperature at a level H_e in the atmosphere where the emission into space is equal to the emission of the planet as such. For the present-day atmosphere $H_e \approx 6$ km. Since an average vertical temperature gradient in the troposphere accounts for 5.5 K/km, then the temperature T_s of the underlying surface must be approximately equal to 288 K. The difference $(T_s - T_e) \approx 33$ K is determined by the greenhouse effect of the atmosphere. According to estimates obtained on the assumption of radiative equilibrium for the atmosphere, an increase δQ in net (short-wave + long-wave) downward radiation at the upper tropospheric boundary for a doubling of the atmospheric CO_2 concentration is about 4 W/m². About one-third of δQ is consumed in Earth's surface heating; the remainder is consumed in tropospheric heating. This increases the long-wave emission by radiatively active atmospheric components (clouds, H_2O, CO_2 and other gases) and thereby enhances warming of the underlying surface. The total increase δT_s of the underlying surface temperature caused by the two above-mentioned factors is $\delta T_s \approx \delta Q/4\sigma T_s^3 \approx 0.75$ K.

This estimate does not take into account the presence of feedbacks; some of these (including those discussed in Section 1.3) are able to change the resulting effect considerably. This circumstance dictates the necessity for more detailed analysis of the greenhouse effect of CO_2. Let us dwell on one such analysis carried out by Spelman and Manabe (1984) using the GFDL model. It covers all subsystems of the climatic system including the deep ocean and is the most exhaustive of all existing analyses.

The spatial distribution of steady state changes in the zonal average temperature of air and water as a result of the quadrupled concentration of atmospheric CO_2 can be judged from Figure 6.3. As can be seen, the stratosphere is cooled down and the troposphere is heated up due to an enhancement of the greenhouse effect. The largest rise in air temperature is noted at high latitudes due to a decrease in albedo of the underlying surface caused by a reduction of snow and sea ice areas. Here a rise in temperature extends up to ~ 12 km, while at low latitudes it covers the whole troposphere up to ~ 18 km. The first circumstance is explained by the influence of stable stratification at high latitudes, the second by intensive convective mixing at low latitudes.

The zonal average water temperature increases at all depths in the ocean. As this takes place the increase in the deep layer temperature amounting to about 7.5 °C turns out to be independent of latitude. The same changes in the sea surface temperature take place in the zone 65–70 °N where an area of cold deep water formation is located and where, as is known, stratification is close to neutral and temperature disturbances penetrate down to the

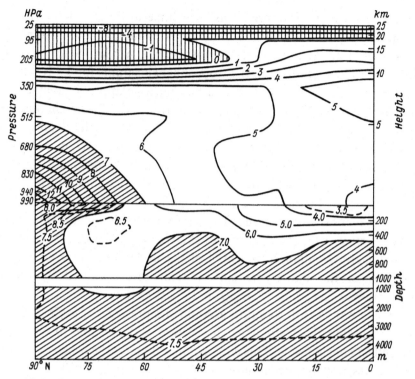

Figure 6.3 Latitude–altitude distribution of changes in the zonal average temperature (°C) for a four-fold increase in the concentration of atmospheric CO_2, according to Spelman and Manabe (1984).

bottom. The increase in temperature of the ocean surface at low latitudes is much lower than that at high latitudes due to less heating of the surface atmospheric layer. Therefore, the total increase in temperature in the deep ocean is greater than in the UML. It is interesting to note that the largest rise in sea surface temperature takes place not in the region of the poles where heating of the atmospheric surface layer is maximum but, rather, in the vicinity of the sea ice edge ($\sim 75\,°N$). Spelman and Manabe associate this result with the influence of the Arctic halocline separating the cold and fresh UML from the warmer and salty deep layer contributing to a decrease in thickness of the UML. But their subsequent experiments carried out at the eight-fold increased concentration of atmospheric CO_2 do not reflect this feature. Thus the question of the maximum rise in sea surface temperature in the vicinity of the sea ice edge remains open.

Changes in the zonal average temperature of air and water presented in Figure 6.3 correspond to the annual mean insolation. Consideration of its seasonal changes must transform the mentioned distribution of the zonally

averaged temperature of air and water, not least because many feedbacks
(say, the feedback between surface temperature and albedo of the snow–ice
cover) are seasonal in nature. As an example we examine the results of
numerical experiments by Wetherald and Manabe (1981) on the seasonal
variability of the climatic system response to doubling of the concentration
of atmospheric CO_2 obtained within the framework of the GFDL model with
the real geography of the continents and the ocean in the form of a 'swamp'
of finite heat capacity. According to Wetherald and Manabe (1981) the changes
in the zonal average surface air temperature T_a at low latitudes turn out to
be small and to vary little during the annual cycle (Figure 6.4). At high latitudes
they are always large and are subject to marked seasonal oscillations especially
in the Northern Hemisphere. The maximum rise in surface air temperature
takes place at the beginning of winter when values of T_a are minimum; the
minimum rise takes place in summer when values of T_a are maximum. In other
words, when the concentration of atmospheric CO_2 increases the amplitude of
seasonal oscillations of T_a at high latitudes decreases considerably.

To explain this feature we recall that the rise in surface air temperature
which accompanies an increase in the concentration of atmospheric CO_2 leads
to a reduction in the area and thickness of sea ice. Accordingly, the albedo
of the underlying surface decreases, and absorption of short-wave radiation

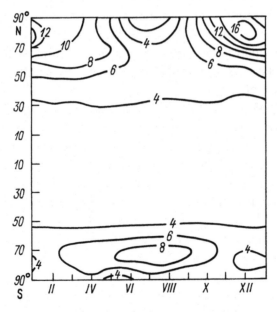

Figure 6.4 Time–space distribution of changes in the zonal average surface air
temperature (°C) for doubling of concentration of atmospheric CO_2, according to
Wetherald and Manabe (1981).

in the UML and its temperature increase. But due to the large heat capacity of the UML, changes in T_a in the absence of ice are considerably less than in its presence; thus, the increase in T_a at high latitudes turns out to be comparatively small. In late autumn when convective heat transfer from the ocean into the atmosphere increases, the excess of heat in the UML results in the delay of the onset of the sea ice formation. This is favoured by an enhancement of the heat exchange between the ocean and the atmosphere and, hence, by a heating of the surface layer in late autumn and early winter.

We have already mentioned that an increase in the concentration of atmospheric CO_2 must be accompanied by intensification of the hydro-logical cycle, that is, by enhancement of evaporation (primarily from the ocean surface) and by an increase in moisture content in the atmosphere and precipitation. All estimates obtained within the framework of three-dimensional climatic models (see Table 6.1), apart from those presented by Gates *et al.* (1981) and by Mitchell (1983), point to an increase in the annual mean global average precipitation. The decrease in precipitation detected by these authors relates to their fixing the sea surface temperature: for constant sea surface temperature and increased concentration of atmospheric CO_2, evaporation remains basically the same, although the air temperature rises a little. Accordingly, the relative humidity of the air and the non-convective precipitation decreases. Thus, if estimates by Gates *et al.* (1981) and Mitchell (1983) are not taken into account it can be affirmed that one of the consequences of increasing the concentration of atmospheric CO_2 is the enhancement of precipitation for the Earth as a whole, and, as a consequence, the intensification of run-off.

A particularly large increase in precipitation and run-off must occur in high latitudes due to the growth of moisture content and enhancement of the meridional transport of water vapour. This conclusion, expressed, probably for the first time, by Manabe and Wetherald (1980) is confirmed by all numerical experiments carried out within the framework of three-dimensional models of the climatic system. It differs for other latitudes, especially for the tropics where modelling results are markedly different.

When taking into account the seasonal cycle of insolation a rise in the concentration of atmospheric CO_2 results in an increase in zonal average values of the difference between precipitation and evaporation at high latitudes and in their decrease at low latitudes throughout the year (see Washington *et al.*, 1980). The same situation occurs when the seasonal cycle of insolation is absent. In other words, the distribution of annual mean zonally averaged disturbances of hydrological cycle characteristics, with and without allowance for the seasonal cycle of insolation, are qualitatively similar to each other. But the presence of seasonal variations of insolation leads to the fact that the annual mean disturbances of hydrological cycle characteristics (in particular, precipitation) turn out to be less in the seasonal model than in the annual mean model.

Seasonal variability of disturbances of hydrological cycle characteristics can be illustrated by calculation of the time–space distribution of changes in the zonal average soil moisture content when the concentration of atmospheric CO_2 is doubled (see Wetherald and Manabe, 1981). The most distinctive feature of this distribution is the increase in soil moisture content at high and temperate latitudes of the Northern Hemisphere (in the zone 30–70 °N) in winter and its decrease in summer. This last circumstance is determined by an increase in the duration of the warm period due to earlier melting of snow.

Note that enhancement of large-scale drying of northern high and mid-latitude continents during summer in the $2 \times CO_2$ experiment is confirmed by most three-dimensional models cited in Table 6.1, and disagrees with the results of the numerical experiments of Hansen *et al.* (1984), Washington and Meehl (1984, 1989) and Mitchell and Warrilow (1987) since the soil in their control experiments is not close to saturation. Therefore, although the summer season increases, nevertheless soil moisture at the beginning of this season increases, too, in consequence of enhanced winter precipitation, so that enhanced summer drying may not be produced.

Now, we turn to analysis of temperature changes caused by an increase in the concentration of atmospheric CO_2. The time–space variability of these changes remains the same for all three-dimensional models of the climatic system in a qualitative, but not in a quantitative, respect. It is essential to understand the causes of such discrepancies. For this purpose we take advantage of available estimates of changes $(\delta \bar{T}_a)_{nX}$ in the annual mean globally averaged surface air temperature $\bar{T}_a$ when the normal (present-day or pre-industrial) concentration X of atmospheric CO_2 increases by n times.

Values of $(\delta \bar{T}_a)_{2X}$ for zero- and one-dimensional thermodynamic models of latitudinal structure are included in the range from 0.6 to 3.3 °C; for one-dimensional models of vertical structure (or, as they are also called, the radiative–convective models), from 0.5 to 4.2 °C; for zonal climatic models, from 1.7 to 3.0 °C; and for three-dimensional models of the climatic system, excluding estimates obtained for fixed sea surface temperature, from 1.3 to 6.5 °C (see Table 6.1). The distinction of estimates $(\delta \bar{T}_a)_{2X}$ is connected with a consideration of different sets of climate-forming factors and with different descriptions of feedbacks in models of different dimension. With increasing

Table 6.1. Annual mean global average of the surface air temperature T_a (°C), precipitation rate P (mm/day) and their changes $\delta\bar{T}_a$ (°C) and $\delta P/P$ (%) for an increase of the concentration of atmospheric CO_2, from results of numerical experiments within the framework of three-dimensional models of the climatic system

| Author | Characteristics of the model | | | | $\bar{T}_a$ | $(\delta\bar{T}_a)_{2x}$ | $(\delta\bar{T}_a)_{4x}$ | $\bar{P}$ | $\dfrac{(\delta P)_{2x}}{P}$ | $\dfrac{(\delta P)_{4x}}{P}$ |
	Distribution of land and ocean	Insolation	Cloudiness	Representation of the ocean						
Manabe and Wetherald (1975)	Idealized (sectorial)	Annual mean	Fixed	'Swamp' of zero heat capacity	21.0	2.9	–	2.55	7.8	–
Manabe and Wetherald (1980)	Same	Same	Calculated	Same	21.4	3.0	5.9	2.58	7.0	11.6
Manabe and Stouffer (1980)	Realistic	Seasonal	Fixed	'Swamp' of finite heat capacity	15.0	–	4.1	2.69	–	6.7
Wetherald and Manabe (1981)	Idealized (sectorial)	Annual mean	Same	Same	16.0	–	6.0	2.35	–	12.8
	Same	Seasonal	Same	Same	16.8	–	4.8	2.40	–	10.0
Schlesinger (1984)	Realistic	Annual mean	Calculated	'Swamp' of zero heat capacity	17.9	2.0	–	2.73	5.1	–
Gates et al. (1981)	Same	Seasonal, accounting for diurnal oscillations	Same	Prescribed time–space distribution of the sea surface temperature	14.8	0.2	0.4	2.69	–1.5	–3.3
Mitchell (1983)	Same	Same	Fixed	Same	12.3	0.2	–	2.83	–2.5	–
Washington and Meehl (1983)	Same	Annual mean	Same	'Swamp' of zero heat capacity	11.7	1.3	2.7	3.67	2.7	6.5
	Same	Same	Calculated	Same	11.7	1.3	3.4	3.66	3.3	6.0
Washington and Meehl (1984)	Same	Seasonal	Same	'Swamp' of finite heat capacity	14.4	3.5	–	3.3	7.1	–
Spelman and Manabe (1984)	Idealized (sectorial)	Annual mean	Fixed	Realistic ocean accounting for heat transport	14.4	3.2	5.4	2.08	7.7	14.4
Manabe and Bryan (1985)	Same	Same	Same	'Swamp' of finite heat capacity	5.2	6.5	13.1	–	–	–

Reference											
Hansen et al. (1984)	Realistic	Seasonal, accounting for diurnal oscillations	Calculated	'Swamp' of finite heat capacity with prescribed heat transport	14.2	4.2	–	3.2	11.0	–
Wetherald and Manabe (1986)	Same	Seasonal	Same	'Swamp' of finite heat capacity	14.8	4.0	–	2.9	8.7	–
Wilson and Mitchell (1987)	Realistic	Seasonal, accounting for diurnal oscillations	Calculated	'Swamp' of finite heat capacity with prescribed heat transport	–	5.2	–	–	15.0	–
Mitchell and Warrilow (1987)	Same	Same	Same	Same	–	5.2	–	–	15.0	–
Wetherald and Manabe (1988)	Same	Seasonal	Fixed	'Swamp' of finite heat capacity	–	4.0	–	–	9.0	–
Schlesinger and Zhao (1989)	Same	Same	Calculated	Same	–	2.8	–	–	8.0	–
Washington and Meehl (1989)	Same	Same	Same	Same	–	4.0	–	–	8.0	–
Gordon and Hunt (see IPCC 90)	Same	Seasonal, accounting for diurnal oscillations	Same	'Swamp' of finite heat capacity with prescribed heat transport	–	4.0	–	–	7.0	–
Mitchell et al. (1989)	Same	Same	Same	Same	–	3.5	–	–	9.0	–
Boer et al. (see IPCC 90)	Same	Same	Same	Same	–	3.5	–	–	4.0	–
Oglesby and Saltzman (1990)	Same	Seasonal	Same	'Swamp' of finite heat capacity	–	3.8	–	–	–	–
McAvaney et al. (see IPCC 90)	Same	Same	Same	Same	–	4.2	–	–	8.3	–
Wang et al. (see IPCC 90)	Same	Same	Same	Same	–	4.2	–	–	8.3	–
Dix et al. (see IPCC 90)	Same	Seasonal, accounting for diurnal oscillations	Same	Same	–	4.8	–	–	10.0	–
Washington and Meehl (1991)	Same	Seasonal	Same	Same	–	4.5	–	–	5.0	–
Wang et al. (see IPCC 92)	Same	Same	Same	Same	–	3.9	–	–	6.9	–
Le Treat et al. (see IPCC 92)	Same	Same	Same	Same	–	5.3	–	–	8.0	–

Note: the observed values T_a and P are equal to 14.2 °C and 2.65 mm/day respectively. Dashes indicate unavailable information.

resolution and sophistication of a model it seems that the accuracy of description of feedback effects should be higher and the range of possible values of $(\delta \bar{T}_a)_{2x}$ should have been less. But this does not occur for the following reasons.

The point is that even three-dimensional models differ from each other in their methods of description of cloudiness and convection in the atmosphere, and of thermal inertia and heat transport in the ocean. They also differ in the size of the domain examined, in the topography of the land surface, in the distribution of land and ocean, in the spatial resolution, and in the prescribed values of external parameters (the solar constant, albedo of land and sea ice, etc.). Finally, they take or do not take into account the seasonal variability of insolation. Let us consider the ways in which the main distinctions affect the estimates of $\delta \bar{T}_a$ obtained within the framework of three-dimensional climatic models.

Cloudiness. An increase in the concentration of atmospheric CO_2 is accompanied by an increase in cloudiness in the low troposphere and in the stratosphere of high latitudes, and by a decrease in cloudiness at all levels in the troposphere of temperate and low latitudes. Such changes in cloudiness detected in all numerical experiments using three-dimensional models are determined by the dependence of cloudiness on the relative air humidity. Indeed, an increase in surface air temperature leads to the intensification of local evaporation and the increase in air absolute humidity. Under conditions of stable stratification the main changes in absolute humidity occur in the low troposphere. Because temperature changes decrease with height, it is obvious that air relative humidity and, hence, cloudiness will increase markedly in the most stable stratified troposphere of high latitudes. Increase in the relative humidity in low layers of the stratosphere is apparently connected with enhancement of moisture transfer through the tropopause due to weakening of hydrostatic stability. As for a decrease in the relative humidity in the troposphere of low and temperate latitudes, this is caused by the intensification of upward, and suppression of downward, vertical motions due to an increase in heat release under phase transitions of water vapour. Since enhancement of relative humidity in an upward flow is limited (relative humidity cannot exceed 100%) and its decrease in a downward flow is not, then the area-weighted average values of relative humidity and fraction of cloudiness decrease.

The explanation given above is valid in those cases where the processes of moisture condensation, cloud formation and convective precipitation are parametrized in terms of the limiting relative humidity. But if cloudiness is determined in another way (say, by using an equation for cloud water) or some more sophisticated parametrizations relating cloud radiative property variations with cloud water content and/or temperature is applied then the qualitative features of cloud distributions are saved but can be explained by other causes (say, by cooling of the upper layers of the troposphere, by subsequent enhancement of convective mixing and then by an increase in water vapour content, relative humidity and cloudiness in the upper part and their decrease in the lower part of the troposphere).

We present available estimates of the influence of cloudiness on the sensitivity of the global average surface air temperature to an increase in the concentration of the atmospheric CO_2. According to Manabe and Wetherald (1980) the value of $(\delta \bar{T}_a)_{2X}$ when considering the feedback between radiation and cloudiness increases by only 0.1 °C. In other words, enhancement of absorption of long-wave radiation in the atmosphere determined by an increase in cloudiness is almost totally compensated by a decrease in short-wave solar radiation due to an increase in the planetary albedo. Washington and Meehl (1983) arrived at the same conclusion, according to which values of $(\delta \bar{T}_a)_{2X}$ for fixed and calculated cloudiness are approximately equal, with values of $(\delta \bar{T}_a)_{4X}$ in the second case being larger by 0.7 °C than those in the first case. A different result was obtained within the framework of the GFDL model using the realistic geography of the continents and presentation of the ocean in the form of a 'swamp' with finite heat capacity. It turned out that subsequent doubling of the concentration of atmospheric CO_2 accompanied by a change in cloudiness are almost exactly the same as those for quadruple increase in the concentration of atmospheric CO_2 and fixed cloudiness: $(\delta \bar{T}_a)_{2X} = 4.0$ °C in the first case, and $(\delta \bar{T}_a)_{4X} = 4.1$ °C in the second case. We note that the estimates by Manabe and Stouffer (1980) and Wetherald and Manabe (1986) were obtained with an allowance for the seasonal variability of insolation whereas those mentioned earlier were not. Because of this distinction the effects of the feedback between radiation and cloudiness can be associated with a more marked increase in cloudiness in the low troposphere of high latitudes in the case where the seasonal variability is not taken into account rather than in the case where it is taken into account.

Seasonal changes in insolation. According to Wetherald and Manabe (1981), an allowance for the seasonal cycle of insolation, all other things being equal, results in a decrease in $(\delta \bar{T}_a)_{4X}$ by 1.2 °C. As has already been mentioned above, this decrease in sensitivity of the seasonal model is associated with the disappearance of snow and sea ice at high latitudes in summer in a period of maximum insolation. Opposite changes in $(\delta \bar{T}_a)_{nX}$ were obtained by Hansen *et al.* (1984) and by Washington and Meehl (1983, 1984), who ascertained that an allowance for the seasonal variations of insolation results in a decrease in $(\delta \bar{T}_a)_{4X}$ by 0.3 °C and 2.2 °C respectively. The inconsistency of the estimates is explained by the different parametrizations of cloudiness and presentation of land and ocean: the estimate by Wetherald and Manabe (1981) complies with fixed cloudiness and the idealized distribution of land and ocean, while estimates by Hansen *et al.* (1984), and Washington and Meehl (1983, 1984) result from calculated cloudiness and use of the realistic distribution of land and ocean. But it is not improbable that the effect of seasonal variability of insolation in the last three works quoted is masked by other distinctions between the models, particularly by the different heat capacities of the ocean.

Thermal inertia of the ocean. It appears that an allowance for this does not affect the equilibrium response of the climatic system, although this statement holds only for the case where the ocean is represented as a 'swamp' with finite heat capacity, and seasonal (and, in general, any other) oscillations of climatic characteristics are absent. Otherwise, an allowance for thermal inertia of the ocean must lead to a change in amplitudes and phase shifts in seasonal oscillations of climatic characteristics, to weakening or enhancement of feedbacks and thereby to the transformation of the equilibrium response of the climatic system to external forcing.

The first estimate of climatic consequences of an increase in the concentration of atmospheric CO_2 when approximating the ocean in the form of a 'swamp' with zero heat capacity was obtained by Manabe and Wetherald (1975) within the framework of zonal and three-dimensional models. According to them and other authors using a similar presentation of the ocean, the value of $(\delta \bar{T}_a)_{2X}$ turned out to vary within the limits from 1.3 to 3.9 °C. In the other case where the ocean heat capacity is assumed to be equal to infinity (this is equivalent to fixing the ocean surface temperature), $(\delta \bar{T}_a)_{2X}$ is about 0.2 °C according to Gates *et al.* (1981) and Mitchell (1983). Such small values of $(\delta \bar{T}_a)_{2X}$ are caused by neglecting the feedback between the sea surface temperature and water vapour content in the atmosphere. When considering the finite heat capacity of the ocean and its prescription as equal to the heat capacity of the UML, the value of $(\delta \bar{T}_a)_{2X}$ changes from 2.2 °C according to McAvaney *et al.* (see IPCC 90) to 6.5 °C according to Manabe and Bryan (1985), and $(\delta \bar{T}_a)_{4X}$ changes from 4.1 °C according to Manabe and Hahn (1977) to 13.1 °C according to Spelman and Manabe (1984). Finally, when the processes in the deep ocean are considered, Manabe and Bryan

(1985) and Spelman and Manabe (1984) recommend $(\delta \bar{T}_a)_{2X} = 3.2\,°C$ and $(\delta \bar{T}_a)_{4X} = 5.4\,°C$.

The estimates presented point to the strong influence of the ocean thermal inertia on the sensitivity of the surface air temperature to an increase in the concentration of atmospheric CO_2. Unfortunately, they were obtained along with simultaneous changes in a number of other climate-forming factors (see Table 6.1) and therefore it is difficult to extract more detailed information from them.

Heat transport in the ocean. It is clear from general considerations that an allowance for the heat transport in the ocean (primarily in the meridional direction) has to be accompanied by a rise in the sea surface temperature at high latitudes, by a reduction in the sea ice area, by an increase in the surface air temperature and by a decrease in the snow-covered areas on the continents. The displacement of boundaries of snow cover and sea ice cover into higher latitudes must result in a weakening of the feedback between surface air temperature and the albedo of the underlying surface. In fact, since insolation and the length of the latitudinal circle decreases with an increase in latitude, then the displacement of boundaries of snow cover and sea ice into higher latitudes must result in a decrease in absorbed solar radiation changes and, hence, in changes in the temperature of the underlying surface and surface atmospheric layer. Thus, an allowance for the meridional heat transport in the ocean must decrease the surface air temperature sensitivity to an increase in the concentration of atmospheric CO_2.

These qualitative arguments are confirmed by estimates by Spelman and Manabe (1984), according to which the values of $(\delta \bar{T}_a)_{4X}$ when the heat transport in the ocean is taken into account turned out to be only 0.58 of that when this was not taken into account (see Table 6.1). A similar result was obtained within the framework of the 0.5-dimensional thermodynamic model of the ocean–atmosphere system described in Section 5.5: the value of $(\delta \bar{T}_a)_{2X}$ equal to 2 °C when approximating the ocean in the form of a 'swamp' of zero heat capacity would decrease by a factor of more than two when prescribing finite heat capacity (the actual depth) and taking into account the meridional heat transport in the ocean. We note that, according to the data of Spelman and Manabe (1984), the influence of the heat transport in the ocean on the change in the zonal average surface air temperature for a quadrupling of the concentration of atmospheric CO_2 manifests itself mostly in high latitudes (Figure 6.5). It is precisely because of this that the changes in the surface air temperature at high and low latitudes differ from each other by much less when the ocean heat transport is considered than

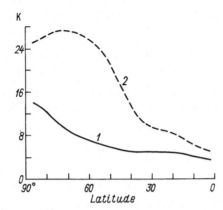

Figure 6.5 Meridional distribution of the zonal average surface air temperature for a four-fold increase in the concentration of atmospheric CO_2 with (1) and without (2) allowance for the heat transport in the ocean, according to Spelman and Manabe (1984).

when it is not. In other words, due to the damping influence of the ocean, smoothing of the latitudinal distribution of changes in the air surface temperature occurs.

Generally speaking, when estimating climatic system sensitivity to an increase in the concentration of atmospheric CO_2, one should keep in mind its dependence on the accuracy of reproduction of the present-day climate. From Figure 6.6 it can be seen that the values of $(\delta \bar{T}_a)_{nX}$ decrease with increased $\bar{T}_a$ as a result of contraction of sea ice and snow cover areas and of the accompanying weakening of feedback between $\bar{T}_a$ and the albedo of the underlying surface. The existence of a strong dependence between $\delta \bar{T}_a$ and $\bar{T}_a$ for the normal concentrations of atmospheric CO_2 suggests that it is precisely this that causes changes in $\delta \bar{T}_a$ in the case where the heat transport in the ocean is taken into account. Indeed, according to Spelman and Manabe (1984), $\bar{T}_a$ for the normal concentrations of atmospheric CO_2, with an allowance for heat transport in the ocean, amounts to 14.4 °C, whereas without allowance for this transport it amounts to only 5.2 °C, that is, the influence of the ocean heat transport on $(\delta \bar{T}_a)_{nX}$ is masked by the change in the annual mean global average surface air temperature for the normal concentrations of atmospheric CO_2.

To estimate the effect of the heat transport in the ocean we turn to Figure 6.6, where the dependence of $\bar{T}_a$ on the concentration of atmospheric CO_2 is shown with and without regard for the factor we are interested in. According to this diagram, the annual mean global average surface air temperature $\bar{T}_a$ is equal to the observed one if the concentration of atmospheric CO_2 amounts to X when the ocean heat transport is taken into account and ~ 2.7X when

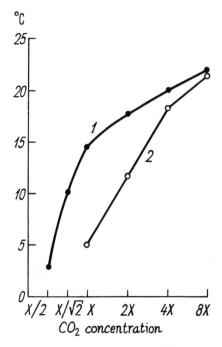

Figure 6.6 Global average surface air temperature $\bar{T}_a$ as a function of the concentration of atmospheric CO$_2$, according to Manabe and Bryan (1985). Curves (1) and (2) were obtained with and without allowance for heat transport in the ocean.

it is not taken into account. A change in $\delta\bar{T}_a$ with an increased concentration of atmospheric CO$_2$ from X to 2X in the first case, and from 2.7X to 5.4X in the second case, is equal to 2.8 and 5.4 °C respectively. The difference (-2.6 °C) between these values serves as a measure of the immediate influence of the ocean heat transport on the sensitivity of $\bar{T}_a$ to doubling of the concentration of atmospheric CO$_2$.

Incidentally, we note one interesting feature of Figure 6.6: the heat transport in the ocean decreases the sensitivity of $\bar{T}_a$ with an increase in the concentration of atmospheric CO$_2$ from X to 8X and increases its sensitivity with a decrease in the concentration of atmospheric CO$_2$ from X to X/2. Manabe and Bryan (1985) explained this rise in sensitivity in the following way: a decrease in the concentration of atmospheric CO$_2$ results in a drop in surface air temperature and sea surface temperature, in a weakening of the heat exchange between the ocean and atmosphere at high latitudes and convective mixing in the area of cold deep water formation, in a decrease in the upwelling velocity and the meridional heat transport in the ocean, and, finally, in an increase in the area of sea ice, that is, an enhancement of feedback between the $\bar{T}_a$ and albedo of the underlying surface.

From the above it is clear that the correct determination of sensitivity is impossible in the absence of an adequate reproduction of the observed annual mean global average surface air temperature. But not all the available estimates satisfy this requirement (see Table 6.1). Moreover, even if it is fulfilled, this does not mean that the model does not distort the average values and time–space distributions of other climatic characteristics. For example, despite small distinctions in the value of the annual mean surface air temperature obtained by Manabe and Stouffer (1980), and by Wetherald and Manabe (1986), and their proximity to the observed value of $\bar{T}_a$ (see Table 6.1), the estimate of $(\delta \bar{T}_a)_{2x}$ in the first of the works mentioned is twice that in the second of the works mentioned (if $(\delta \bar{T}_a)_{2x}$ is determined as $(\delta \bar{T}_a)_{4x}/2$). This can be caused not only by rejection of an assumption on fixing cloudiness but also by an increase in the sea ice area in the Southern Hemisphere.

Thus any model intended to examine sensitivity must reproduce (and quite precisely) all the major features of the present-day climate. Otherwise, hopes of obtaining reliable estimates will be dashed whichever sophisticated model forms its basis.

6.4 Equilibrium response to a change in land surface albedo

As is well known, land makes up 30% of the Earth's surface, and only 50% of the land is covered with vegetation (meadows, forests, pastures and agricultural lands), the remainder comprising deserts (25–30%), continental ice sheets (11%), tundra (6–9%), and lakes, rivers and marshes (2–3%). Different types of land surface have different reflectance power (albedo) and react differently to incoming solar radiation. Albedo, in its turn, depends on the Sun's altitude, the type of vegetation, the nature and age of the snow–ice cover and the wavelength of the solar radiation.

Cloudiness has a pronounced effect on the albedo of the Earth's surface. Clouds not only absorb solar radiation but scatter it, causing an increase in the effective zenith angle of the Sun for small ($<60°$), and a decrease at large ($>60°$), altitudes of the Sun. The influence of solar radiation wavelength has considerable impact on the value of the albedo of vegetative cover. This is connected with the fact that the absorption bands of chlorophyll are concentrated in the wavelength band from 0.4 to 0.7 μm. Within this band plants reflect three times less solar radiation than in the neighbouring IR band covering wavelengths from 0.7 to 4.0 μm. For snow and soil the ratio between albedo values in indicated bands amounts to on average 0.5 and 2.0 respectively.

The different nature of attenuation in two spectral intervals for almost equal incoming solar radiation at the upper atmospheric boundary determines the time–space variability of the net (integral over the spectrum) albedo of the underlying surface. Specifically, a decrease in the net albedo in winter compared with summer occurs as the result of weakening absorption and enhancement of scattering of solar radiation in the first interval, whereas a decrease in the net albedo in the tropics as compared with the midlatitudes in summer when the diurnal insolation is essentially independent of latitude occurs as the result of enhancement of absorption of solar radiation by water vapour in the IR band.

We also mention the *trap effect* (a decrease in albedo due to multiple reflection), which is created by undulations of the underlying surface and by elements of the vegetative cover. So, in changing vegetative height from 0.2 to 10 m, its albedo decreases from 0.25 to 0.1; for Scots fir woods this decrease is twice that for pine forests. It should be borne in mind that the albedo of the vegetative cover depends on the temperature of the environment controlling the intensity of evaporation from plants and, therefore, their colour. Perhaps it is these properties that explain the fact that the albedo of the vegetative cover at high latitudes is less than that at low latitudes.

Troubles in determination of ice and snow albedo are no less than in the case of vegetative cover. Indeed, according to Dickinson (1983) ice without air bubbles has the same albedo as sea water and, because of this, the ice albedo is controlled first of all by the size and distribution of air bubbles. Similarly, the snow albedo is determined by the size and distribution of ice crystals in air. The albedo of snow also depends on its age (the effect of 'snow ageing'), non-homogeneity of the underlying surface, and on cloudiness when related to high latitudes.

The albedo of soil and sand surfaces varies within the limits from 0.1 for dark organic soils to 0.5 and more for white sands, and also depends on the size and colour of their constituent particles (when the latter increases, the albedo decreases). But the moisture of soils, their aggregate state, the degree of roughness, and, especially, the relative content of absorbing organics and mineral components completely overrule this effect. The dependence of albedo on the determining parameters is perhaps best known for the ocean surface, for which the net albedo for different values of the Sun's altitude and degree of roughness (waves) changes from 0.05 to 0.15. In addition, the influence of waves manifests itself more with decreasing altitude of the Sun.

At present, two methods for the determination of the surface albedo have gained wide acceptance: satellite and inventory methods. The first is based on the use of multichannel scanning radiometers recording the radiation field in different spectral ranges; the second is based on the thorough analysis

of data from ground-based measurements of albedo with consideration given
to the area of every type of underlying surface. The introduction of both
methods has led to the construction of global maps of the surface albedo
which have turned out to differ because of errors in the separation of the
useful signal from atmospheric noise (cloudiness, water vapour and aerosols)
in the first case, and subjectivism of estimates of the Sun's altitude, roughness
of the underlying surface, soil humidity and density of vegetative cover in the
second case. This results in a considerable scatter of estimates even for the
zonal average values of the surface albedo obtained by different authors:
maximum differences amounting to 0.3 occur in high latitudes and result from
non-identical prescription of the sea ice area, melted water, leads and
hummocks. But even in temperate and low latitudes the differences between
available estimates of zonal average surface albedo can reach 0.06. It is clear
that local peculiarities of spatial fields can differ even more.

We now direct our attention to a discussion of the results of numerical
experiments on the equilibrium response of the climatic system to a change
in the land surface albedo. In so doing we restrict ourselves to presentation
of only the main facts. Discussion of this issue within the framework of
three-dimensional models was initiated by Charney et al. (1977), who carried
out three experiments. In the first experiment the albedo of the snowless land
surface was assumed to be equal to 0.14; in the region of deserts in the
Northern Hemisphere it was assumed to be 0.35. In the second experiment,
values of the surface albedo inherent in deserts were extended to the region
of the Sahel (the southern periphery of the Sahara desert), Rajputan (India)
and the western part of Grand Plains (USA) which imitate the process of
desertification. Finally, in the third experiment a similar change in albedo
was realized in three other regions: Central Africa, the region of Bangladesh
and in the Mississippi valley which imitated the process of destruction of
vegetation. In all cases sea surface temperature and soil moisture were
assumed to be constant.

The most remarkable result of this series of numerical experiments was the
fact that, regardless of expectation, an increase in the surface albedo on
average for the six regions examined led not to a decrease but to an increase
in solar radiation absorbed by the underlying surface. This last circumstance
is due to a decrease in cloudiness, which is caused by a decrease in the
local evaporation and horizontal transport of water vapour from neighbouring
regions. In turn, a decrease in the local evaporation is determined by
attenuation of downward long-wave radiation and, accordingly, by a decrease
in the value of radiation balance and temperature of the underlying surface;
and a decrease in the horizontal transport of water vapour is caused by the

formation of a circulation cell of the monsoon type, superimposed on the large-scale atmospheric circulation.

By the formation of a circulation cell of the monsoon type we mean the following. A decrease in the temperature of the underlying surface and the surface atmospheric layer results in a rise in atmospheric pressure and amplification of the downward vertical motions over the region of increased albedo if the background temperature in this region is less than that in the neighbouring region or, otherwise, in the attenuation of upward vertical motions over the region of increased albedo. Opposite changes in the intensity of vertical motions occur in the neighbouring region. This favours intensification of the process of cloud formation and condensation. Accordingly, the horizontal transport of water vapour into the area of increased albedo decreases.

Thus, both sources of water vapour for the atmosphere over the region of increased albedo – local evaporation and horizontal transport – decrease, that is, an increase in albedo favours the enhancement of aridity of the climate. This is the essence of the *effect of self-amplification of deserts*, discovered by Charney (1976). An increase in the recurrence of droughts in the Sahel region as a result of pasture spoiling serves as a manifestation of the above.

Further, since the horizontal temperature contrast determining the intensity and direction of circulation in a cell of the monsoon type depends on the location and extent of the area of increased albedo, then, naturally, the ratio between the change in local evaporation from the underlying surface and the change in the horizontal transport of water vapour in the atmosphere does not remain equal everywhere. So, according to Charney *et al.* (1977), in regions of the Sahel, Rajputan, and Central Africa a decrease in the horizontal transport of water vapour dominates over a decrease in local evaporation; in the western part of the Grand Plains the opposite is true; and in the Mississippi valley and in Bangladesh a decrease in local evaporation is even accompanied by an increase in the horizontal transport of water vapour.

Similar numerical experiments within the framework of a three-dimensional general circulation model were carried out by Carson (1982). In his work, as in the work of Charney *et al.* (1977), the sea surface temperature and soil moisture were fixed; in return, the albedo was changed from 0.1 to 0.3, not in restricted regions but, rather, over the whole surface of continents free from snow and ice. As a result it was established that an increase in albedo occurring everywhere leads to a decrease in evaporation, horizontal transport of water vapour and precipitation over continents, and to some increase in precipitation (due to the fixing of sea surface temperature and, to some extent, evaporation) over oceans.

It is clear that the assumption of constancy of soil moisture when the surface albedo is changed is not fulfilled in actuality. Thus the next step had to be its rejection. This step was taken by Potter *et al.* (1981) and Chervin (1979). In the first-mentioned work a zonal model was used as the basis; in the second, a three-dimensional general circulation model with a fixed sea surface temperature formed the basis. In both works the changes in the land surface albedo were prescribed as local: in the first work, the albedo increased from 0.16 to 0.35 for a region of area 9×10^6 km² located in the vicinity of the 20° N parallel, and from 0.07 to 0.16 in two other regions (each of area 7×10^6 km²) in the vicinity of the equator and the 10 °S parallel; in the second work, the actual distribution of albedo in North Africa (from the Mediterranean Sea to the 7.5 °N parallel) and in the western part of the Grand Plains in the USA was replaced by a constant value equal to 0.45.

The data analysis presented by Potter *et al.* (1981) points to the fact that an increase in albedo in the vicinity of the 20 °N parallel determines a local decrease in absorbed solar radiation, temperature of the underlying surface, evaporation and precipitation. But the matter does not end there. It also causes an increase in the meridional temperature gradient, enhancement of evaporation in the northern Hadley cell and its displacement to the south. The latter leads to a fall in temperature and moisture content in the atmosphere, and, hence, to a decrease in the meridional sensible and latent heat transport to the pole, which in turn involves an increase in sea ice area and a further fall in temperature and moisture content in the atmosphere of the Northern Hemisphere. In the Southern Hemisphere changes in climatic characteristics become apparent by a shift of the southern Hadley cell and its associated decrease in cloudiness, increase in the absorbed solar radiation and a rise in temperature of the underlying surface and the atmosphere in the vicinity of the 30 °S parallel. As a result the temperature and moisture content of the Southern Hemisphere even increase a little. The global fall in surface air temperature amounts to 0.2 °C; the global decrease in moisture content of the atmosphere is equal to 0.04 g/kg.

Numerical experiments carried out by Chervin (1979) confirm Charney *et al.*'s (1977) conclusion that there is a change in the velocity of vertical motions and a decrease in temperature of the underlying surface and precipitation in the region of increased surface albedo. According to Chervin (1979), the velocity of upward vertical motions in North Africa decreases up to 2 mm/s; the temperature of the underlying surface decreases up to 2 °C; the precipitation decreases up to 5 mm/day, and the soil moisture content decreases up to 5 cm (for a soil moisture capacity equal to 15 cm). Simultaneously, in the southern part of the area examined (in the zone between the 12.5° and 7.5 °N parallels) the temperature of the underlying surface does not fall but rises by about 0.5 °C. In other words, here, instead of the negative correlation between albedo and temperature of the underlying surface, a positive correlation takes place. The cause of such a change in sign is the suppression of the effect of evaporation due to a sharp decrease in soil moisture content.

But the changes in climatic characteristics discussed above are inherent only in local regions of increasing albedo. In neighbouring regions they have an opposite sign. As a result it turns out that, for the Earth as a whole, changes in the temperature of the underlying surface and of the surface atmospheric layer amount to only −0.25 °C. This estimate of temperature drop is close to the one obtained by Potter *et al.* (1981). But it should be remembered that the above-mentioned estimate corresponds to the local increase in land albedo, and the estimate of Potter *et al.* (1981) complies with a global increase in land albedo.

It is no mere chance that we focus our attention on the approximate character of the available estimates. By this we emphasize that these do not necessarily have to coincide with each other if for no other reason than the differences between the models and the steady states corresponding to them. But then would it be better, instead of comprehensive three-dimensional models, to use simple thermodynamic models which describe the meridional heat transport in the ocean explicitly? The fact that such a question is not unfounded is attested to by the results of a numerical experiment that were obtained within the framework of a 0.5-dimensional seasonal model of the ocean–atmosphere system (see Section 5.5), in increasing the albedo of the snowless land surface from 0.19 to 0.25.

Judging from the results of this numerical experiment an increase in albedo leads to a decrease in absorbed solar radiation at the underlying surface, as well as a decrease in temperature of the land surface, evaporation, precipitation and run-off. Accordingly the moisture content of the atmosphere and heat release due to phase transitions of water vapour are reduced. A decrease in the temperature of the land surface and heat sources in the atmosphere is accompanied by a drop in air temperature and, hence, by a weakening of downward long-wave radiation. It would have caused an increase in net long-wave radiative flux at the underlying surface but this flux decreases due to the increase in upward long-wave radiation flux.

A decrease in the net flux of long-wave radiation and in the absorbed solar radiation at the underlying surface do not counterbalance each other completely: the latter dominates. As a result the value of net radiation flux at the underlying surface decreases. Latent heat flux at the underlying surface changes in a similar way that determines a decrease in heat transfer from the ocean into the atmosphere in the area of cold deep water formation, and an increase in heat transfer from the atmosphere into the ocean in temperate regions and at low latitudes. The heat transport from these regions into the area of cold deep water formation also decreases. This is accompanied not by a rise but, rather, by a fall in temperature of the UML at temperate and low latitudes of the ocean. This last circumstance is the consequence of displacement towards the south of the boundary between the northern and the southern boxes, causing a reduction of ocean area in the southern box (see Section 5.5).

It should also be noted that a decrease in the meridional heat transport from temperate and low latitudes of the ocean results in a fall in temperature in the area of cold deep water formation, and this, in turn, results in weakening of the heat transport into the polar ocean, increase of sea ice area, and further enhancement of the effects of positive feedback between the albedo of the underlying surface and surface air temperature.

6.5 Equilibrium response to a change in soil moisture content

Nowadays, it is assumed that the major storage of soil moisture is contained in the upper two-metre layer, the moisture content of which amounts to 0.2 m. If one excludes a part of the land occupied by glaciers and permanent snow cover ($\sim 10\%$), permanent frost ($\sim 14\%$), and by arid and semiarid regions ($\sim 21\%$), where soil moisture is present during relatively short periods, then the area of the moisture-containing soil cover will be about 55% of the land area or 82×10^{12} m^2. The total storage of soil moisture is thus 16.4×10^{12} m^3.

Soil moisture exists in two forms: liquid (water) and gaseous (water vapour). Water transport is determined by the motion of water in capillaries and non-capillary pores, and water vapour transport is due to molecular diffusion. Because of this, maximum soil moisture content (moisture capacity) depends, generally speaking, on the distribution of capillary and non-capillary porosity (fraction of capillaries and pores in a unit volume of soil), the level of underground water, the value of underground run-off, water surface tension in capillaries, and on the coefficient of molecular diffusion. In the presence of vegetation it will also be determined by the depth of penetration of the plant root system, its vertical distribution, and by the capacity to absorb moisture from soil. Information about these parameters is extremely fragmentary, which fact would not have given cause for concern were the soil moisture content not to affect the state of the climatic system. But actually the case is somewhat different and the results of numerical experiments for excessively moist and dry soil are the most convincing evidence of the impact of soil moisture content on the climatic system state. Note that under conditions of excessive moistening, the soil water content is equal to its critical value for which evaporation is equal to evaporability; in dry soil it is equal to zero.

In this discussion of the results of numerical experiments we begin with those carried out by Kagan *et al.* (1990) on the basis of the 0.5-dimensional seasonal model of the ocean–atmosphere system. Comparison of results of the first and control (complying with the present-day climate) experiments shows that excessive soil moistening results in an increase in evaporation and precipitation, and in a decrease in run-off. An increase in latent heat flux at the land surface leads to a rise in temperature and moisture content in the atmosphere, and to an increase in absorption of short-wave solar radiation in the atmosphere and its decrease at the underlying surface. The net radiative flux at the land and ocean surface increases at temperate and low latitudes, and decreases in the area of cold deep water formation. At the same time, heat transfer from the ocean into the atmosphere in this area increases. Nevertheless, the sea water temperature in the area of cold deep water formation increases as a result of a diminution of this area because of a displacement toward the north of the boundary between the northern and southern boxes. A rise in temperature in the area of cold deep water formation promotes warming of the deep ocean and a reduction of the sea ice area and snow mass at the sea ice surface. The temperature of the snow–ice cover surface, as well as that of the sea surface, increases, while the surface air temperature remains constant, which is explained by the compensating effect of a fall in the land surface temperature.

A decrease in soil water content down to zero causes opposite (as compared to the first experiment) changes of climatic characteristics in the southern box, as could be expected. But in the northern box the following events occur. Here, because of the abrupt rise in the land surface temperature, the surface air temperature also increases. This results in a decrease in the meridional temperature contrast, in a displacement of the boundary between boxes toward the north, and in a decrease in land fraction in the northern box and, hence, in a water–air temperature difference in the area of cold deep water formation in autumn and winter. As a result, the heat transfer into the atmosphere decreases in this area, its temperature rises, the heat transport into the polar ocean increases, and sea ice disappears. An increase in temperature in the area of cold deep water formation is accompanied by warming of the deep layer at temperate and low latitudes, by a decrease in the temperature difference between these areas, and by a weakening of upwelling. The latter causes an increase in thickness of the UML and, hence, a decrease in the sea surface temperature at temperate and low latitudes. Finally, we note a strong (of almost two orders of magnitude) decrease in the meridional latent heat transport in the atmosphere determined by an appropriate decrease in contrast of air humidity between the northern and southern boxes.

Similar limiting situations were also reproduced within the framework of three-dimensional general circulation models. As this takes place, the sea surface temperature and sea ice area were assumed to be fixed and to comply with the present-day climate. The last suggestion, in the light of the data presented above, looks somewhat artificial. Nevertheless to estimate tendencies to climate changes appearing, say, when operating large irrigation or drainage systems the simulation of such a situation makes sense even in such a formulation. From this viewpoint, for the case of dry soil the most significant consequences are, on the one hand, a rise in the land surface temperature that is connected with the exclusion of the cooling effect of evaporation, a decrease in the surface atmospheric pressure, amplification of upward vertical motions, and, despite all of this, a decrease in cloudiness and precipitation (due to a reduction in air humidity) over continents, and, on the other hand, attenuation of downward vertical motions and, hence, a decrease in cloudiness and precipitation over the ocean. In the case of moist soil the changes in climatic characteristics over continents and the ocean are opposite to those given above. It is interesting to note that maximum changes in precipitation–evaporation difference occur in the tropics in this case, too, due to the displacement of the northern and southern Hadley cells with a consequent change in the vertical and horizontal transport of water vapour in the atmosphere.

The strong dependence of the climatic system state on changes in soil moisture content produces rather stringent requirements as to the accuracy of the chosen parametrization of the moisture transport in soil. At present, three methods, proposed by Budyko (1956), Hansen *et al.* (1984), and Deardorff (1977), have gained widest acceptance. The first of these is based on the presentation of the active soil layer in the form of a single layer with finite thickness and on fixing a critical value of moisture content and moisture capacity (a modified version of this method, called the 'bucket' method, is presented in Section 5.5); the second is based on singling out two layers of finite thickness which interact by means of diffusive moisture transfer; and the third is based on the separation of two weakly interacting layers – the upper 0.5-cm layer which responds to precipitation and evaporation, and the lower 50-cm layer which serves as a source of moisture for the upper layer. In the latter case it is assumed that diffusive transfer is always directed from the low layer to the upper one, and moisture influx into the low layer occurs due to instantaneous infiltration of water from the land surface.

The test of these three parametrizations carried out within the framework of a one-dimensional radiative-convective model showed (see Hunt, 1985) that the first two methods entail undesirably long periods to establish maximum values of temperature, and a short period to establish minimum values of soil moisture content. In this sense the third method (the so-called force–restore method) is preferable, in any event, in the absence of vegetation. In the presence of vegetation the deep soil layer provides moisture for evapotranspiration. But this method also has its own disadvantages, the main one of which is an excessive sensitivity to the choice of characteristic time of diffusive moisture transfer in soil and the thickness of layers.

6.6 Equilibrium response to a change in vegetative cover

As is known, in the presence of vegetation, not all precipitated moisture reaches the underlying surface. Part of it is detained by the vegetation (the *effect of interception*). The evaporation of the moisture detained by vegetation increases the water vapour content in the atmosphere and (in the case of condensation of water vapour) the intensity and duration of precipitation. The moisture is extracted from the soil by the roots of plants and, by means of transpiration via leaf stomata, returns to the atmosphere. The rate of transpiration depends on density, height and the type of vegetative cover. For example, for grass cover transpiration is approximately equal to evaporation, while for a tall wood transpiration can be more than evaporation by an order of magnitude. Moreover, it can happen (as in a dense, deciduous forest) that the latent heat flux at the surface of the vegetative cover and soil is directed into the atmosphere, and the sensible heat flux if soil is in shadow is directed to the underlying surface. It is clear that such a situation occurs only in limited areas of continuous forest tracts. When increasing spatial scales, the effects of the horizontal sensible and latent heat transport and of precipitation dominate.

The mechanism of redistribution of heat and moisture between soil and vegetative cover mainly affects the formation of diurnal oscillations of meteorological parameters. But this does not exclude the fact that vegetation can affect processes with long time scales. We illustrate this by the following example. Wood usually decreases the maximum run-off during snow melting and torrential rains, and at the same time prevents rivers becoming shallow in rainless periods. According to Baumgartner (1979) the ratio between run-off and precipitation in forest tracts amounts to 0.18; in the open country it amounts to 0.42.

It is established that vegetation in droughty regions tends to minimize water loss by soil, and in regions with excessive moisture it tends to maximize biomass at the expense of increase in vegetation density. Different kinds of vegetation show this tendency in different ways: in arid regions plants turn their leaves in the direction of the sun's rays, and in regions with excessive moisture they turn their leaves perpendicular to the sun's rays, thereby decreasing transpiration and evaporation in the first case and increasing them in the second. There are cases of more complicated organization where the upper plant leaves shield the lower ones without regard to the degree of their moisture or lack of moisture in the overlying air layer. The estimations of tranpiration in arid regions and in regions with excessive moisture point to the fact that the tendency mentioned above is observed over a wide range of spatial scales from 10 to 100 km. This gives hope for a proper description of the effects of large plant associations in global climatic models.

Until recently the processes of heat and moisture exchange between the vegetative cover and the atmosphere have been described by the introduction of different kinds of dependencies of exchange intensity on the humidity and roughness of the underlying surface and/or on horizontal coordinates. The biospheric models proposed by Dickinson *et al.* (1986), and Sellers *et al.* (1986) take into account the influence of the physiological characteristics of plants on heat and moisture exchange. In these models, the temperature and mass of drop-shaped moisture at the surface of the vegetative cover, soil moisture content at certain levels and the mass of water at the soil surface are used as dependent variables. The transpiration and water transport via plant roots are parametrized in terms of the resistance law according to which the flux of the substance examined is assumed to be proportional to the difference in its values at boundaries, with the proportionality factor characterizing an exchange rate. The latter is considered to be a function of the coefficient of diffusive resistance for the underlying surface and vegetative cover and of the rate of moisture exchange inside individual plants and at their surfaces. For transpiration, the resistance coefficient is represented as the sum of the leaf diffusive resistance and stoma resistance depending on the value of solar radiation, season and amount of water extracted by the plant from the soil. The root resistance is determined by the degree of soil moistening, by the size of the root system, and by the highest possible (for a given type of plant) transpiration. Dickinson *et al.* (1986) singled out 18 types of underlying surface differing by the size of the root system as well as by the physiological, aerodynamic and radiative properties of vegetation. In turn soil is divided into twelve classes according to its mechanical structure, and into eight classes according to its colour. The main restrictions of the proposed models are related to determination of the average properties of plant associations and the effect of their interaction. Overcoming these restrictions is a matter for the future.

Since the time of its appearance, humanity has been continuously perfecting ways to destroy the vegetative cover. From the beginning this was by gathering roots and the fruits of wild plants, then when man had learned to make fire he deliberately provoked fires, and nowadays we see both the replacement of the natural vegetation by cultivated vegetation and the deterioration in the composition and structure of flora (the *synanthropization of the vegetable kingdom*). The first considerable changes in vegetative cover started as far back as the transition from hunting and fishing to cattle-breeding and agriculture. Since then these changes have continued with increasing rate. The scales of the changes can be judged from the following figures. According to Sagan *et al.* (1979), since the first appearance of human beings the area of the deserts has increased by $9 \times 10^6 \, \text{km}^2$; on the other hand the area of the forests has decreased by $8 \times 10^6 \, \text{km}^2$ at temperate latitudes and by $7 \times 10^6 \, \text{km}^2$ at tropical latitudes; urbanization has resulted in the total destruction of the vegetation of an area of about $1 \times 10^6 \, \text{km}^2$; the area of lands with excessive salt content has increased by $0.6 \times 10^6 \, \text{km}^2$. Thus, anthropogenic changes have affected a total of $26 \times 10^6 \, \text{km}^2$, or 17%, of the land area.

The most intense destruction of forest occurs in the tropics. According to data from Woodwell *et al.* (1983) the rate of decrease in the area of the tropical forests is from 0.16×10^6 to $0.19 \times 10^6 \, \text{km}^2$ per year. The new estimates based on the FAO 1988 and 1990, Tropical Forest Assessments show that tropical deforestation increased from $0.132 \times 10^6 \, \text{km}^2$ in 1980 to $0.193 \times 10^6 \, \text{km}^2$ in 1990. If such deforestation continues after 1990, with the rate depending on growth of population in the countries of the tropical belt, then according to IPCC (1992), from 73% ($14.47 \times 10^6 \, \text{km}^2$) to 91% ($16.86 \times 10^6 \, \text{km}^2$) of tropical forests will be cleared by the end of the next century. All this points to the fact that, at present, anthropogenic changes in the vegetative cover have acquired a global character. What climatic consequences can this lead to? To answer this question we examine two extreme situations: the destruction of forest tracts and the total destruction of the vegetative cover (see Kagan *et al.*, 1990).

Destruction of forest tracts. This is accompanied by changes in the albedo of the underlying surface, soil moisture and the concentration of CO_2 in the atmosphere. In their turn, these changes depend on the form of secondary vegetation and the measures undertaken for soil humus conservation. We consider the following scenarios of the economic activities of mankind: (1) wood is burned, vegetation on the cleared territories is not recovered, soil humus is not preserved; (2) wood is burned, vegetation on the cleared territories is recovered, soil humus is not preserved; (3) wood is burned, cleared lands are used for agriculture, appropriate measures are undertaken for soil humus conservation.

We define the change $\delta\alpha_s$ in the albedo of the underlying surface owing to the destruction of forests as $(\alpha_{nf} - \alpha_f)s_f/s$ where α_{nf} is the albedo of cleared territory; α_f and s_f are the albedo and area of forests; s is the land area. According to Whittaker and Likens (1975) the total area of forests in 1950 amounted to 48.5×10^6 km^2, of which 24.5×10^6 km^2 formed a share of tropical forests (tropical jungles and seasonal tropical forests); 5.0×10^6 km^2 of evergreen subtropical forests; 7.0×10^6 km^2 of seasonal forests at temperate latitudes, and 12×10^6 km^2 of boreal forests. When assuming that the albedo of tropical jungles and boreal forests is equal to 0.10, and that of seasonal tropical forests, evergreen subtropical forests and seasonal forests at temperate latitudes is equal to 0.15, then the area-weighted average value of α_f is equal to 0.12. Assuming now that the albedo of cleared lands, meadow vegetation and cultivated areas is equal to 0.14, 0.20 and 0.25, respectively, we obtain $\delta\alpha_s \approx 0.01$, 0.03 and 0.04. We note for comparison that the maximum albedo change over the last 25–30 years caused by the effects of desertification, salinization, temperate and tropical deforestation and urbanization is between 3.3×10^{-4} and 6.4×10^{-4} (Henderson-Sellers and Gornitz, 1984).

Further, according to data from Whittaker and Likens (1975), the biomass density (in units of carbon) in tropical jungles amounts to 20.25 kgC/m^2, in seasonal tropical and evergreen subtropical forests it amounts to 13.5 kgC/m^2, and in boreal forests it amounts to 9.0 kgC/m^2. Therefore, having available information on the area of forests (see above), we can speculate that, when burning all the forests, about 0.744 TtC (1 Tt $= 10^{12}$ t) is transferred into the atmosphere. Destruction of forests is also accompanied by enhancement of soil breathing. Assuming that the soil biomass density in tropical jungles is equal to 13.0 kgC/m^2, that in seasonal tropical and evergreen subtropical forests is equal to 7.0 kgC/m^2, that in seasonal forests of temperate latitudes is equal to 10.7 kgC/m^2 and that in boreal forests is equal to 20.6 kgC/m^2, then the total carbon content in the humus of forest soils amounts to 0.631 TtC. If soil breathing results in a 15% reduction in the carbon content of humus we obtain an additional gain of carbon in the atmosphere that will be equal to 0.095 TtC. By this measure the total input of carbon into the atmosphere amounts to 0.839 TtC.

An excess amount of carbon in the atmosphere has to be partly absorbed by the ocean and partly assimilated by the terrestrial biota. We will suppose that the production of primary vegetation on lands not occupied by forest is limited, not by the CO_2 content in the atmosphere, but by other factors (for example, by the flux of short-wave solar radiation and soil moisture content). In this case the sink of excessive amounts of atmospheric carbon has to be determined by its absorption by the ocean and secondary vegetation on cleared lands. We assume that about half of the CO_2 excess is absorbed by the ocean and that the sink of atmospheric carbon, owing to its assimilation by secondary vegetation, can be evaluated by the biomass density of grassland and cultivated lands which, according to Whittaker and Likens (1975), is equal to 0.72 and 0.45 kgC/m^2. As a result we arrive at the conclusion that the transformation of cleared lands into grasslands and cultivated lands entails the removal from the atmosphere of 0.035 and 0.022 TtC, respectively, and that the total increase in the concentration of atmospheric CO_2 for the three scenarios listed above is equal to 196, 188 and 169 ppm respectively.

The results of calculations for different scenarios of human economic activity obtained within the framework of the 0.5-dimensional seasonal thermodynamic model (see Section 5.5) point to the fact that when burning forest and failing to renew vegetation (the first scenario) the increase in the concentration of atmospheric CO_2 dominates, and because of this the changes in the climatic characteristics coincide qualitatively with those caused by doubling of the atmospheric CO_2 concentration whereas they differ quantitatively by several tens of percent. The effect of an increase in albedo manifests itself only in a small decrease in the land surface temperature in the southern box. The transformation of forest tracts into grasslands and cultivated lands (the second and third scenarios) leads to opposite changes in climatic characteristics, due to the domination of the effect of the land surface albedo increase over the effect of increased concentration of atmospheric CO_2.

Total destruction of the vegetative cover. The series of numerical experiments described above leads to a somewhat unexpected result: destruction of forest tracts entails an increase in soil moisture content and a decrease in run-off. In reality, the destruction of vegetative cover (say, when transforming forests into savannas or savannas into deserts) is accompanied by a decrease in soil moisture content. Obviously, the result obtained is a consequence of the constancy of the hydrological cycle parameters (soil moisture capacity W_0 and the fraction γ of precipitation spent on the formation of run-off), which, generally speaking, depend on the type and state of the vegetative cover. Since such dependencies for the Earth as a whole are unknown there is only one thing to do: to examine an extreme situation where the usual landscape is replaced by a surface devoid of any vegetation at all and reminiscent, by its properties, of dry grey soil.

Let the albedo of this surface be equal to 0.25 and the moisture be zero. We also take into account that in the absence of terrestrial vegetation the carbon in it and in the soil humus has to be redistributed between the atmosphere and the ocean. According to Whittaker and Likens (1975), the carbon content in terrestrial vegetation is 0.828 TtC, in soil humus it is 2.9 TtC, and from the 2.9 TtC contained in soil humus only 15% of 2.0 TtC (the carbon amount in the labile part of soil humus which takes part in exchange with the atmosphere) is added to the atmosphere. Because of this, the net incoming carbon in the ocean–atmosphere system is equal to 1.128 TtC. Assuming, as before, that this is redistributed between the ocean and the atmosphere equally, we have that the concentration of atmospheric CO_2 has to increase by 264 ppm compared with to its present-day value. Thus, the limiting values of the albedo of the land surface, the soil moisture content and the concentration of atmospheric CO_2 are known.

It is clear that the climatic consequences of the total destruction of the vegetative cover can be established only in the case where all three listed factors are considered, not separately, but in combination. Their combined effect demonstrates results close to those obtained in an experiment with zero soil moisture content. This indicates that the first two factors, in the sense of their impact on the separate characteristics of the climatic system, compete between themselves and, for the majority of them, practically compensate each other. The only exception is the annual mean temperature of the land surface: the predominant effect of increased albedo leads to a decrease in the land surface temperature (compared with its value corresponding to the zero soil moisture content), followed by a decrease in the surface air temperature.

Overall, the total destruction of the Earth's vegetation causes an increase in planetary albedo, a decrease in radiative cooling of the atmosphere and in the net radiative flux at the underlying surface, attenuation of the meridional sensible and latent heat transport in the atmosphere and ocean, a fall in the mass-weighted average air temperature and a rise in the surface air temperature (as a whole for the hemisphere), a decrease in evaporation and precipitation at temperate and low latitudes and their increase at high latitudes, an increase in run-off, a displacement of the boundary between the boxes to the north, a rise in temperature in the area of cold deep water formation, the disappearance of sea ice, a rise in temperature in the deep ocean at temperate and low latitudes, weakening of upwelling, an increase in the thickness of the UML, and, finally, a decrease in temperature in the upwelling area. As this takes place, the amplitudes of the seasonal oscillations of the mass-weighted average temperature and humidity of the atmosphere, as well as precipitation and evaporation, decrease, while the amplitude of the seasonal oscillations of the surface air temperature increases at temperate and low latitudes and decreases sharply at high latitudes.

In summary, we present some estimates of the equilibrium response of the climatic system to anthropogenic changes in the vegetative cover. According to Sagan *et al.* (1979) the changes in the land surface albedo related to the already-mentioned increase in desert area, the destruction of forests at tropical and temperate latitudes, the salinization of soil and urbanization result in a global fall in the surface air temperature by 1 °C. According to estimates by Potter *et al.* (1981), an increase in the land surface albedo at the expense of continuing (during the last millennia) desertification and destruction of tropical forests leads to a global reduction in the surface air temperature and precipitation by 0.2 °C and 1.6% respectively (in the Northern Hemisphere the surface air temperature decreases by 0.6 °C). Finally, most estimates characterizing an increase in global average values of the surface air temperature and precipitation due to doubling of the concentration of atmospheric CO_2 are found in intervals from 1.2 to 4.5 °C, and from 0.06 to 0.20 mm/day. Thus, despite different initial prerequisites the available estimates of consequences of anthropogenic changes in vegetative cover are qualitatively similar.

6.7 Transient response to a change in the concentration of atmospheric CO_2

The problem contained in the title of this section is immediately related to the problem of prediction of potential climatic changes, the importance of which for mankind's economic activities it is impossible to overestimate. It is not without reason that this problem has drawn the attention not only of separate groups of scientists and the scientific community in general, but also of government organizations who have to plan and take decisions about the economic, social and technical development of human societies.

According to estimates by Keeling (1973), Marland (1984) and Rotty (1987), who analysed data on the consumption of coal, oil and natural gas in all regions of the world, the total incoming CO_2 has, for 125 years since the dawn of the industrial revolution, amounted to 187×10^9 tC. Of this 98% is related to the burning of fossil fuel, and 2% to the production of cement. Since the industrial revolution the annual mean emission of CO_2 has increased from 0.1×10^9 tC in 1864 to 5.3×10^9 tC in 1984. Judging by the most recent estimates of Marland (1989), and Marland and Boden (see IPCC, 1992), the cumulative release of carbon from fossil fuel use and cement production between 1850 and 1987 amounts to 200 ± 20 GtC, and the global annual mean CO_2 industrial emission equals 5.7 GtC/year in 1987 and 6.0 ± 0.5 GtC/year in both 1989 and 1990. An additional release of carbon to the atmosphere is due to impacts on the terrestrial biota (deforestation and changes in land use). It is important to understand that the role of each of the above-mentioned anthropogenic factors, and primarily of the terrestrial biota, especially that enhancing the emission of CO_2 when burning fossil fuel and increasing the rate of photosynthesis and biomass are competing.

There are several methods for estimating changes in biomass of the terrestrial biota. The most representative is an inventory one based on the use of historical data on the nature of land tenure. Having systematized such data Houghton *et al.* (1983) singled out six types of anthropogenic changes in terrestrial ecosystems: (1) clearing of land for agriculture; (2) the use of natural ecosystems as pastures; (3) the growth of secondary forests after clearing; (4) halting of cultivation of agricultural lands; (5) halting of exploitation of pastures; and (6) reforestation (planting of trees in previously cleared areas). The state of each ecosystem was characterized by a change in its area, age and carbon content in the biomass and organic matter of soil. As has been shown by Houghton and Skole (see IPCC, 1992), the annual emission of CO_2 due to the transformation of terrestrial ecosystems exceeds the intensity of industrial effluents up to 1960, and the total emission of CO_2 from 1860 to 1985 amounts to 115×10^9 tC with a standard deviation of $\pm 35 \times 10^9$ tC determined by uncertainty in assignment of initial values of forest biomass and inorganic matter of soil, and by inaccurate information on the dynamics of land use. This estimate, together with other inventory estimations of CO_2 exchange between the atmosphere and terrestrial biota is presented in Table 6.2.

All other methods of estimating the anthropogenic impact on the terrestrial biota are indirect. We mention, in particular, the method based on the use of data on changes of ratio between concentrations of stable carbon isotopes ^{13}C and ^{12}C in the annual rings of perennial wood plants and corals. As is well known, the isotope ^{13}C is contained not only in the atmosphere but also in the terrestrial biomass and in fossil fuel, and in the latter its relative content is about 20% less than that in the atmosphere. Because of this, burning of fossil fuel or destruction of biota must result in a reduction in $^{13}C/^{12}C$ by a value $\delta^{13}C$, where $\delta^{13}C = 10^3[(^{13}C/^{12}C) - (^{13}C/^{12}C)_{PDB}]/ (^{13}C/^{12}C)_{PDB}$ is the relative departure of $^{13}C/^{12}C$ from the so-called PDB-standard, defined as the relative content of ^{13}C and ^{12}C, in shells of the fossil mollusc *Belemnitella americana* discovered in Cretaceous sedimentary

Table 6.2. *Carbon input into the atmosphere due to anthropogenic impact on terrestrial ecosystems according to various researchers*

Author	Observed period	Total input (10^9 tC)	Annual mean input (10^9 tC year^{-1})
Revelle and Munk (1977)	1860–1970	70–80	0.6–0.7
Bolin (1978)	1800–1975	40–120	0.2–0.7
Siegenthaler *et al.* (1978)	1860–1974	133–195*	1.2–1.7*
Stuiver (1978)	1850–1950	120*	1.2*
Freyer (1979)	1860–1974	70	0.6
Hampicke (1979)	1860–1980	180	1.5
Chan and Olson (1980)	1860–1970	150	1.4
Moore *et al.* (1981)	1860–1970	148	1.3
Houghton *et al.* (1983)	1860–1980	180	1.5
Peng *et al.* (1983)	1850–1976	240*	1.9*
Peng (1985)	1800–1980	144*	0.8*
Kobak and Kondrashova (1987)	1860–1983	60–85	0.7
Houghton and Skole (see IPCC, 1992)	1860–1985	115 ± 35	0.85 ± 0.30

* Estimates obtained from data of isotopic analysis.

deposits. Estimates of the intensity of the biotic source of CO$_2$ obtained by this method (they are marked by asterisks) are also presented in Table 6.2. As can be seen they do not contradict the inventory estimates: both point to the fact that the carbon input into the atmosphere for the last 120 years, because of the anthropogenic impact on the terrestrial biota, must amount to (60–90) $\times$ 10^9 tons, that is, of the order of half the emission of CO$_2$ at the expense of burning fossil fuel.

If one makes use of Marland's (1989) estimate for fossil fuel combustion and cement manufacturing and Houghton and Scoles's (see IPCC, 1992) estimate for the release of carbon from land-use changes, then the total release of carbon to the atmosphere during the period from 1850 to 1986 must be 312 ± 40 GtC. This release of carbon must increase the concentration of atmospheric CO$_2$ by 147 ppm (1 ppm CO$_2$ of the global atmosphere equals 2.12 GtC). Thus, if the preindustrial concentration of atmospheric CO$_2$ was 288 ppm as reconstructed from ice core analyses (IPCC, 1990), then the global annual mean average concentration of atmospheric CO$_2$ in 1986 should be 435 ppm. However, measurements available from that time show a concentration of atmospheric CO$_2$ of 348 ppm corresponding to 41 ± 6% of the cumulative release. The remaining fraction must be redistributed between ocean and terrestrial biota. In other words, this means that the global net uptake of CO_2 by the ocean and productivity of vegetation have increased since the beginning of the industrial revolution. The former is explained by increasing CO_2 partial pressure difference between ocean and atmosphere, while an increase in the latter is not so evident. It may be associated with the so-called *fertilization effect* (enhanced vegetative growth with increasing CO_2 levels).

Direct measurements show that the inter-hemispheric CO_2 concentration difference (currently about 8 ppm) is smaller than one would expect if nearly all of the fossil releases occurred in the Northern Hemisphere. This suggests that there is an unexpectedly large sink in the Northern Hemisphere equivalent to more than half of CO_2 release of the fossil fuel. Tans *et al.* (1990) partitioned the northern sink of carbon between ocean and terrestrial biota, using data on the CO_2 partial pressure in surface waters, whereas Keeling *et al.* (1989) presupposed that a large fraction of this sink was due to a natural imbalance in the Northern Hemisphere carbon cycle consisting of a net transport of carbon from the Northern to the Southern Hemisphere in the ocean and a return transport of carbon in the atmosphere. The last statement is equivalent to an assumption about the existence of a countergradient flux of CO_2 in the atmosphere: the inter-hemispheric CO_2 concentration difference mentioned above implies a continuous flux of CO_2 from the Northern to the Southern Hemisphere.

The second feature is a relatively small net input of carbon to the atmosphere in the tropics because of both outgassing of CO_2 from warm tropical waters and deforestation. According to Keeling *et al.* (1989), this may be explained by a significant (about 50%) reduction of the net flux of CO_2 from the tropics as a consequence of the fertilization effect. However, this conclusion, as well as the conclusions of Tans *et al.* (1990) and Keeling *et al.* (1989) concerning the origin of the large carbon sink in the Northern Hemisphere, needs to be tested.

In discussing forthcoming changes in the concentration of atmospheric CO_2 we note that to predict these it is necessary to know, first, the prediction of the development of power engineering, and, second, that fraction of CO_2 emission which remains in the atmosphere. The solution of the first problem is inseparably linked with the prospects for the development of society with its complex of economic, social, demographic and ecological problems. It is clear that such predictions or scenarios, as they are usually called, can only be approximate. Nevertheless, we introduce two of them, proposed by Legasov and Kuzmin (1981) and Working Group III of the Intergovernmental Panel on Climate (IPCC, 1990). The following considerations form the basis of the first prediction.

Energy consumption should provide a sufficiently high living standard for a population. When proposing that, by the end of next century, the population of the Earth will amount to 10–12 billion, and that by that time the specific energy consumption (consumption of energy per capita) will have reached the present-day level of 10 KW·year/year per capita in all the developed countries, then the total energy consumption will be equal to 100–120 TW·year/year (1 TW·years = 10^9 KW·year = 31.54×10^{18} J). But the growth in population and the process of industrialization in many of the developing countries tends to increase in energy expenses per unit of Gross National Product (GMP). Industrialization also contributes to the additional consumption of materials (metals, plastics, etc.) per capita and, accordingly, increases the specific energy capacity of production. Additional energy will be required for production of foodstuff for the increased population and this in turn will require the cultivation of lands with low fertility and new energy consumption to produce vast amounts of fertilizers for the amelioration of lands. It will be necessary to increase energy consumption to exploit new deposits of mineral resources, for the recovery of industrial waste, and to provide humanity with fresh water. Additional energy consumption will be required for the preservation of the environment (sewage disposal, air cleaning, etc.). All this leads to the fact that the specific energy consumption will increase up to 10– 20 KW·year/year per capita, or up to 100–200 TW· year/year. On the other hand, a transition from the policy of producing even more energy to one of increasing the efficiency of the use of energy and its maximum economy is inevitable. The resulting energy consumed by the Earth's population by the end of the twenty-first century may amount to 60 TW·year/year. Such a prediction was assumed by Legasov and Kozmin (1981) as the basis when estimating the intensity of the net emission of CO_2 of an industrial origin.

The IPCC scenario (the so-called Business-as-Usual (BaU) scenario) covers the emissions of carbon dioxide (CO_2) and other greenhouse gases (methane (CH_4), nitrous oxide (N_2O), carbon monoxide (CO), nitrogen oxides (NO_3) and chlorofluorocarbons (CFCs)) from the present up to the year 2100. Population is assumed to approach 10.5 billion in the second half of the next century. Growth of the economy is taken to be 2–3% annually in the coming decade in the OECD countries and 3–5% in the Eastern European and devyeloping countries. The economic growth levels are considered to decrease thereafter. The energy supply is coal and on the demand side only modest efficiency increases are achieved. CO controls are modest, deforestation continues until the tropical forests are depleted and agricultural emissions of CH_4 and NO_3 are uncontrolled.

Since completion of the 1990 IPCC scenarios, new information has become available. It includes revision of the population growth rate, re-estimation of sources and sinks of greenhouse gases and improved data on tropical deforestation and forest biomass, as well as the consequences of recent political events in the former USSR, Eastern Europe and the Middle East, affecting the level of economic activity and the efficiency of energy production and use. This requires an update of the 1990 IPCC scenarios. As a result, we have six alternative 1992 IPCC scenarios differing among themselves as to the population and economic growth rates, assumptions about use of fossil fuel and biotic carbon emissions.

In the first of the scenarios referred to as the IS92a scenario, the update population assumptions are about 10% higher than those in the 1990 IPCC BaU scenario: global population increases from 5.25 billion in 1990 to 8.41 billion in 2025 and to 11.31 billion in 2100, with about 94% of the growth taking place in developing countries.

The assumptions about economic growth from 1990 to 2000 are also higher than those used in the BaU scenario. In accordance with them, the annual average rate of increase of Gross National Product (GNP) is assumed to be equal to 2.9% from 1990 to 2025 and 2.3% from 1990 to 2100. The estimates of the GNP growth rate for the initial 35 years, from 1990 to 2024, are substantially lower than that experienced by most world regions in the past 35 years, from 1955 to 1989. Over the last 75 years of the coming century, GNP is assumed to be slowing due to expected decrease in the population growth. The future GNP per capita is assumed to be rising everywhere throughout the next century, but most rapidly in the developing countries where even in 2100 the income per capita will still remain below levels in the developed countries.

Other assumptions concern primary energy consumption, resource availability, land-use changes and emissions. In particular, it is suggested that the annual average decline in total primary energy requirements per unit of GNP will be 0.8% for 1990–2025 and 1.0% for 1990–2100, while the cumulative net fossil C emissions, total tropical deforestation and related cumulative net C emissions from deforestation will accordingly be equal to 28.5 GtC, $6.78 \times 10^6 \text{ km}^2$ and 42 GtC for 1990–2025 and 1386 GtC, $14.47 \times 10^6 \text{ km}^2$ and 77 GtC, for 1990–2100. Finally, the CO_2 emission rate from energy, cement production and deforestation is taken to be 7.4 GtC/year in 1990, 12.2 GtC/year in 2025 and 20.3 GtC/year in 2100.

The remaining 1992 IPCC scenarios were designed to examine the sensitivity of future greenhouse gas emissions to a wide range of alternative input assumptions regarding population and economic growth rates, oil and gas resource availability, resulting in higher or lower prices and promoting expansion of nuclear and renewable energy, additional use of coal-mine methane for energy supply, improvements in regional pollution control (by means of a 30% environmental surcharge on fossil energy use), changes in deforestation rates and carbon stored within the biomass, etc. Overall, the scenarios indicate that the emissions of CO_2 and other greenhouse gases might rise substantially in the absence of stronger control measures.

Now that the net CO_2 emissions of industrial or industrial plus biotic origin have been determined, the next task must be to estimate the fraction of these emissions which remains air-borne. The simplest way is to assume that the same fraction remains air-borne as was observed during the last decade, i.e. $46 \pm 7\%$. *A propos*, it is this method that has been used to produce estimates of climate changes consequent on BaU and IS92a scenarios. One may also use the results of analyses of air samples in ice cores from Greenland and Antarctic ice sheets and data on variations of the annual average CO_2 concentration at monitoring stations, on the one hand, and estimates of industrial emissions, on the other hand. In the latter case, the average value of α for the period from 1760 to 1984 is found to be equal to 56% in the absence and $(39 \pm 5)\%$ in the presence of the biotic source with an emission rate of (1.6 ± 0.8) GtC/year (see Bjutner, 1986) and, again, difficulties arise with the prediction of future atmospheric CO_2 concentrations. In practice the assumption of the constancy of α is not fulfilled: even for the relatively short period from 1963 to 1978 the value of α calculated without consideration of the biotic source increases from 49 to 60% (see Bjutner, 1986). It remains only to use the global models of the carbon cycle where the parameter α is not fixed but, rather, is determined, together with all other variables sought.

First of all let us examine a model situation: a step-wise increase in the concentration of atmospheric CO_2. The appropriate transient response of the climatic system is described within the framework of a zero-dimensional model by the equation

$$c \, d\delta T/dt = \delta Q - G_f^{-1}\delta T, \qquad (6.6.1)$$

where δT is the disturbance of the global average surface air temperature; $G_f = \partial T/\partial Q$ is the parameter describing the sensitivity of the climatic system to variations of heat input δQ; c is the heat capacity of the climatic system.

For a step-wise change of δQ, and δT being equal to zero at the initial time instant, the solution of Equation (6.6.1) has the form $\delta T = \delta T_{eq}[1 - \exp(-t/\tau_e)]$, where $\delta T_{eq} = G_f \delta Q$ is the equilibrium temperature disturbance and $\tau_e = cG_f$ is the relaxation time. As can be seen, the latter depends not only on the heat capacity and, hence, on whether the most inertial link of the climatic system (ocean) is taken into account or not, but also on its sensitivity.

An analysis of the response of the ocean–atmosphere system to an instantaneously changing CO_2 concentration within the framework of the three-dimensional global circulation models (GCMs) was carried out by Hansen *et al.* (1984), Schlesinger *et al.* (1985), Bryan *et al.* (1982), Spelman and Manabe (1984), Washington and Meehl (1989) and Stouffer *et al.* (1989). In the first of the works mentioned the ocean current velocity in the UML was derived from empirical data, and to describe the vertical heat transport in the deep ocean a one-dimensional (in the vertical direction) diffusive model was used which did not consider heat transport by ordered motions. With such assumptions the reorganization of the ocean circulation under the influence of temperature disturbances and accompanying changes in the meridional heat transport are not taken into account. In the other works mentioned above, these restrictions were excluded. But in the second of them the integration of equations of a coupled ocean–atmosphere GCM was carried out only for the first 16 years, and then the calculation was continued within the framework of a zero-dimensional thermodynamic model of the atmosphere and a diffusive box model of the ocean (in such a model the UML is represented in the form of a well-mixed box, and the deep ocean is considered to be at rest so that the heat transport in it is produced only by the vertical eddy diffusion). Thus, the model is able to estimate the equilibrium response of global average values of surface air temperature and sea water temperature at certain depths, but does not permit the detection of local effects. Bryan *et al.* (1982), and Spelman and Manabe (1984) ran the GFDL model for 50 years after instantaneously doubling CO_2. Washington and Meehl (1989) performed a similar experiment over a 30-year period with the NCAR model in which the atmospheric submodel was represented by a spectral GCM, and the ocean submodel was represented by a coarse-grid GCM. Stouffer *et al.* (1989) also used a spectral atmospheric GCM coupled to a coarse-grid ocean GCM with isopycnal diffusion and heat and fresh water flux corrections for a 60-year period. We discuss their results in detail.

To compare non-equilibrium and equilibrium responses we introduce, according to Bryan *et al.* (1982), the relative temperature deviation $r = (T - T_0)/(T_{eq} - T_0)$, where T_0 and T_{eq} are equilibrium values of temperature for current and increased concentrations of atmospheric CO_2. Latitude–altitude distributions of the relative temperature deviation in the troposphere and in the upper one-kilometre ocean layer after 25 years following quadrupling of the concentration of atmospheric CO_2 are shown in Figure 6.7. It can be seen that the distribution of r in the troposphere and UML remains homogeneous

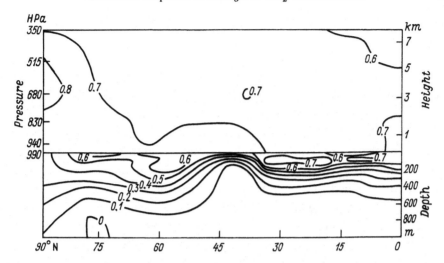

Figure 6.7 Latitude–altitude distribution of the relative deviation r of the zonal average temperature in the troposphere and in the upper 1 kilometre ocean layer after 25 years from the beginning of the integration, according to Spelman and Manabe (1984).

with latitude. The exception is the polar area where sea ice shields the ocean from the atmosphere and the water temperature is close to freezing point. In the vicinity of the sea ice southern boundary ($\sim 75\,°N$) the temperature disturbances are localized in the thin UML separated from the deep ocean by a sharp halocline. The existence of the sea ice and the halocline in the polar ocean and a related decrease in heat exchange between the atmosphere and deep ocean favours a rise in heating of the surface atmospheric layer in high latitudes. To the south of 65 °N the temperature disturbances penetrate to greater depths, with the maximum depth of their penetration being limited to the area of cold deep water formation (about 60 °N). To the south of it the value of r does not depend on latitude at the ocean surface and decreases rapidly with depth, approaching zero in the deep layer. Thus, the meridional distribution of the zonal average temperature after 25 years from an instantaneous increase in the concentration of atmospheric CO$_2$ closely resembles the equilibrium distribution in the atmosphere and UML but differs markedly from it in the deep ocean, from which it follows that the use of estimates of the equilibrium response when predicting the zonal average surface air temperature is justified only in the case where the characteristic time scale of the external forcing is more than 25 years.

All instantaneous CO$_2$ experiments performed within the framework of the coupled GCMs agree with the earlier mixed-layer model experiments. In particular, they show a larger warming at higher latitudes, as well as enhanced drying in the mid-continental regions in summer and an increase in the soil moisture in winter.

Let us now turn to a discussion of the transient response of the climatic system to a gradual increase in the concentration of atmospheric CO_2. First we note that the basis of most available estimates of potential climate changes, caused by an increase in the concentration of atmospheric CO_2, is formed by the *a priori* assumption of weak interaction between the carbon and thermodynamic cycles in the ocean–atmosphere system. There is no conclusive proof of this assumption. Moreover, simple qualitative arguments point to exactly the contrary. Indeed, if the burning of fossil fuels and the destruction of vegetation is accompanied by an increase in the concentration of atmospheric CO_2 and a temperature rise in certain subsystems of the climatic system, this must lead to a shift in the chemical equilibrium between carbon dioxide dissolved in the surface ocean layer, on the one hand, and bicarbonate and carbonate ions, on the other. The shift in chemical equilibrium must imply a change in intensity of absorption of anthropogenic CO_2 by the ocean and, hence, the redistribution of CO_2 between the atmosphere and ocean. Thus, the interaction between carbon and thermodynamic cycles is not in doubt.

This consideration was taken into account by Kagan *et al.* (1990) when simulating the evolution of the ocean–atmosphere climatic system from the dawn of the industrial revolution until the end of the twenty-first century. The equations of the 0.5-dimensional model (see Section 5.5) used in this work were supplemented by the appropriate equations for the carbon dioxide budget in atmospheric boxes, and for inorganic carbon in ocean boxes, as well as by expressions for the carbon eddy flux at the low boundary of the UML and for the carbon equivalent flux at the upper boundary of the deep layer, by expressions for the CO_2 flux at the ocean–atmosphere interface, and by hydrochemical relationships describing the behaviour of CO_2 in solution. Moreover, it was assumed that the boundary between the northern and southern atmospheric boxes is fixed (coincident with a circle of latitude $60\,°N$) and that the source R_f of atmospheric CO_2 is determined only by the burning of fossil fuels, and the influence of the biotic sources and sinks of CO_2 is manifested only by means of an increase in the sink R_p of atmospheric CO_2 due to a rise in production of terrestrial biomass, and does not have an effect on the vital functions of the terrestrial and marine biota. Then, assuming that all carbon dioxide emission is concentrated in the southern box we obtain $R_{f1} = (\mu_{CO_2}/\mu_C)(dN/s_1^A\,dt)$, $R_{f2} = 0$, $R_{pi} = (R_{pi}^A)_0[(1 + \beta \ln(c_i^A/c_{i0}^A)]$, $i = 1, 2$, where $dN/dt = rN_\infty((N_\infty/N_0) - 1)^{-1}\,e^{rt}/[1 + ((N_\infty/N_0) - 1)^{-1}\,e^{rt}]^2$, $r = 0.03$ l/year, $N_0 = 4.5 \times 10^9$ tC, $N_\infty = 5000 \times 10^9$ tC, c_{i0}^A and $(R_{pi}^A)_0$ are the concentration and sink of CO_2 in the pre-industrial period subject to the influence of the natural seasonal variability, and $\beta = 0.3$. We note that the expression for the production, dN/dt, of anthropogenic CO$_2$ was chosen only from considerations of convenience of its realization and the magnitude of the factor β to provide satisfactory agreement between calculated and observed (during the monitoring period) changes in the annual mean global average concentration of atmospheric CO$_2$.

We begin the discussion of the calculated results with an indication of a method of determining the initial conditions. The initial values of the CO$_2$ concentration in the atmosphere and the total carbon in the ocean on 1 January 1860 were found from the steady-state time-periodic solution in the absence of an anthropogenic source of CO$_2$ in the atmosphere. The initial values of the mass-weighted average air temperature in the northern and southern boxes and sea water temperature in the polar ocean, in the area of cold deep water formation, and in the area of upwelling were prescribed on the same basis. The annual mean values of these and all other characteristics of the carbon and thermodynamic cycles corresponding to time-periodic solution are shown in the first column of Table 6.3. Calculated deviations of the annual mean climatic characteristics from their initial values relating to different years of the period examined (1860–2100) are indicated in the other columns of the table.

As can be seen, an increase in the concentration of atmospheric CO$_2$ causes an enhancement of the absorption of long- and short-wave radiation, a rise in mass-weighted average values of the air temperature and humidity, a decrease in short-wave radiation and an increase in net long-wave radiation at the underlying surface, a domination of the latter over the former and, as a result, a rise in temperature of the surface of land, snow–ice cover and ocean in the area of upwelling, accompanied by an increase in local evaporation and precipitation. As for the temperature in the area of cold deep water formation, this decreases irregularly: sea water temperature falls up to the year 2025, then rises and in the last quarter of the coming century it starts to fall again.

Further, a rise in air temperature in the polar ocean results in a decrease in sensible and latent heat fluxes. This, together with an increase in the net long-wave radiation flux, causes a weakening of the heat transport from the lower to the upper surface of the snow–ice cover, followed by a decrease in the thickness and area of sea ice and the planetary albedo. Snow mass at the sea–ice surface decreases and at the land surface in the northern atmospheric box it increases. The former is connected with a reduction in the sea ice area, the latter with an increase in the precipitation–evaporation difference. An increase in precipitation favours the intensification of run-off during the whole period examined, and growth of the soil moisture content in

Table 6.3. *Annual mean values of the climatic characteristics at the beginning of the industrial revolution (1860) and deviations from these in 1985 and as forecast for the twenty-first century*

Characteristic	Year						
	1860	1985	2000	2025	2050	2075	2100
Radiation balance at the upper atmospheric boundary (W/m²):							
northern box	−74.8	−0.3	−0.4	−0.6	0.9	−1.5	−1.0
southern box	11.5	−0.1	−0.2	−0.3	0.4	0.5	0.0
Absorbed short-wave radiation (W/m²) in the atmosphere:							
northern box	45.7	0.2	0.3	0.4	0.5	0.9	1.7
southern box	91.8	0.3	0.4	0.8	1.1	1.8	2.4
Net long-wave radiation flux (W/m²) in the atmosphere:							
northern box	−134.6	−1.5	−2.2	−3.1	−3.6	−6.4	−9.0
southern box	−148.6	−1.3	−2.0	−3.1	−4.4	−7.5	−10.4
Planetary albedo (%)	31.40	−0.05	−0.07	−0.08	−0.07	−0.15	−0.29
Sea ice area (10⁶ km²)	12.45	−0.66	−0.99	−1.14	−0.53	−1.50	−3.47
Sea ice thickness (m)	2.76	−0.06	−0.10	−0.11	−0.05	−0.15	−0.37
Heat balance (W/m²) of the snow–ice cover in the polar ocean:							
short-wave radiation flux	31.2	−0.1	−0.1	−0.2	0.4	0.5	0.2
long-wave radiation flux	−24.4	0.9	1.4	2.1	1.5	2.6	4.2
sensible heat flux	−0.1	0.0	0.0	0.0	−0.1	−0.2	−0.2
latent heat flux	−16.3	−0.2	−0.3	−0.5	−0.8	−1.6	−2.7
heat exchange between upper and lower surfaces	7.0	−0.3	−0.5	−0.6	−0.7	−1.5	−2.3
heat release due to ice melting phase transitions	2.6	−0.3	−0.5	−0.8	−0.3	0.2	0.8
Heat balance (W/m²) of the ocean surface in the area of cold deep water formation:							
short-wave radiation flux	92.6	−1.3	−2.0	−2.4	−1.4	−3.4	−6.4
long-wave radiation flux	−81.7	3.8	5.7	7.6	7.0	13.2	21.6
sensible heat flux	−189.5	14.0	21.0	26.1	20.9	43.0	73.9
latent heat flux	−35.7	−1.0	−1.5	−1.7	−1.6	−3.6	−5.1
heat exchange with lower layers	214.3	−15.5	−23.2	−29.6	−24.9	−49.2	−84.0

Heat balance (W/m²) of the ocean surface in the upwelling area:

short-wave radiation flux	168.4	−0.2	−0.3	−0.7	−1.0	−1.7	−2.2
long-wave radiation flux	−87.1	1.3	1.9	3.7	5.4	8.5	11.2
sensible heat flux	−0.8	0.0	0.0	0.0	0.0	0.0	0.0
latent heat flux	−73.9	−0.9	−1.4	−2.7	−4.0	−6.5	−8.8
heat exchange with lower layers	−6.6	−0.2	−0.2	−0.3	−0.4	−0.3	−0.2

Gas exchange (gC/m²/year) between the ocean and atmosphere:

the domain of cold deep water formation area	40.1	43.1	65.5	129.0	227.4	323.2	280.6
the domain of upwelling area	−1.4	5.0	7.3	13.1	21.5	27.4	30.9

Heat balance (W/m²) of the land surface in the northern box:

short-wave radiation flux	73.7	−0.1	−0.1	−0.2	−0.3	−0.4	0.2
long-wave radiation flux	−55.8	0.5	0.7	1.2	1.4	2.7	2.7
sensible heat flux	−0.6	0.0	0.0	0.0	0.0	−0.1	−0.2
latent heat flux	−15.5	−0.4	−0.6	−0.9	−1.0	−2.0	−2.4
heat release due to snow melting	1.8	0.0	0.0	0.1	0.1	0.2	0.3

Heat balance (W/m²) of the land surface in the southern box:

short-wave radiation flux	146.9	−0.2	−0.3	−0.6	−0.9	−1.5	−2.0
long-wave radiation flux	−99.7	0.8	1.2	2.4	3.5	5.8	7.8
sensible heat flux	−1.3	0.0	0.0	0.0	0.0	0.0	0.0
latent heat flux	−45.9	−0.6	−0.9	−1.8	−2.6	−4.3	−5.8

Air temperature (K) at the middle level in the atmosphere:

northern box	240.67	0.51	0.75	0.87	0.51	1.26	2.30
southern box	258.50	0.23	0.35	0.56	0.80	1.36	1.87

Air humidity (g/kg) at the middle level in the atmosphere:

northern box	0.95	0.01	0.02	0.03	0.04	0.07	0.15
southern box	3.28	0.04	0.06	0.12	0.18	0.29	0.40

Partial pressure (ppm) of CO_2 in the atmosphere:

northern box	280.3	65.4	100.5	203.4	397.8	696.5	1031.5
southern box	280.6	66.3	101.8	205.9	401.6	702.3	1038.0

Precipitation rate (mm/day):

northern box	0.96	0.02	0.04	0.06	0.07	0.14	0.20
southern box	2.14	0.03	0.04	0.08	0.12	0.19	0.26

(continued)

349

Table 6.3. (*continued*)

Characteristic	Year						
	1860	1985	2000	2025	2050	2075	2100
Soil water content (cm):							
northern box	3.10	0.00	0.00	0.01	0.06	0.08	0.24
southern box	5.35	−0.02	−0.02	−0.04	−0.06	−0.09	−0.10
Run-off (cm/year):							
northern box	17.45	0.44	0.70	1.00	1.30	2.44	4.36
southern box	18.84	0.22	0.33	0.54	0.84	1.43	1.99
Snow mass (10^{14} kg) at the underlying surface:							
sea ice	4.74	−0.02	−0.02	0.05	−0.11	−0.27	−0.72
land in the northern box	8.87	0.26	0.40	0.56	0.71	1.22	1.74
Surface air temperature (K):							
northern box	261.88	0.82	1.23	1.55	1.49	3.00	5.10
southern box	292.78	0.22	0.34	0.59	0.95	1.54	2.00
hemisphere as a whole	288.64	0.31	0.46	0.72	1.02	1.73	2.43
Land surface temperature (K):							
northern box	264.72	0.82	1.23	1.58	1.49	2.93	5.05
southern box	294.01	0.27	0.40	0.70	1.13	1.81	2.35
Snow–ice cover surface temperature (°C) in the polar ocean	−14.12	0.71	1.10	1.35	1.50	2.99	4.80
Sea water temperature (°C):							
subice layer in the polar ocean	−1.80	0.0	0.0	0.0	0.0	0.0	0.0
deep layer in the polar ocean	0.50	−0.09	−0.13	−0.20	−0.06	−0.02	−0.07
cold deep water formation area	0.42	−0.10	−0.14	−0.19	−0.04	−0.01	−0.06
UML in the upwelling area	19.14	0.20	0.30	0.57	0.84	1.35	1.81
deep layer in the upwelling area	3.73	−0.01	−0.02	−0.04	−0.07	−0.08	−0.09

Total carbon concentration (10^{-3} mole CO_2/l) in the ocean:							
polar area	2.17	0.02	0.02	0.04	0.08	0.13	0.20
cold deep water formation area	2.17	0.02	0.03	0.05	0.09	0.14	0.21
UML in the upwelling area	2.05	0.04	0.05	0.10	0.16	0.22	0.26
deep layer in the upwelling area	2.28	0.01	0.01	0.01	0.02	0.03	0.05
pH of sea water:							
cold deep water formation area	8.33	−0.03	−0.04	−0.09	−0.19	−0.35	−0.56
upwelling area	8.31	−0.07	−0.10	−0.18	−0.30	−0.44	−0.56
UML thickness (m) in the upwelling area	50.78	−0.03	−0.05	−0.09	−0.14	−0.17	−0.19
Heat flux (W/m²) at the upper boundary of the deep layer in the upwelling area	16.22	0.08	0.13	0.25	0.59	1.05	1.33
Carbon fluxes (gC/m² year) at the UML/deep layer interface in the upwelling area:							
diffusive	−31.3	4.3	6.5	11.7	18.8	24.9	28.3
equivalent	−26.7	3.8	5.7	10.2	16.6	22.2	25.4
Upwelling velocity (10^{-7} m/s)	0.73	0.02	0.03	0.04	0.00	−0.01	0.00
Meridional heat transport (10^{14} W):							
heat transport from the upwelling area into the area of cold deep water formation	9.22	0.38	0.54	0.80	0.33	0.43	0.78
sensible heat transport from the southern into the northern atmospheric box	12.73	−0.43	−0.63	−0.51	0.27	0.05	−0.69
latent heat transport from the southern into the northern atmospheric box	3.51	−0.02	−0.03	0.04	0.24	0.31	0.21

Note: positive values of heat due to phase transitions correspond to sea ice formation; negative values correspond to sea ice melting; positive values of gas exchange indicate the absorption of CO_2 by the ocean; negative values of gas exchange indicate the emission of CO_2 into the atmosphere.

the northern atmospheric box in the second half of this period. At the end of the current century and in the first quarter of the coming century the soil moisture content in the northern atmospheric box remains practically constant, which is explained by compensation of the effects of evaporation and precipitation. On the other hand, the soil moisture content in the southern atmospheric box reduces due to the predominance of evaporation over precipitation. We also note the tendency to a decrease in temperature of the deep ocean in the area of upwelling and a simultaneous increase in heat transport from the UML into the deep layer. This result, contradicting estimates of the equilibrium response (see Section 6.2), is connected with an increase in temperature contrast between the UML and the deep ocean over time.

The calculated results for the characteristics of the carbon cycle are not unexpected. It can be seen from Table 6.3 that an increase in the concentration of atmospheric CO_2 is accompanied by an enhancement of gas exchange between the atmosphere and ocean; this, in turn, causes an increase in the total carbon content and a decrease in the pH in the upper ocean layer, as well as an increase in transfer of carbon excess from the UML to the deep layer and, eventually, an increase in the carbon concentration in the deep layer.

Turning back to the discussion of the characteristics of the thermodynamic cycle we will attempt to explain the above-mentioned change in sign of the temperature variations in the area of cold deep water formation. From Figure 6.8, it follows that, from the end of the first half of the coming century, the sea water temperature in this area will be subject to oscillations. Similar

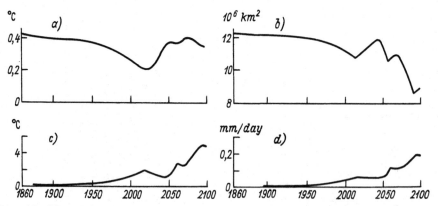

Figure 6.8 Change in the annual mean temperature in the area of cold deep water formation (*a*); sea ice area (*b*); and deviations of the surface air temperature (*c*); and of precipitation (*d*) in the northern atmospheric box from the beginning of the industrial revolution to the end of the twenty-first century, according to Kagan *et al.* (1990).

oscillations manifest themselves in the secular changes in the sea ice area, surface air temperature and precipitation. Furthermore they precede sea water temperature oscillations by about 4–9 years, depending on whether the sea ice area increases or decreases. These features can be interpreted in the following way. Let us recall that, when fixing the boundary between boxes, the extent of the area of cold deep water formation was defined as the difference between the ocean area in the northern box and the sea ice area. Because of this a decrease in the sea ice area, along with an increase in the concentration of atmospheric CO_2, must result in an increase in the extent of the area of cold deep water formation and, hence, to a fall in sea water temperature in this area owing to trapping of colder waters from the polar ocean. But the fall in sea water temperature favours a decrease in the heat transfer from the area of cold deep water formation into the subice ocean layer and thereby enhances stabilization or even an increase in the sea ice area. With an increase in the sea ice area (which occurs when the heat transfer in the subice ocean layer is less than the vertical heat flux in the snow–ice cover), trapping of colder water from the polar ocean into the area of cold deep water formation terminates. This entails a rise in temperature in the area of cold deep water formation, an increase in the heat transport in the subice ocean layer and, as a result, a further decrease in the sea ice area. Then the cycle of decrease and increase in the sea ice area is repeated.

It is likely that this is the nature of autooscillations appearing in the system of the polar ocean and the area of cold deep water formation at the moment in time when the sea ice area reaches a certain value. Allowing for changes in the secular changes in climatic characteristics generated by these autooscillations has an important consequence: it protects the sea ice from total disappearance. This distinguishes the results obtained here from previous ones, according to which an increase in the concentration of atmospheric CO_2 determines the total disappearance of northern sea ice in the second quarter of the twenty-first century. The estimates presented do not agree with this conclusion. According to them the annual mean sea ice area in the second quarter of twenty first century is reduced to 1.1×10^7 km^2 after which it varies against the background of a slower secular trend. As a result it turns out (see Figure 6.8(b)) that at the end of the twenty-first century the annual mean sea ice area will amount to about 9×10^6 km^2, that is, it will decrease as compared with its preindustrial value by 3.4×10^6 km^2.

The reliability of the estimates can be judged on the basis of a comparison of predicted and actual changes in the concentration of atmospheric CO_2 during the monitoring period (from 1958 until the present time). As can be seen from Figure 6.9 they show satisfactory agreement between themselves, and their

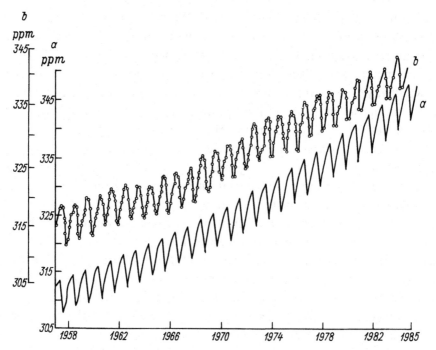

Figure 6.9 Change in the partial pressure of atmospheric CO_2, according to Kagan *et al.* (1990). (*a*) calculated results in the southern atmospheric box; (*b*) observational data from the Mauna Loa observatory.

agreement extends not only to inter-annual changes but also to seasonal oscillations of the characteristics examined.

Detailed information on the spatial structure of the climatic system response to a gradual increase in the concentration of atmospheric CO_2 may be obtained only on the basis of experiments with coupled ocean–atmosphere GCMs. To date, four such experiments have been performed. The first was realized with the NCAR model (Washington and Meehl, 1989); the second, with the modified version of the GFDL model (Stouffer *et al.*, 1989); the third, with the UKMO model (Murphy, 1990); and the fourth, with the MPI model (Cubasch *et al.*, 1991). We note, first, that the equilibrium response of each of these models to doubling CO_2 is different: when the ocean is represented as a specified-depth mixed layer with no heat transport, the equilibrium sensitivity is 4.5 °C for the NCAR model, 4.0 °C for the GFDL model, and 2.6 °C for the MPI model. Further, these models have different CO_2 doubling times (100 years for the NCAR model, 70 years for the GFDL and UKMO models, and 60 years for the MPI model). Finally, these models differ among themselves in parametrizations of various physical processes, spatial resolution and the use of some unphysical devices (e.g. correction of heat, fresh water and momentum fluxes at the ocean–atmosphere interface) when coupling the ocean and atmospheric submodels.

But despite these differences all of the coupled ocean–atmosphere GCMs exhibit a number of similar overall features in their transient and spatial responses. So, in all cases the annual mean globally averaged increase in surface air temperature is approximately 60% of the equilibrium warming due to thermal inertia of the deep ocean. All of the models demonstrate markedly smaller warming over oceans than over land in the Southern Hemisphere, compared to the Northern Hemisphere. That could be a result of the stronger heat uptake by the deep ocean. In the Northern Hemisphere the largest warming is found over middle continental areas in summer, and over the Arctic Ocean in winter. Accordingly, this leads to an increase in seasonality over the southern half of North America and south-east Europe and to a reduction in seasonality over the Arctic region. All of the models predict pronounced warming in the deep ocean adjacent to Antarctica and Greenland where the deep convective mixing occurs. In the other parts of the ocean the warming is largely confined to the upper 0.5 km layer. As a consequence, sea surface temperature changes in the northern North Atlantic and around Antarctica are small and even may be negative during the first decade of the simulation.

There is no consensus about changes in the deep water production and ocean circulation regime. Thus, Stouffer *et al.* (1989), and Washington and Meehl (1989) concluded that increasing surface temperature and precipitation would be accompanied by a decrease in the deep water production as the increased precipitation plus runoff exceed the increased evaporation thereby freshening the water. This does not take place in the Southern Ocean where the deep water formation is governed by a wider set of determining factors. In addition, Stouffer *et al.* (1989) noted significant weakening of the North Atlantic thermohaline circulation, while Washington and Meehl (1989) did not find this. Both models disagree on the ocean thermohaline circulation response for the Southern Hemisphere.

As regards the hydrological cycle changes, they are similar to those in comparable equilibrium experiments using models with a specified-depth upper mixed layer ocean. In winter, precipitation is generally higher over land areas north of 50 °N and is little changed or decreased further south. In summer, precipitation is lower over most mid-latitude continental areas and increases over Central America and south-east Asia. Correspondingly, soil moisture is greater over mid- to high-latitude northern continents in winter and is lower in summer. Exceptions are the Indian subcontinent and the Mediterranean region. They are described by a summer increase and a winter decrease in soil moisture respectively. The pattern of changes in the Southern Hemisphere is less well defined because of these changes, and their seasonal variations are small, but decreases in precipitation over most of the southern subtropical oceans in winter are apparent.

Assessing the results of the transient CO_2 experiments with coupled ocean–atmosphere GCMs it should be remembered that three of these (GFDL, MPI and UKMO models) use the procedure of flux correction lest the ocean temperature and salinity be essentially different from those presently observed. This correction causes substantial changes of fluxes comparable in magnitude to the fluxes themselves. Moreover, these changes tend to be of opposite signs to the signs of the fluxes (IPCC, 1992). Therefore, the use of the flux correction entails, in principle, significant distortions in the model response to small perturbations and particularly to perturbations with inter-annual and decadal timescales created by a gradual increase of the atmospheric CO_2. On the other hand, if the flux correction is not applied (as in the NCAR model) the solution drift introduces large systematic errors in the coupled ocean–atmosphere simulation. Again, we are facing the choice: which of the evils is the lesser? The answer to this question did not wait: as has been shown by Manabe *et al.* (1991, 1992) within the framework of the GFDL model with and without flux corrections, the simulated changes in the ocean–atmosphere climate induced by a gradual increase of atmospheric CO_2 are similar and may not be substantially affected.

We now compare available estimates of the increase in the global average surface temperature during the coming century obtained with an allowance for the thermal inertia of the climatic system. Analysis of the data presented in Table 6.4 shows that these estimates do not differ very much from each other, despite the differences in the models and in the scenarios for the production of industrial CO_2. This, and the fact that the detected long-term tendency to change in climatic characteristics agrees in general with empirical estimates are encouraging. In particular, judging from the analyses of instrumental record data (IPCC, 1990, 1992), the linear trend in land surface air temperature for the period 1881–1989 gives a rate of warming of 0.53 °C/100 year in the Northern and 0.52 °C/100 year in the Southern Hemisphere. According to ship data the increase in sea surface temperature between the late nineteenth century and the latter half of the twentieth century was less than 0.3 °C in the Northern and 0.5 °C in the Southern Hemisphere, so that the overall increase in global average sea surface temperature during the period indicated amounts to about 0.4 °C (0.43 °C between the periods 1861–1900 and 1981–1990). Combined land and marine observation data yield an increase in global average temperature of 0.47 °C between the 20-year period 1881–1900

Table 6.4. *Possible changes in the annual mean global average surface air temperature (°C) in the twenty-first century relative to its preindustrial value, generated by the increase in the concentration of atmospheric CO_2*

Author	Year				
	2000	2025	2050	2075	2100
Schneider and Thompson (1981)	0.45–0.80	0.90–1.50	1.40–2.25	2.20–3.45	–
Hansen *et al.* (1981)	0.2–0.3	0.5–1.0	0.7–2.4	0.9–3.4	1.2–4.4
Dickinson (1982)	0.40	0.65	1.05	1.50	–
Schlesinger (1983)	0.30–0.55	0.65–0.95	1.05–1.45	1.45–1.80	1.80–2.0
Hansen *et al.* (1984)	0.75–0.90	–	–	–	–
Harvey and Schneider (1985)	0.7	1.3	2.1	3.1	–
Kagan *et al.* (1986)	0.46	0.72	1.02	1.73	2.43
Peng *et al.* (1987)	0.55	0.90	1.40	1.90	–
Budyko and Israel (1987)	1.0	1.5	1.75–2.25	–	–
IPCC (1990)	1.1	1.75	2.6	3.25	4.25

and the latest decade 1981–1990. Comparable increases in the Northern and Southern Hemispheres are 0.47 and 0.48 °C respectively. There is also evidence, though not yet conclusive on global or even hemispheric scales due to limitations in the quantity and quality of the available information, of an enhancement of precipitation in high and middle latitudes and an increase in the frequency of droughts (a decrease in soil moisture) in lower latitudes of the continents. The estimates given in Table 6.3 do not contradict these conclusions. Moreover, if we take into account that the observed changes in the climatic characteristics are determined not only by an increase in the concentration of the atmospheric CO_2 but also by an increase in the concentration of other greenhouse gases, then the discrepancies between calculated and observed estimates can be recognized as acceptable, or at least explainable. Of course, this does not exclude the necessity for further perfection of prognostic models and improvement in the quality of forecasts.

References

Akerblom, F. (1908). Recherches sur les courants les plus bas de l'atmosphère au-dessus du Paris. *Nova Acta Req. Soc. Sc. Uppsala*, **Ser. 4**, (2), 1–45.

Ariel, N. Z., Bjutner, E. K. and Strokina, L. A. (1991). Time-space variability of the CO_2 exchange between the ocean and the atmosphere. *Meteorologiya i gidrologiya*, no. 2, pp. 54–9.

Augstein, E., and Krugermeyer, L. (1974). Errors of temperature, humidity and wind speed measurements on ships in tropical latitudes. *Meteor. Forsch. Reihe B*, **9**, 1–10.

Barron, E. J., Thompson, S. E. and Hay, W. W. (1984). Continental distribution as a forcing factor for global scale temperature. *Nature*, **310**, 574–5.

Baumgartner, A. (1979). *Climatic Variability and Forestry*. WCC, Overview Paper, no. 21, 27 pp.

Baumgartner, A. and Reichel, E. (1975). *The World Water Balance*. Elsevier, Amsterdam, NY, 179 pp.

Benilov, A. Y., Gumbatov, A. I., Zaslavskii, M. M. and Kitaigorodskii, S. A. (1978). Non-stationary model of the marine boundary layer development with generating surface waves. *Izv. AN. SSSR, fizika atmosfery i okeana*, **14**, 1177–87.

Benton, G. S., Fleagle, R. G., Leipper, D. F., Montgomery, R. B., Rakestraw, N., Richardson, W. S., Riehl, H. and Snodgrass, J. (1963). Interaction between the atmosphere and the ocean. *Bull. Amer. Meteorol. Soc.*, **44**, 4–17.

Bjutner, E. K. (1978). *Dynamics of the Surface Air Layer*. Gidrometeoizdat, Leningrad, 157 pp.

Bjutner, E. K. (1986). *The Planetary Gas Exchange of O_2 and CO_2*. Gidrometeoizdat, Leningrad, 240 pp.

Blanc, T. V. (1983). An error analysis of profile flux, stability and roughness length measurements made in the marine atmospheric surface layer. *Boundary-Layer Meteorol.*, **26**, 243–67.

Blanc, T. V. (1985). Variation of bulk-derived surface flux, stability and roughness results due to the use of different transfer coefficient schemes. *J. Phys. Oceanogr.*, **15**, 650–69.

Blanc, T. V. (1987). Accuracy of bulk-method determined flux, stability and sea surface roughness. *J. Geophys. Res.*, **92**, 3867–76.

Bolin, B. (1978). Modelling the global carbon cycle. In *Carbon Dioxide, Climate and Society*. Pergamon Press, Oxford, NY, pp. 41–3.

Bortkovskii, R. S. (1983). *Ocean and Atmosphere Heat and Moisture Exchange Under Storm Conditions*. Gidrometeoizdat, Leningrad, 158 pp.

Bortkovskii, R. S., Bjutner, E. K., Malevskii-Malevich, S. P. and Preobrazhenskii, L. Yu. (1974). *Transfer Processes Near the Ocean–Atmosphere Interface.* Gidrometeoizdat, Leningrad, 239 pp.

Bryan, F. (1986). High latitude salinity effects and interhemispheric thermohaline circulations. *Nature*, **323**, 301–4.

Bryan, F. and Oort, A. H. (1984). Seasonal variation of the global water balance based on aerological data. *J. Geophys. Res.*, **89**, 11717–30.

Bryan, K. (1982). Poleward heat transport by the ocean: Observations and models. *Ann. Rev. Earth Planet. Sci.*, **10**, 15–38.

Bryan, K., Komro, F. G., Manabe, S. and Spelman, M. J. (1982). Transient response to increasing atmospheric carbon dioxide. *Science*, **215**, 56–8.

Bryan, K., Manabe, S. and Pacanowsky, R. C. (1975). A global ocean–atmosphere climate model. Part II. The ocean circulation. *J. Phys. Oceanogr.*, **5**, 30–46.

Budyko, M. I. (1956). *Heat Balance of the Earth's Surface.* Gidrometeoizdat, Leningrad, 255 pp.

Budyko, M. I. (1969). The effect of solar radiation variations on the climate of the Earth. *Tellus*, **21**, 611–19.

Budyko, M. I. (1974). *Climate Change.* Gidrometeoizdat, Leningrad, 280 pp.

Budyko, M. I. (1980). *Climate of Past and Future.* Gidrometeoizdat, Leningrad, 350 pp.

Budyko, M. I. and Israel, Yu. A. (eds.) (1987). *Anthropogenic Changes in Climate.* Gidrometeoizdat, Leningrad, 406 pp.

Budyko, M. I., Ronov, A. B. and Yanshin, A. L. (1985). *History of the Atmosphere.* Gidrometeoizdat, Leningrad, 207 pp.

Busch, N. E. (1977). Fluxes in the surface boundary layer over the sea. In *Modelling and Prediction of the Upper Layer of the Ocean*, E. B. Kraus (ed.), Pergamon Press, Oxford, pp. 72–91.

Cacuci, D. G. and Hall, M. G. (1984). Efficient estimation of feedback effects with application to climate model. *J. Atmos. Sci.*, **41**, 2063–8.

Cahalan, R. F. and North, G. R. (1979). A stability theorem for energy-balance climate models. *J. Atmos. Sci.*, **36**, 1205–16.

Carissimo, B. C., Oort, A. H. and Vonder-Haar, T. H. (1985). Estimating the meridional energy transport in the atmosphere and ocean. *J. Phys. Oceanogr.*, **15**, 82–91.

Carson, D. J. (1982). Current parametrization of land surface processes in atmospheric general circulation models. *Proc. Conf. on Processes in Atmospheric General Circulation Models.* Greenbelt, 5–10 January 1981. WMO, Geneva, pp. 67–108.

Chalikov, D. V. (1982). Zonal models of the atmosphere. *Izv. AN SSSR, fizika atmosfery i okeana*, **18**, 1247–55.

Chamberlin, T. C. (1906). On a possible reversal of deep-sea circulation and its influence on geologic climates. *J. Geol.*, **14**, 363–73.

Chan, J. H. and Olson, J. S. (1980). Limits of the organic storage of carbon from burning fossil fuels. *J. Environ. Management*, **11**, 147–63.

Charney, J. G. (1969). What determines the thickness of the planetary boundary layer in the neutrally stratified atmosphere? *Okeanologia*, **9**, 143–5.

Charney, J. G. (1975). Dynamics of desert and drought in the Sahel. *Quart. J. Roy. Meteorol. Soc.*, **101**, 193–202.

Charney, J. G., Quirk, W. J., Chow, S. H. and Kornfield, J. (1977). A comparative study of the effects of albedo change on drought in semiarid regions. *J. Atmos. Sci.*, **34**, 1366–85.

Chervin, R. M. (1979). Response of the NCAR general circulation model to changed land surface albedo. *Report of the JOC Study Conf. on Climate Models, 1.* GARP Publ. Ser., no. 22, pp. 563–81.

Ching, J. K. S. (1976). Ships' influence on wind measurements determined from Bomex mast and boom data. *J. Appl. Meteorol.,* **15,** 102–6.

Coantic, M. (1986). A model of gas transfer across air–water interface with capillary waves. *J. Geophys. Res.,* **91,** 3925–43.

Corby, G. A., Gilchrist, A. and Rowntree, P. R. (1977). United Kingdom Meteorological Office five-level general circulation model. In General circulation models of the atmosphere (J. Chang, ed.), *Methods in Comput. Physics,* **17,** 67–110.

Cubasch, U. (1989). A global coupled atmosphere–ocean model. *Phil. Trans. R. Soc., Lond.,* **A329,** 263–73.

Cubasch, U., Hasselmann, K., Höck, H., Maier-Reimer, E., Mikolajewicz, U., Santer, B. D. and Sausan, R. (1991). *Time Dependent Greenhouse Warming Computations with a Coupled Ocean–Atmosphere Model.* Max Planck Inst. Meteorol., Report 67, Hamburg.

Deacon, E. L. (1977). Gas transfer to and across an air–water interface. *Tellus,* **29,** 363–74.

Deacon, E. L. and Webb, E. K. (1962). Small-scale interactions. In *The Sea,* M. N. Hill (ed.), Vol. 1. Interscience, NY, London, pp. 43–87.

Deardorff, J. W. (1977). A parametrization of ground-surface moisture content for use in atmospheric prediction models. *J. Appl. Meteorol.,* **16,** 1182–5.

Denman, K. L. and Miyake, M. (1973). Behaviour of the mean wind, drag coefficient, and the wave field in the open ocean. *J. Geophys. Res.,* **78,** 1917–31.

Dickinson, R. E. (1981). Convergence rate and stability of ocean–atmosphere coupling schemes with a zero dimensional climate model. *J. Atmos. Sci,* **38,** 2112–20.

Dickinson, R. E. (1982). Modelling climate changes due to carbon dioxide increases. In *Carbon Dioxide Review 1982,* W. C. Clark (ed.). Oxford Univ. Press, London, NY, pp. 101–33.

Dickinson, R. E. (1983). Land–surface processes and climate–surface albedos and energy balance. *Adv. in Geophys.,* **25,** 305–53.

Dickinson, R. E., Henderson-Sellers, A., Kennedy, P. J. and Wilson, M. F. (1986). *Biosphere–Atmosphere Transfer Scheme (BATS) for the NCAR Community Climate Model.* NCAR Techn. Note NCAR/TN-275 STR, 69 pp.

Donelan, M. A. (1982). The dependence of the aerodynamic drag coefficient on wave parameters. *First International Conference on Meteorology and Air–Sea Interaction of the Coastal Zone.* Amer. Meteorol. Soc., Boston, pp. 381–7.

Dunckel, M., Hasse, L., Krügermeyer, L., Schrievers, D. and Wucknitz, J. (1974). Turbulent fluxes of momentum, heat and water vapour in the atmospheric surface layer at sea during ATEX. *Boundary-Layer Meteorol.,* **6,** 81–106.

Dyer, A. J. (1961). Measurements of evaporation and heat transfer in the lower atmosphere by an automatic eddy-correlation technique. *Quart. J. Roy. Meteorol. Soc.,* **87,** 401–12.

Dyer, A. J. (1974). A review of flux-profile relationships. *Boundary-Layer Meteorol.,* **7,** 363–72.

Dymnikov, V. P., Perov, V. L. and Lykosov, V. N. (1979). A hydrodynamic zonal model of the atmospheric general circulation. *Izv. AN SSSR, fizika atmosfery i okeana,* **15,** 484–97.

Ekman, V. W. (1902). Om jordrotationens inverkan pa vindströmmar i havet. *Nyt. Mag. für Naturvid*, **40**, 37–63.

Ekman, V. W. (1905). On the influence of the Earth's rotation on ocean currents. *Arkiv Math., Astron., Fysik.* Uppsala, Stockholm, **2**, 11–52.

Ellis, J. S., Vonder-Haar, T. H., Levitus, S. and Oort, A. H. (1978). The annual variation in the global heat balance of the Earth. *J. Geophys. Res.*, **83**, 1958–62.

Emmanuel, C. B. (1975). Drag and bulk aerodynamic coefficients over shallow water. *Boundary-Layer Meteorol.*, **8**, 465–74.

Esbensen, S. K. and Kushnir, V. (1981). *The Heat Budget of the Global Ocean: An atlas based on estimate from surface marine observations.* Clim Res. Inst., Oregon State Univ., Corvallis, 27 pp.

Ewing, G. and McAlister, E. D. (1960). On the thermal boundary layer of the ocean. *Science*, **131**, 1374–6.

Exner, F. M. (1912). Zur Kenntnis der intersten Winde über Land und Wasser. *Ann. Hydrogr. und Marit. Meteorol.*, **5**, 226–39.

Faegre, A. (1972). An interactive model of the earth–atmosphere–ocean system. *J. Appl. Meteorol.*, **11**, 4–6.

Foreman, S. J., Grahame, N. S., Naskell, K. and Roberts, D. L. (1988). *Feedbacks and Error Mechanisms in a Global Coupled Ocean–Atmosphere–Sea Ice Model. Modelling the Sensitivity and Variations of the Ocean–Atmosphere System.* Report of a Workshop at the European Centre for Medium Range Weather Forecasts (11–13 May 1988), WCRP-15, pp. 271–9.

Freyer, H. D. (1979). Variations in the atmospheric CO_2 content. In *The Global Carbon Cycle*, B. Bolin, E. T. Degens, S. Kempe and P. Ketner (eds.), SCOPE 13, John Wiley and Sons, NY, pp. 79–99.

Friehe, C. A. and Schmitt, K. F. (1976). Parametrization of air–sea interface fluxes of sensible heat and moisture by the bulk aerodynamic formulas. *J. Phys. Oceanogr.*, **6**, 801–9.

Garratt, J. R. and Hyson, P. (1975). Vertical fluxes of momentum, sensible heat and water vapour during the air mass transformation experiment (AMTEX-1974). *J. Meteorol. Soc. Jap.*, **53**, 149–60.

Gates, W. L., Cook, K. H. and Schlesinger, M. E. (1981). Preliminary analysis of experiments of the climatic effects of increased CO_2 with an atmospheric general circulation model and a climatological ocean. *J. Geophys. Res.*, **86**, 6385–93.

Gates, W. L., Han, Y. J. and Schlesinger, M. E. (1985). The global climate simulated by a coupled ocean–atmosphere general circulation model: preliminary results. In *Coupled Ocean–Atmosphere Models.* J. C. J. Nihoul (ed.). Elsevier, Amsterdam, pp. 131–51.

Geernaert, G. L., Katsaros, K. B. and Richter, K. (1986). Variation of the drag coefficient and its dependence on sea state. *J. Geophys. Res.*, **91**, 7667–9.

Golitsyn, G. S. (1973). *Introduction to Dynamics of Planetary Atmospheres.* Gidrometeoizdat, Leningrad, 103 pp.

Gongwer, C. A. and Finkle, E. (1960). Turbulence meter as an instrument of oceanographic research. *Proc. Instrum. Soc. Amer.*, **15**, 88–93.

Graham, N. E. and White, W. B. (1988). The El Niño cycle: a natural oscillator of the Pacific ocean–atmosphere system. *Science*, **240**, 1293–1302.

Grassel, H. (1976). The dependence of the measured cool skin of the ocean on wind stress and total heat flux. *Boundary-Layer Meteorol.*, **8**, 465–74.

Green, J. S. (1970). Transfer properties of the large-scale eddies and the general circulation of the atmosphere. *Quart. J. Roy. Meteorol. Soc.*, **96**, 157–85.

Hampicke, V. (1979). Net transfer of carbon between the land biota and the atmosphere, induced by man. In *The Global Carbon Cycle*, B. Bolin, E. T. Degens, S. Kempe and P. Ketner (eds.). SCOPE 13, John Wiley and Sons, NY, pp. 219–36.

Han, Y. J., Schlesinger, M. E. and Gates, W. L. (1985). An analysis of the air–sea ice interaction simulated by the OSU coupled atmosphere–ocean general circulation model. In *Coupled Ocean–Atmosphere Models*. J. C. J. Nihoul (ed.). Elsevier, Amsterdam, pp. 167–82.

Hansen, J., Johnson, D., Lacis, A., Lebedeff, S., Lee, P., Rind, D. and Russell, G. (1981). Climate impact of increasing atmospheric carbon dioxide. *Science*, **213**, 957–66.

Hansen, J., Lacis, A., Rind, D., Russell, G., Stone, P., Fung, I., Ruedy, R. and Lerner, J. (1984). Climate sensitivity: analysis of feedback mechanisms. In *Climate Processes and Climate Sensitivity*, J. E. Hansen and T. Takahashi (eds.). *Geophys. Monogr.*, **29**, 130–63.

Harvey, L. D. D. and Schneider, S. H. (1985). Transient climate response to external forcing on 10^0–10^4 year time scales. Part 1: Experiments with globally averaged coupled atmosphere and ocean energy balance models. *J. Geophys. Res.*, **90**, 2191–205.

Hasse, L. O. (1968). On the determination of the vertical transports of momentum and heat in the atmosphere boundary layer at sea. *Hamburg, Geophys. Einzelschriften*, **11**, 70.

Hasse, L., Grünewald, M., Wucknitz, J., Dunckel, M. and Schriever, D. (1978). Profile derived turbulent fluxes in the surface layer under disturbed and undisturbed conditions during GATE. '*Meteor*' *Forschungsergeb*, **13**, 24–40.

Henderson-Sellers, A. and Gornitz, V. (1984). Possible climatic impacts of land cover transformations with particular emphasis on tropical deforestation. *Clim. Change*, **6**, 231–57.

Hesselberg, T. (1954). The Ekman spirals. *Arch. für Meteorol., Geophys. und Bioklim.*, **A7**, 329–43.

Hicks, B. B. and Dyer, A. L. (1970). Measurements of eddy fluxes over the sea. *Quart. J. Roy. Meteorol. Soc.*, **96**, 523–8.

Hill, R. H. (1972). Laboratory measurements of heat transfer and thermal structure near the air–water interface. *J. Phys. Oceanogr.*, **2**, 190–8.

Houghton, R. A., Hobbie, J. E., Mellilo, J. M., Moore, B., Peterson, B. J., Shaver, G. R. and Woodwell, G. M. (1983). Changes in the carbon content of terrestrial biota and soils between 1860 and 1980: a net release of CO_2 to the atmosphere. *Ecological Monographs*, **53**, 235–65.

Hsu, S. A. (1974). A dynamic roughness and its application to wind stress determination at the air–sea interface. *J. Phys. Oceanogr.*, **4**, 116–20.

Hughes, P. A. (1956). A determination of the relation between wind and sea-surface drift. *Quart. J. Roy. Meteorol. Soc.*, **82**, 499–502.

Hunjua, G. G. and Andreyev, E. G. (1974). Experimental study of heat exchange between sea and atmosphere with a small-scale impact. *Izv. AN SSSR, fizika atmosfery i okeana*, **10**, 1100–10.

Hunt, B. G. (1985). A model study of some aspects of soil hydrology relevant to climate modelling. *Quart. J. Roy. Meteorol. Soc.*, **111**, 1071–6.

Intergovernmental Panel on Climate Change, 1990 (IPCC 90). *Climate Change*, J. T. Houghton, G. J. Jenkins and J. J. Ephraums (eds.). Cambridge University Press, 364 pp.

Intergovernmental Panel on Climate Change, 1992 (IPCC 92). *Climate Change 1992*.

The Supplementary Report to the IPCC Scientific Assessment. J. T. Houghton, B. A. Callander and S. K. Varney (eds.). Cambridge University Press, 200 pp.

Jaeger, L. (1983). Monthly and areal patterns of mean global precipitation. In *Variations in the Global Water Budget.* D. Reidel, p. 129–40.

Kagan, B. A. (1971). Ocean–Atmosphere Interaction (theoretical models of a boundary layer ocean–atmosphere system). *Itogy nauki. Okeanologia,* 1970. VINITI AN SSSR, Moskva, pp. 37–69.

Kagan, B. A., Laikhtman, D. L., Oganesian, L. A. and Piaskovskii, R. V. (1974). Two-dimensional thermodynamic model of the World Ocean. *Izv. AN SSSR, fizika atmosfery i okeana,* 10, 1118–22.

Kagan, B. A. and Maslova, N. B. (1991). Non-uniqueness of the thermohaline circulation in a three-layer ventilated ocean. *Fluid Dyn. Res.,* 6, 287–95.

Kagan, B. A., Ryabchenko, V. A. and Safray, A. S. (1986). Simulation of the transient response of the ocean–atmosphere system to increasing atmospheric CO_2 concentration. *Izv. AN SSSR, fizika atmosfery i okeana,* 22, 1131–41.

Kagan, B. A., Ryabchenko, V. A. and Safray, A. S. (1990). *The Ocean–Atmosphere System Response to External Forcings.* Gidrometeoizdat, Leningrad, 303 pp.

Kagan, B. A. and Tsankova, I. S. (1986). Time–space variability of heat content in the World Ocean. *Meteorologiya i gidrologiya,* no. 11, pp. 111–14.

Kagan, B. A. and Tsankova, I. S. (1987). Time–space variability of the meridional heat transport in the World Ocean. *Meteorologiya i gidrologiya,* no. 4, pp. 66–71.

Kahma, K. K. and Leppäranta, M. (1981). On errors in wind speed observation on R/V 'Aranda'. *Geophysica,* 17, 155–65.

Katsaros, K. S., Liu, W. T., Businger, J. A. and Tillman, J. E. (1977). Heat transport and thermal structure in the interfacial boundary layer measured in an open tank of water in turbulent free convection. *J. Fluid Mech.,* 83, 334–55.

Kazanskii, A. B. and Monin, A. S. (1961). On dynamic interaction between the atmosphere and the Earth's surface. *Izv. AN SSSR, ser. geofiz.,* no. 5, 786–8.

Keeling, C. D. (1973). Industrial production of CO_2 from fossil fuels and limestone. *Tellus,* 25, 174–98.

Keeling, C. D., Piper, S. C. and Heimann, M. (1989). A three-dimensional model of atmospheric CO_2 transport based on observed winds: 4. Mean annual gradients and interannual variations. In *Aspects of climate variability in the Pacific and the Western Americas.* D. H. Peterson (ed.). *Geophys. Monograph,* 55, 305–63.

Kerman, B. R. (1984). A model of interfacial gas transfer for a well-roughened sea. *J. Geophys. Res.,* 89, 1439–46.

Kidwell, K. B. and Seguin, W. R. (1978). *Comparison of Mast and Boom Wind Speed and Direction Measurements on US GATE B-Scale Ships.* NOAA Tech. Rep. EDS 28, 41 pp.

Killworth, P. D. (1983). Deep convection in the World Ocean. *Rev. Geophys. Space Phys.,* 21, 1–26.

Kitaigorodskii, S. A. (1970). *Physics of the Ocean–Atmosphere Interaction.* Gidrometeoizdat, Leningrad, 284 pp.

Kitaigorodskii, S. A. (1984). On the fluid dynamical theory of turbulent gas transfer across an air–surface interface in the presence of breaking wind-waves. *J. Phys. Oceanogr.,* 14, 960–72.

Kitaigorodskii, S. A., Kuznetsov, O. A. and Panin, G. N. (1973). Coefficients of drag, sensible heat, and evaporation in the atmosphere over the sea surface. *Izv. Akad. Nauk SSSR, fizika atmosfery i okeana,* 9, 1135–41.

Klauss, E., Hinzpeter, H. and Müller-Glewe, J. (1970). Messungen zur Temperaturstruktur im Wasser an der Grenzfläche. *Ozean–Atmosphere. Meteorol. Forschungsergeb.*, **Ser. B**, no. 5, pp. 90–4.

Kobak, K. I. (1988). *Biotic Components of the Carbon Cycle.* Gidrometeoizdat, Leningrad, 248 pp.

Kobak, K. I. and Kondrashova, N. Yu. (1987). Anthropogenic impacts on the soil reservoir of carbon and on the carbon cycle. *Meteorologiya i gidrologiya*, no. 5, pp. 39–46.

Kolmogorov, A. N. (1941). Local structure of turbulence in incompressible fluids at very large Reynolds numbers. *Doklady AN SSSR*, **30**, 299–303.

Kondo, J. Y. (1975). Air–sea bulk transfer coefficients in diabatic conditions. *Boundary-Layer Meteorol.*, **9**, 91–112.

Korzun, V. I. (ed.). (1974). *The World Water Balance and Water Resources of the Earth.* Gidrometeoizdat, Leningrad, 638 pp.

Koschmider, G. (1938). *Dynamical Meteorology.* Gosud. Nauchno-Tekhnich. Izd., Moskva, 376 pp.

Kraus, E. B. (1972). *Atmosphere–Ocean Interaction.* Clarendon Press, Oxford.

Kraus, E. B. and Turner, J. S. (1967). A one-dimensional model of the seasonal thermocline II. The general theory and its consequences. *Tellus*, **19**, 98–106.

Krügermeyer, L. (1976). Vertical transports of momentum, sensible and latent heat from profiles at the tropical Atlantic during ATEX. '*Meteor*' *Forschungsergeb.*, **11**, 51–77.

Krügermeyer, L., Grünewald, M. and Dunckel, M. (1978). The influence of sea waves on the wind profile. *Boundary-Layer Meteorol.*, **14**, 403–14.

Kuharets, V. P., Tsvang, L. R. and Yaglom, A. M. (1980). On correlation of surface and boundary atmospheric layers characteristics. In *Fizika atmosfery i problemy klimata*, Golitsyn, G. S. and Yaglom, A. M. (eds.). Nauka, Moskva, pp. 162–93.

Kuznetsov, O. A. (1970). Results of experimental study of the air flow over the sea surface. *Izv. AN SSSR, fizika atmosfery i okeana*, **4**, 798–803.

L'vovich, M. I. (1973). The global water balance. *Trans. Am. Geophys. Un.*, **54**, 28–42.

Laikhtman, D. L. (1958). On Wind Drift of Ice. *Trudy Leningr. Gidrometeorol. Inst.*, issue 7, pp. 129–37.

Laikhtman, D. L. (1966). Dynamics of the boundary atmospheric and ocean layers with an allowance for their interaction and non-linear effects. *Izv. AN SSSR, fizika atmosfery i okeana*, **2**, 1017–25.

Laikhtman, D. L. (1970). *Physics of the Boundary Layer of the Atmosphere.* Gidrometeoizdat, Leningrad, 341 pp.

Laikhtman, D. L. (1986). Non-linear theory of the wind ice drift. *Izv. AN SSSR, fizika atmosfery i okeana.* **4**, 1220–3.

Laikhtman, D. L., Kagan, B. A. and Timonov, V. V. (1968). Ways of studying the ocean–atmosphere interaction. *Meteorol. issledovanya*, no. 16, pp. 287–300.

Large, W. G. and Pond, S. (1981). Open ocean momentum flux measurements over the ocean. *J. Phys. Oceanogr.*, **11**, 324–36.

Large, W. G. and Pond, S. (1982). Sensible and latent heat flux measurements over the ocean. *J. Phys. Oceanogr.*, **12**, 464–82.

Launiainen, J. (1983). Parametrization of the water vapour flux over a water surface by the bulk aerodynamic method. *Annales Geophys.*, **1**, 481–92.

Lebedev, A. A. (1986). Ice Balance of the World Ocean. *Problemy Arktiki i Antarktiki*, issue 62, pp. 18–28.

Lemke, P., Trinkl, E. W. and Hasselmann, K. (1980). Stochastic dynamic analysis of polar sea ice variability. *J. Phys. Oceanogr.*, **10**, 2100–20.

Levitus, S. (1982). *Climatological Atlas of the World Ocean*. NOAA Prof. Paper 13. Rockville, MD, 171 pp.

Liu, W. T. (1983). Estimation of latent heat flux with Seasat-SMMR, a case study in N. Atlantic. In *Large-scale Oceanographic Experiments and Satellites*, C. Gautier and M. Fieux (eds.). D. Reidel, Dordrecht e.a., pp. 205–22.

Lorenz, E. N. (1967). *The Nature and Theory of the General Circulation of the Atmosphere*. World Meteorological Organization, Geneva, 161 pp.

Lorenz, E. N. (1968). Climatic determinism. *Meteorol. Monogr.*, **8**, 1–3.

Lorenz, E. N. (1970). Climatic change as a mathematical problem. *J. Appl. Meteorol.*, **9**, 325–9.

Lorenz, E. N. (1975). Climate predictability. *The Physical Bases of Climate and Climate Modelling*. GARP Publ. Ser., no. 16, World Meteorological Organization, pp. 132–6.

Lorenz, E. N. (1979). Forced and free variations of weather and climate. *J. Atmos. Sci.*, **36**, 1367–76.

Lueck, R. and Reid, R. (1984). On the production and dissipation of mechanical energy in the ocean. *J. Geophys. Res.*, **89**, 3439–45.

Lumley, J. L. (1980). Second-order models for turbulent flows. In *Prediction Method for Turbulent Flows*. W. Kollman (ed.). Hemisphere, pp. 7–34.

Makova, V. I. (1968). Resistance coefficient and roughness parameter for large wind velocities. *Trudy Gos. Okeanograph. Inst.*, issue 93, pp. 173–90.

Malevskii-Malevich, S. P. (1970). Influence of the cool skin on heat exchange between the ocean and atmosphere. *Trudy Glav. Geofiz. Observ.*, issue 257, pp. 35–45.

Malkevich, M. S. (1973). *Satellite Optical Researches of the Atmosphere*. Nauka, Moskva, 303 pp.

Manabe, S. and Bryan, K. (1969). Climate and the ocean circulation. *Mon. Wea. Rev.*, **97**, 739–827.

Manabe, S. and Bryan, K. (1985). CO_2-induced change in a coupled ocean–atmosphere model and its paleoclimatic implications. *J. Geophys. Res.*, **90**, 11689–707.

Manabe, S., Bryan, K. and Spelman, M. J. (1975). A global ocean–atmosphere climate model. Part I. The atmosphere circulation. *J. Phys. Oceanogr.*, **5**, 3–29.

Manabe, S., Bryan, K. and Spelman, M. J. (1979). A global ocean–atmosphere climate model with seasonal variation for future studies of climate sensitivity. *Dyn. Atmos. Oceans*, **3**, 395–426.

Manabe, S. and Hahn, D. G. (1977). Simulation of the tropical climate of an ice age. *J. Geophys. Res.*, **82**, 3889–911.

Manabe, S., Spelman, M. J. and Stouffer, R. J. (1992). Transient responses of a coupled ocean–atmosphere model to gradual changes of atmospheric CO_2. Part II: Seasonal response. *J. Climate*, **5**, 105–26.

Manabe, S. and Stouffer, R. J. (1980). Sensitivity of a global climate model to an increase of CO_2 concentration in the atmosphere. *J. Geophys. Res.*, **85**, 5529–54.

Manabe, S. and Stouffer, R. J. (1988). Two stable equilibria of a coupled ocean–atmosphere model. *J. Clim.*, **1**, 841–66.

Manabe, S., Stouffer, R. J., Spelman, M. J. and Bryan, K. (1991). Transient responses of a coupled ocean–atmosphere model to gradual changes of atmospheric CO_2. Part I: Annual mean response. *J. Climate*, **4**, 785–818.

Manabe, S. and Wetherald, R. T. (1975). The effects of doubling the CO_2 concentration on the climate of a general circulation model. *J. Atmos. Sci.*, **32**, 3–15.

Manabe, S. and Wetherald, R. T. (1980). On the distribution of climate changes resulting from an increase in CO_2 content of the atmosphere. *J. Atmos. Sci.*, **37**, 99–118.

Marchuk, G. I., Dymnikov, V. P., Zalesnyi, V. B., Lykosov, V. N. and Galin, V. Ya. (1984). *Mathematical Modelling of the General Circulation of the Atmosphere and Ocean*. Gidrometeoizdat, Leningrad, 319 pp.

Marland, G. (1989). *Fossil Fuels CO_2 Emissions: Three countries account for 50% in 1988*. CDIAC Communications, Winter 1989, 1–4, Carbon Dioxide Information Analysis Center, Oak Ridge National Laboratory, USA.

Marland, G. and Rotty, R. M. (1984). Carbon dioxide emissions from fossil fuels: a procedure for estimation and results for 1950/1982. *Tellus*, **36B**, 232–61.

McCreary, J. R. (1986). Coupled ocean–atmosphere models of El Niño and the Southern Oscillations. In *Large-scale Transport Processes in Oceans and Atmosphere*. J. Willebrand and D. L. T. Anderson (eds.). D. Reidel, Dordrecht, pp. 247–80.

Memery, L. and Merlivat, L. (1985). Modelling of gas flux through bubbles at the air–water interface. *Tellus*, **37B**, 272–85.

Minnet, P. J., Zadovy, A. M. and Llewellyn-Jones, D. T. (1983). Satellite measurements of sea-surface temperature for climate research. In *Large-scale Oceanographic Experiments and Satellites*. C. Gautier and M. Fieux (eds.). D. Reidel, Dordrecht, e.a., pp. 57–86.

Mitchell, J. F. B. (1983). The seasonal response of a general circulation model to changes in CO_2 and sea temperature. *Quart. J. Roy. Meteorol. Soc.*, **109**, 113–52.

Mitchell, J. F. B., Senior, C. A. and Ingram, W. J. (1989). CO_2 and climate: a missing feedback? *Nature*, **341**, 132–4.

Mitchell, J. F. B. and Warrilow, D. A. (1987). Summer dryness in northern mid-latitudes due to increased CO_2. *Nature*, **330**, 238–40.

Mitchell, J. M. (1976). An overview of climatic variability and its causal mechanisms. *Quaternary Res.*, **6**, 1–13.

Miyake, M., Donelan, M., McBean, G., Paulson, C., Badgly, F. and Leavitt, E. (1970). Comparison of turbulent fluxes over water determined by profile and eddy correlation techniques. *Quart. J. Roy. Meteorol. Soc.*, **96**, 132–7.

Monin, A. S. (1950). Dynamical turbulence in the atmosphere. *Izv. AN SSSR, ser. geogr. and geofiz.*, **14**, 232–54.

Monin, A. S. (1965). On symmetry properties for turbulence in the air surface layer. *Izv. AN SSSR, fizika atmosfery i okeana*, **1**, 45–54.

Monin, A. S. (1969). *Forecast as a Problem of Physics*. Nauka, Moskva, 184 pp.

Monin, A. S. (1982). *Introduction to the Climate Theory*. Gidrometeoizdat, Leningrad, 245 pp.

Monin, A. S. and Obukhov, A. M. (1954). Main regularities of the turbulent mixing in the surface layer of the atmosphere. *Trudy Geofiz. Inst. AN SSSR*, no. 24 (151), pp. 163–87.

Monin, A. S. and Yaglom, A. M. (1965). *Statistical Hydromechanics. Part I. Mechanics of Turbulence*. Nauka, Moskva, 639 pp.

Moore, B., Boone, R. D., Hobbie, J. E., Houghton, R. A., Mellilo, J. M., Peterson, B. J., Shaver, G. R., Vörösmarty, C. Y. and Woodwell, G. M. (1981). A simple model for analysis of the role of terrestrial ecosystems in the global carbon budget. In *Carbon Cycle Modelling*, B. Bolin (ed.). SCOPE 16, John Wiley and Sons, NY, pp. 365–85.

Munk, W. H. (1966). Abyssal recipes. *Deep-sea Res.*, **13**, 707–30.

Murphy, J. M. (1990). Prediction of the transient response of climate to a gradual increase in CO_2 using a coupled ocean atmosphere model with flux correction. In *Research Activities in Atmospheric and Oceanic Modelling*. CAS/JSC Working Group on Numerical Experimentation, Report no. 14, WMO/TO, no. 376, 9.7–9.8.

Nakamura, N. and Oort, A. H. (1988). Atmospheric heat budgets of the polar regions. *J. Geophys. Res.*, **93**, 9510–24.

Neumann, G. and Pirson, W. J. (1966). *Principles of Physical Oceanography*. Prentice-Hall.

Nieuwstadt, F. T. M. (1981). The steady-state height and resistance laws of the nocturnal boundary layers: theory compared with Cabaw observations. *Boundary-Layer Meteorol.*, **20**, 3–17.

North, G. R. (1975). Theory of energy balance climate models. *J. Atmos. Sci.*, **32**, 2033–43.

North, G. R., Cahalan, R. F. and Coakley, J. A. (1981). Energy balance climate models. *Rev. Geophys. Space Phys.*, **19**, 91–121.

Oglesby, R. J. and Saltzman, B. (1990). Sensitivity of the equilibrium surface temperature of a GCM to systematic changes in atmospheric carbon dioxide. *Geophys. Res. Letters*, **17**, 1089–92.

Oort, A. H. (1983). *Global Atmospheric Circulation Statistics, 1958–1973*. NOAA Prof. Paper 14, Rockville, MD, 180 pp.

Oort, A. H. (1985). Balance conditions in the Earth's climate system. *Adv. in Geophys.*, **28A**, 75–98.

Oort, A. H., Ascher, S. C., Levitus, S. and Peixóto, J. P. (1989) New estimates of the available potential energy in the World Ocean. *J. Geophys. Res.*, **94**, 3187–200.

Oort, A. H. and Peixóto, J. P. (1983). Global angular momentum and energy balance requirements from observations. *Adv. in Geophys.*, **25**, 355–490.

Oort, A. H. and Rasmusson, E. M. (1971). *Atmospheric Circulation Statistics*, NOAA Prof. Paper W5. Rockville, MD, 323 pp.

Pandolfo, J. P. (1969). A numerical model of the atmosphere–ocean planetary boundary layer. *Proc. WMO/JUGG Sympos. Numerical Weather Prediction*. World Meteorological Organization, Tokyo, pp. 1131–40.

Panin, G. N. (1985). *Heat and Mass Exchange Between Water Basin and Atmosphere Under Natural Conditions*. Nauka, Moskva, 295 pp.

Panteleyev, N. A. (1960). Study of turbulence in the surface water layer of the Antarctic part of the Indian and Pacific Oceans. *Trudy Okeanograph. Komissii AN SSSR*, **10**, 137–40.

Paulson, C. A. and Parker, T. W. (1972). Cooling of a water surface by evaporation, radiation and heat transfer. *J. Geophys. Res.*, **77**, 491–5.

Paulson, C. A. and Simpson, J. J. (1981). The temperature difference across the cool skin of the ocean. *J. Geophys. Res.*, 86, 11044–54.

Peixóto, J. P. and Oort, A. H. (1983). The atmospheric branch of the hydrological cycle and the climate. In *Variations in the Global Water Budget*. D. Reidel, Dordrecht, e.a. 5–65.

Peixóto, J. P. and Oort, A. H. (1984). Physics of climate. *Rev. Modern Phys.*, **56**, 365–429.

Peng, T. H. (1985). Atmospheric CO_2-variations based on the tree-ring ^{13}C record.

In *The Carbon Cycle and Atmospheric CO_2 Natural Variations Archean to Present. Geophys. Monogr.*, **32**, 123–31.

Peng, T. H., Broecker, W. S., Freyer, H. D. and Trumbore, S. (1983). A deconvolution of the tree ring based on $\delta^{13}C$ record. *J. Geophys. Res.*, **88**, 3609–20.

Peng, L., Chou, M. D. and Arking, A. (1987). Climate warning due to increasing atmospheric CO_2: simulations with a multilayer coupled atmosphere–ocean seasonal energy balance model. *J. Geophys. Res.*, **92**, 5505–21.

Petuhov, V. K. (1989). A zonal climatic model of the heat and moisture exchange in the atmosphere over the ocean. In *Physics of the Atmosphere and the Problem of Climate*, G. I. Golytzyn and A. M. Yaglom (eds.). Nauka, Moskva, pp. 8–41.

Petuhov, V. K. and Manuilova, N. I. (1984). Estimation of some climate-generating factors in a simple thermodynamic model of climate. *Meteorologiya i gidrologiya*, no. 18, pp. 31–7.

Philander, S. G. H. (1985). El Niño and La Niña. *J. Atmos. Sci.*, **42**, 3652–62.

Philander, S. G. H. and Rasmusson, E. M. (1985). Southern Oscillation and El Niño. *Adv. in Geophys.*, **28**, Part A. Climate dynamics, pp. 461–77.

Phillips, O. M. (1977). *The Dynamics of the Upper Ocean.* Cambridge University Press, 336 pp.

Pond, S. D., Fissel, B. and Paulson, C. A. (1974). A note on bulk aerodynamic coefficients for sensible heat and moisture fluxes. *Boundary-Layer Meteorol.*, **6**, 333–9.

Pond, S., Phelps, G., Paquin, J. E., McBean, G. and Stewart, R. W. (1971). Measurements of the turbulent fluxes of momentum, moisture and sensible heat over the ocean. *J. Atmos. Sci.*, **28**, 901–11.

Potter, G. L., Ellsaesser, H. W., MacCracken, M. C. and Ellis, J. S. (1981). Albedo change by man: test of climate effects. *Nature*, **291**, 47–9.

Radikevich, V. M. (1968). Study of some characteristics of the atmosphere–ocean boundary layer interaction based on a new theoretical model. *Trudy Leningr. Gidrometeorol. Instit*, Issue 32, pp. 3–15.

Report of the Joint Scientific Committee ad hoc Working Group on Radiative Flux Measurements. World Climate Programme, 1987, no. 136, 76 pp.

Revelle, R. and Munk, W. (1977). The carbon dioxide cycle and the biosphere. In *Energy and Climate*. Natl. Acad. Sci., Wash., DC, pp. 140–58.

Romanov, Ju. A., Fedorova, I. B., Chervyakov, M. S. and Shapiro, G. I. (1983). On the increase in accuracy of wind speed and direction ship measurements on the basis of aerodynamic testing of a ship model. *Okeanologia*, **23**, 355–60.

Rooth, C. (1982). Hydrology and ocean circulation. *Progr. Oceanogr.*, **11**, 131–49.

Rossby, C. G. (1932). A generalization of the theory of the mixing length with application to atmospheric and oceanic turbulence. *Mass. Inst. Technol., Meteorol.*, Pap., **1**, 1–36.

Rotta, J. C. (1951). Statistische Theorie Nichthomogener Turbulenz, 1. *Zs. Phys.*, **129**, 547–72; Statistische Theorie Nichthomogener Turbulenz, 2. *Zs. Phys.*, **132**, 51–77.

Rotty, R. M. (1987). Estimates of seasonal variation in fossil fuel CO_2 emissions. *Tellus*, **39B**, 184–202.

Sagan, C., Toon, O. B. and Pollack, J. B. (1979). Anthropogenic albedo changes and earth's climate. *Science*, **206**, 1363–8.

Saltzman, B. (1983). Climate system analysis. *Adv. in Geophys.*, **25**, 173–233.

Sattinger, D. H. (1980). Bifurcation and symmetry breaking in applied mathematics. *Bull. Amer. Math. Soc.*, **3**, 779–820.

Sausen, R., Barthel, K. and Hasselmann, K. (1988). Coupled ocean–atmosphere model with flux correction. *Climate Dynamics*, **2**, 145–63.

Schlesinger, M. E. (1983). A review of climate models and their simulation of CO_2-induced warming. *Intern. J. Environ. Studies*, **20**, 103–14.

Schlesinger, M. E. (1984). Climate model simulations of CO_2-induced climatic change. *Adv. in Geophys.*, **26**, 141–235.

Schlessinger, M. E., Gates, W. L. and Han, Y.-J. (1985). The role of the ocean in CO_2-induced climate change: preliminary results from the OSLI coupled atmosphere–ocean general circulation model. In *Coupled Ocean–Atmosphere Models*. J. C. J. Nikoiel (ed.). Elsevier, Amsterdam, pp. 447–78.

Schlesinger, M. E. and Zhao, Z. C. (1989). Seasonal climatic change induced by doubled CO_2 as simulated by the OSU atmospheric GCM/mixed-layer ocean model. *J. Climate*, **2**, 429–95.

Schneider, S. H. and Thompson, S. L. (1981). Atmospheric CO_2 and climate: importance of the transient response. *J. Geophys. Res.*, **86**, 3135–47.

Schooley, A. H. (1967). Temperature differences near the sea–air interface. *J. Mar. Res.*, **25**, 60–8.

Schumann, U. (1977). Realizability of Reynolds-stress turbulent models. *Phys. Fluids*, **20**, 721–5.

Sellers, P. J., Minz, Y., Sud, Y. C. and Dalcher, A. (1986). A simple biosphere model (SiB) for use within general circulation models. *J. Atmos. Sci.*, **43**, 505–31.

Sheppard, P. A. (1958). Transfer across the Earth's surface and through the air above. *Quart. J. Roy. Meteorol. Soc.*, **84**, 205–24.

Sheppard, P. A., Tribble, D. T. and Garratt, J. R. (1972). Studies of turbulence in the surface layer over water (Lough Neagh), part 1. Instrumentation, programme, profiles. *Quart. J. Roy. Meteorol. Soc.*, **98**, 627–41.

Shukla, J. (1985). Predictability. *Adv. in Geophys.*, **28**, Part B. Weather Dynamics, pp. 87–122.

Shvetz, M. E. (1939). To the theory of ocean currents. *Meteorologiya i gidrologiya*, no. 3, pp. 3–7.

Siegenthaler, U., Heimann, M. and Oeschger, H. (1978). Model responses of the atmospheric CO_2-level and $^{13}C/^{12}C$ ratio to biogenic CO_2 input. In *Carbon Dioxide, Climate and Society*. Pergamon Press, Oxford, NY, pp. 79–88.

Slingo, A. and Pearson, D. W. (1987). A comparison of the impact of an envelope orography and of a parametrization of orographic gravity wave-drag on model simulations. *Quart. J. Roy. Meteorol. Soc.*, **113**, 847–70.

Smith, S. D. (1967). Thrust-anemometer measurements of wind velocity spectra and Reynolds stress over a coastal inlet. *J. Mar. Res.*, **25**, 239–62.

Smith, S. D. (1970). Thrust-anemometer measurements of wind turbulence, Reynolds stress, and drag coefficient over the sea. *J. Geophys. Res.*, **75**, 6758–70.

Smith, S. D. (1974). Eddy flux measurements over Lake Ontario. *Boundary-Layer Meteorol.*, **6**, 235–55.

Smith, S. D. (1980). Wind stress and heat flux over the ocean in gale force winds. *J. Phys. Oceanogr.*, **10**, 709–26.

Smith, S. D. and Anderson, R. J. (1984). Spectra of humidity, temperature and wind over the sea at Sable Island, Nova Scotia. *J. Geophys. Res.*, **88**, 2029–40.

Smith, S. D. and Banke, E. G. (1975). Variation of the sea surface drag coefficient with wind speed. *Quart. J. Roy. Meteorol. Soc.*, **101**, 665–73.

Smith, S. D. and Jones, E. P. (1985). Evidence for wind-induced pumping of air–sea gas exchange based on direct measurements of CO_2 fluxes. *J. Geophys. Res.*, **90**, 869–75.

Smith, W. L., Bishop, W. P., Dvorak, V. F., Hayden, C. M., McElroy, J. H., Mosher, F. R., Oliver, V. J., Purdom, J. F. and Wark, D. Q. (1986). The meteorological satellite: overview of 25 years of operation. *Science*, **231**, 455–62.

Spelman, M. J. and Manabe, S. (1984). Influence of oceanic heat transport upon the sensitivity of a model climate. *J. Geophys. Res.*, **89**, 571–86.

Stepanov, V. N. (1974). *World Ocean*. Znanie, Moskva, 256 pp.

Stephens, G. L., Campbell, G. G. and Vonder-haar, T. H. (1981). Earth radiation budgets. *J. Geophys. Res.*, **86**, 9739–60.

Stewart, R. B. and Grant, H. L. (1962). Determination of the rate of dissipation of turbulent energy near the sea surface in the presence of waves. *J. Geophys. Res.*, **67**, 3177–80.

Stommel, H. (1958). *The Gulf Stream. A Physical and Dynamical Description.* London Univ. Press, Berkeley – Los Angeles, 202 pp.

Stommel, H. (1961). Thermohaline convection with two stable regimes of flow. *Tellus*, **13**, 224–30.

Stone, P. H. (1972). A simplified radiative-dynamical model for the static stability of rotating atmosphere. *J. Atmos. Sci.*, **29**, 405–18.

Stone, P. H. and Miller, D. A. (1980). Empirical relations between seasonal changes in meridional temperature gradients and meridional fluxes of heat. *J. Atmos. Sci.*, **37**, 1708–21.

Stouffer, R. J., Manabe, S. and Bryan, K. (1989). On the climate change induced by a gradual increase of atmospheric carbon dioxide. *Nature*, **342**, 660–2.

Strokina, L. A. (1982). Zonally averaged values of sea-water and air temperature for the World Ocean. *Meteorologia i gydrologiya*, no. 4, pp. 50–5.

Strokina, L. A. (1989). *Heat Balance of the Sea Surface (Handbook).* Gidrometeoizdat, Leningrad, 447 pp.

Stuiver, M. (1978). Atmospheric carbon dioxide and carbon reservoir changes. *Science*, **199**, 253–8.

Takahashi, T., Broecker, W. S. and Bainbridge, A. E. (1981). The alkalinity and total carbon dioxide concentration in the World oceans. In *Carbon Cycle Modelling*. B. Bolin (ed.), SCOPE 16. NY: John Wiley and Sons, pp. 271–86.

Tans, P. P., Fung, I. Y. and Takahashi, T. (1990). Observational constraints on the global atmospheric carbon dioxide budget. *Science*, **247**, 1431–8.

Taylor, P. K. (1983). The determination of surface fluxes of heat and water by satellite microwave radiometry and *in situ* measurements. In *Large-scale Oceanographic Experiments and Satellites*. C. Gautier and M. Fieux (eds.). D. Reidel, Dordrecht, pp. 223–46.

Taylor, P. K. (1985). TOGA Surface Fluxes of Sensible and Latent Heat by *in situ* Measurements and Microwave Radiometry. *Report of the Third Session of the JSC/CCCO TOGA Scientific Steering Group*. WMO/TD, no. 81, 30 pp.

Taylor, P. K. (1986). Validation of ERS-1 wind data using observations from research and voluntary observing ships. *Proc. Workshop on ERS-1 Wind and Wave Calibration*. FRG, 2–6 June, 1986, ESA SP-262, pp. 69–75.

Tibaldi, S. (1984). On the relationship between the systematic error of the ECMWF forecast model and orographic forcing. In *Predictability of Fluid Motions*. G. Holloway and B. J. West (eds.). Amer. Inst. of Physics, pp. 397–418.

Toba, Y. (1991). *Sea-surface Roughness Length Fluctuating in Concert with Wind and Waves*. Collected Papers. Faculty of Sciences, Tohoku University, Sendai, Japan.

Tomczak, G. (1963). Neuere Untersuchungen mit Treibkörpern zur Bestimmung des Windeinflusses auf Oberflächenströmungen. *Meer. Ber. Dtsch. Wetterdienstes*, **91**, 18–25.

Tsukamoto, O., Hayashi, T., Monji, N. and Mitsuta, Y. (1975). Transfer coefficients and turbulence-flux relationship as directly observed over the ocean during the AMTEX-74. *Scient. Report, 4th AMTEX Study Conf., Tokyo, Sept. 1975*, pp. 109–12.

Untersteiner, N. (1961). On the mass and heat budget of Arctic sea ice. *Arch. Meteorol. Geophys. Bioklimatol.*, **Ser. A, 12**, 151–82.

Verbitskii, M. Y. and Chalikov, D. V. (1986). *Modelling a Glacier–Ocean–Atmosphere System*. Gidrometeoizdat, Leningrad, 131 pp.

Volkov, Yu. A. and Koprov, B. M. (1974). On methods for measurement of heat, moisture, and momentum from aboard ship. *Tropeks*-72. Gidrometeoizdat, Leningrad, pp. 313–18.

Walsh, J. E. and Johnson, C. M. (1979). An analysis of Arctic sea ice fluctuations 1953–77. *J. Phys. Oceanogr.*, **9**, 580–91.

Warren, S. G. and Schneider, S. H. (1979). Seasonal simulation as a test for uncertainties in the parametrization of Budyko–Sellers zonal climate models. *J. Atmos. Sci.*, **36**, 1377–91.

Washington, W. M. and Meehl, G. A. (1983). General circulation model experiments on the climatic effects due to a doubling and quadrupling of carbon dioxide concentration. *J. Geophys. Res.*, **88**, 6600–10.

Washington, W. M. and Meehl, G. A. (1984). Seasonal cycle experiment on the climate sensitivity due to a doubling of CO_2 with an atmospheric general circulation model coupled to a simple mixed-layer ocean model. *J. Geophys. Res.*, **89**, 9475–503.

Washington, W. M. and Meehl, G. A. (1989). Climate sensitivity due to increased CO_2: Experiments with a coupled atmosphere and ocean general circulation model. *Climate Dyn.*, **4**, 1–38.

Washington, W. M. and Meehl, G. A. (1991). Characteristics of coupled atmosphere–ocean CO_2 sensitivity experiments with different ocean formulations. In *Greenhouse–Gas–Induced Climatic Change: A Critical Appraisal of Simulations and Observations*. Elsevier, Amsterdam, pp. 79–110.

Washington, W. M., Semtner, A. J., Meehl, G. A., Knight, D. J. and Mayer, T. A. (1980). A general circulation experiment with a coupled atmosphere, ocean and sea-ice model. *J. Phys. Oceanogr.*, **10**, 1887–908.

Washington, W. M. and Williamson, D. L. (1977). Description of the global circulation models of the National Centre for Atmospheric Researches. In *Methods in Computational Physics*, **17**, General circulation models of the Atmosphere, J. Chang (ed.), pp. 111–73.

Welander, P. (1986). Thermohaline effects in the ocean circulation and related simple models. In *Large-scale Transport Processes in Oceans and Atmosphere*. J. Willebrand and D. L. T. Anderson (eds.). D. Reidel, Dordrecht, pp. 125–46.

Weller, H. S. and Burling, R. W. (1967). Direct measurements of stress and spectra in the boundary layer over the sea. *J. Atmos. Sci.*, **24**, 653–64.

Wetherald, R. T. and Manabe, S. (1981). Influence of seasonal variations upon the sensitivity of a model climate. *J. Geophys. Res.*, **86**, 1194–204.

Wetherald, R. T. and Manabe, S. (1986). An investigation of cloud cover change in response to thermal forcing. *Clim. Change*, **8**, 5–26.

Wetherald, R. T. and Manabe, S. (1988). Cloud feedback processes in a general circulation model. *J. Atmos. Sci.*, **45**, 1397–1415.

Whittaker, R. H. and Likens, G. E. (1975). The biosphere and man. Primary productivity of the biosphere. *Ecological Studies*, **14**, Springer-Verlag, NY, pp. 305–28.

Wieringa, J. (1974). Comparison of three methods for determining strong wind stress over Lake Flevo. *Boundary-Layer Meteorol.*, **7**, 3–19.

Wiin-Nielsen, A. (1975). Predictability and climate variation illustrated by a low-order system. *Seminars on Scientific Foundation of Medium Range Weather Forecasts*, 1–12 September 1975. ECMWF, Reading, UK.

Williams, G. P. and Davis, D. R. (1965). A mean motion model of the general circulation. *Quart. J. Roy. Meteorol. Soc.*, **91**, 471–89.

Wilson, C. A. and Mitchell, J. F. B. (1987). Simulated climate and CO_2 induced climate change over western Europe. *Clim. Change*, **10**, 11–42.

WMO (1984). *International List of Selected, Supplementary and Auxiliary Ships.* WMO-47, WMO, Geneva.

Woodcock, A. H. and Stommel, H. (1947). Temperature observed near the surface of a fresh-water pond at night. *J. Meteorol.*, **4**, 102–3.

Woodwell, C. M., Houghton, R. A., Hobbie, J. E., Melillo, J. M., Moore, B., Peterson, B. J. and Shaver, G. R. (1983). Global deforestation and the carbon dioxide problem. *Science*, **222**, 1081–6.

Wu, J. (1969). Wind stress and surface roughness at air–sea interface. *J. Geophys. Res.*, **74**, 444–56.

Wu, J. (1980). Wind-stress coefficients over the sea surface near neutral conditions. A revisit. *J. Phys. Oceanogr.*, **10**, 727–40.

Wylie, D. P. and Hinton, B. B. (1983). A summary of the wind data available from satellites from the past history to future sensors. In *Large-scale Oceanographic Experiments and Satellites*. C. Gautier and M. Fieux (eds.). D. Reidel, Dordrecht, pp. 125–46.

Yaglom, A. M. and Kader, B. A (1974). Heat and mass transfer between rough wall and turbulent fluid flow at high Reynolds and Peclét numbers. *J. Fluid Mech.*, **62**, 601–23.

Yamada, T. (1976). On the similarity functions A, B and C of the planetary boundary layer. *J. Atmos. Sci.*, **33**, 781–93.

Yoshihara, H. (1968). Sea–air interaction: a simplified model. *J. Atmos. Sci.*, **25**, 729–35.

Zilitinkevich, S. S. (1970). *Dynamics of the Atmospheric Boundary Layer.* Gidrometeoizdat, Leningrad, 290 pp.

Zilitinkevich, S. S. (1972). On the determination of the height of the Ekman boundary layer. *Boundary-Layer Meteorol.*, **3**, 144–5.

Zilitinkevich, S. S. and Laikhtman, D. L. (1965). On closing of the equation system of eddy motion for the atmospheric boundary layer. *Trudy Glav. Geofiz. Observ.*, issue 167, 44–8.

Zilitinkevich, S. S., Laikhtman, D. L. and Monin, A. S. (1967). Dynamics of the boundary atmospheric layer. *Izv. AN SSSR, fizika atmosfery i okeana*, **3**, 297–333.

Zilitinkevich, S. S. and Monin, A. S. (1977). *The Global Atmosphere–Ocean Interaction.* Gidrometeoizdat, Leningrad, 23 pp.

Zilitinkevich, S. S., Monin, A. S. and Chalikov, D. V. (1978). Ocean–atmosphere interaction. *Fizika okeana, 1*, 1, *Gidrofizika okeana*,Nauka, Moskva, pp. 208–339.

Zubkovskii, S. L. and Kravchenko, T. K. (1967). Direct measurements of some characteristics of atmospheric turbulence in the marine surface layer. *Izv. AN SSSR, fizika atmosfery i okeana*, **3**, 127–35.

Index